SOCIETY FOR EXPERIMENTAL BIOLOGY
SEMINAR SERIES: 35

THE DIVISION AND SEGREGATION
OF ORGANELLES

SOCIETY FOR EXPERIMENTAL BIOLOGY SEMINAR SERIES

A series of multi-author volumes developed from seminars held by the Society for Experimental Biology. Each volume serves not only as an introductory review of a specific topic, but also introduces the reader to experimental evidence to support the theories and principles discussed, and points the way to new research.

THE DIVISION AND SEGREGATION OF ORGANELLES

Stephen A. Boffey

The Hatfield Polytechnic, Hatfield

David Lloyd

University College, Cardiff

The right of the
University of Cambridge
to print and sell
all manner of books
was granted by
Henry VIII in 1534.
The University has printed
and published continuously
since 1584.

CAMBRIDGE UNIVERSITY PRESS

Cambridge

New York   New Rochelle   Melbourne   Sydney

CAMBRIDGE UNIVERSITY PRESS
Cambridge, New York, Melbourne, Madrid, Cape Town,
Singapore, São Paulo, Delhi, Tokyo, Mexico City

Cambridge University Press
The Edinburgh Building, Cambridge CB2 8RU, UK

Published in the United States of America by Cambridge University Press, New York

www.cambridge.org
Information on this title: www.cambridge.org/9780521282116

First published 1988
First paperback edition 2011

A catalogue record for this publication is available from the British Library

Library of Congress Cataloguing in Publication data
The division and segregation of organelles.

(Seminar series / Society for Experimental Biology ; 35)
1. Cell organelles.   2. Cell division.   I. Boffey,
Stephen A.   II. Lloyd, David, 1940–
III. Series: Seminar series (Society for Experimental
Biology (Great Britain)) ; 35.
QH581.2.D58   1988   574.87'62   87–32014

ISBN 978-0-521-33436-5 Hardback
ISBN 978-0-521-28211-6 Paperback

CONTENTS

CONTRIBUTORS

Brian J. Baumgartner,
Program in Biological Science, University of Houston–Clear Lake, Houston, TX 77058–1098, USA.

S.A. Boffey,
School of Natural Sciences, The Hatfield Polytechnic, PO Box 109, College Lane, Hatfield, Herts. AL10 9AB, UK.

J.A. Bryant,
Department of Biological Sciences, University of Exeter, Washington Singer Laboratories, Perry Road, Exeter EX4 4QG, UK.

Th. Butterfass,
Botanisches Institut der Universität, Postfach 11 19 32, D-6000 Frankfurt, a.M. 11, Federal Republic of Germany.

H.G. Dickinson,
Department of Botany, Plant Sciences Laboratories, The University of Reading, Whiteknights, PO Box 221, Reading RG6 2AS, UK.

W.D. Donachie,
Department of Molecular Biology, University of Edinburgh, King's Buildings, Mayfield Road, Edinburgh, EH9 3JR, UK.

C.J. Duncan,
Department of Zoology, University of Liverpool, PO Box 147, Liverpool L69 3BX, UK.

E.B. Gingold,
Department of Biological Sciences, The Hatfield Polytechnic, PO Box 109, Hatfield, Herts. AL10 9AB, UK.

H. Hashimoto,
Commonwealth Scientific and Industrial Research Organization, Division of Horticultural Research, GPO Box 350, Adelaide, South Australia 5001.

R.M. Leech,
Department of Biology, University of York, Heslington, York YO1 5DD, UK.

F.L. Li,
Beijing Forestry College, Box 120, Beijing, China.

D. Lloyd,
Department of Microbiology, University College, Newport Road, Cardiff CF2 1TA, UK.

W.R. Mills,
Program in Biological Science, University of Houston—Clear Lake, Houston, TX 77058—1098, USA.

J. Oross,
Commonwealth Scientific and Industrial Research Organization, Division of Horticultural Research, GPO Box 350, Adelaide, South Australia 5001.

J.V. Possingham,
CSIRO Division of Horticultural Research, GPO Box 350, Adelaide, South Australia 5001.

K.A.Pyke,
Department of Biology, University of York, Heslington, York, YO1 5DD, UK.

R.J. Rose,
Department of Biological Sciences, The University of Newcastl , New South Wales, 2308, Australia.

R. Rosenberger,
National Institute for Medical Research, Mill Hill, London NW7 1AA, UK.

R.A.E. Tilney-Bassett,
School of Biological Sciences, University College of Swansea, Singleton Park, Swansea SA2 8PP, UK.

J.M. Whatley,
Department of Plant Sciences, University of Oxford, South Parks Road, Oxford OX1 3RA, UK.

INTRODUCTION

In the early years of this century, ideas of partial genetic autonomy for chloroplasts and mitochondria were first postulated by cytologists and geneticists. The initial recognition of cytoplasmic inheritance by Correns and by Baur in 1908 was a description of non-Mendelian inheritance of a factor influencing chloroplast development in leaves. Subsequent work, especially by Renner, Michaelis & Rhodes, established the principles of maternal inheritance of plastid mutations and of plastid autonomy.

Parallel developments in mitochondrial research awaited the discovery of the *poky* mutations in *Neurospora* by Mitchell & Mitchell in 1952 and *petite* mutations in yeast by Ephrussi in 1953. Another decade elapsed before the identification of chloroplast DNA in *Chlamydomonas* by Sager, and the clear demonstration of DNA in mitochondria by Nass and Nass. Since the early 1960s the pace of discovery has quickened; detailed genetic maps and complete sequences have been elucidated, organelle gene products have been identified, mechanisms of protein import and membrane assembly have been probed. A huge area of research has opened up.

The chapters in this book are based on lectures presented as a Seminar at a meeting of the Society for Experimental Biology, held in York during Easter 1987. In this Seminar we restricted ourselves to what is, after all, *the* fundamental problem in the transmission and continuity of organelle-encoded information: the orderly division and segregation of organelles and their genomes during growth and development. A major feature of this Seminar was that it brought together contributions related to animals, plants, yeast and bacteria, and so presented a unique opportunity to look for common themes and illuminating differences between these systems.

Several of the chapters review and reinforce a considerable body of evidence which indicates that chloroplasts and the mitochondria of yeast or plants arise only by the division of pre-existing organelles. However, such evidence is not yet available for the mitochondria of animal cells, and, in view of the existence of complete copies of the mitochondrial genome within the nucleus of human cells (Rosenberger), it would be rash to assume that *de novo* synthesis of mammalian mitochondria never occurs.

What do we know of the physical mechanisms of organelle division? Recently Rhodamine 123 vital staining has become a popular way of tracing mitochondrial movements. The fluorescent images show fast shape changes (< 0.3 s) and often predominantly Brownian movement, but also active locomotion at velocities up to 100 μm/min has been noted. Generation of the necessary forces is mysterious, but attachment to cytoskeletal elements seems a possibility. Similarly, migration and segregation at cell division, or the more precise partitioning required, for instance, during germ tube formation in the dimorphic yeast *Candida albicans*, as studied by Kuroiwa's group, may involve the active participation of contractile elements. Whatley, on the basis of microscopic studies, gives a full description of the visible changes associated with chloroplast division, and provides evidence for the involvement of a contractile system and for control of the polarity of division. Although the nature of the contractile system and the pattern of membrane growth are not yet known, there is the exciting prospect that developments in technology have now reached the point where they should be able to throw light on these important areas.

Information on the dynamics of organelle behaviour is hard-earned. Those working on the division of organelles could be excused for feeling embarrassed by their inability to demonstrate, on demand, an organelle going through the separation step of division. There is no longer any need for such embarrassment, since Leech & Pyke have shown that the separation step is extremely brief for chloroplasts, and so will be observed very rarely even amongst a population of chloroplasts which is actively dividing. Presumably a similar argument could be applied to mitochondria. Even in a system which may appear simple at first sight complexities abound. For instance, the mitochondria of yeast appear in thin sections as discrete, sausage-shaped organelles, many to a cell. This textbook concept was unquestioned until 1973, when Hoffman & Avers showed, in extensive sequences of serial sections, that baker's yeast sometimes has a single, highly branched, giant organelle. Similar three-dimensional reconstructions by Stevens confirmed this complex morphology, as did high voltage electron microscopy of yeasts. Similar findings were reported for other organisms such as Trypanosomes. Cyclic changes between sausages and large networks were shown in *Euglena* and *Chlamydomonas*. These dynamic complexities necessitated reaffirmation that some cell types do in fact contain large numbers of small mitochondria, e.g. the alga *Gonyostomum*, the ciliate *Tetrahymena*, and mammalian liver.

The regulation of chloroplast division has been the subject of intensive study. There is much evidence (Leech & Pyke) that chloroplast growth and division is regulated in such a way that cells maintain a fixed (presumably optimum) 'cover' of chloroplasts in relation to the cell size. This would result in a consistent ability to trap incident radiation, and would be expected to have been favoured by evolution. However, it is not known how this regulation occurs. The mechanisms that regulate mitochondrial numbers per cell can be probed by the use of drugs that disturb the control loops operating between nuclear and mitochondrial genomes, e.g. chloramphenicol treatment gives *Tetrahymena* with a greatly increased population of tiny mitochondria. Although such results do not allow us to deduce the precise mechanism for regulation of mitochondrial division, they demonstrate that this experimental system has the potential to allow us to identify key steps in the regulatory pathway.

For those who work on organelles there is a reminder of their probable ancestry in the chapter by Donachie, in which he describes the division of bacterial cells. Genes have been identified responsible for almost every aspect of prokaryotic division, and it seems likely that many such genes are also involved in the division of organelles. A major aim of research into organelle division and segregation must be to characterise these genes in as much detail as has been achieved for bacteria.

A recurrent theme of the chapters in this book is the importance of the nuclear genome in the division of organelles and the replication of their DNA. Butterfass shows a strong correlation between nuclear DNA content and the number of chloroplasts per guard cell, and suggests this could be a reflection of the mechanism of regulation of proplastid division (though presumably not of chloroplast division). There is much evidence for the idea that replication of organelle DNA depends entirely on the products of nuclear genes, as in the cases of Albostrians barley (Possingham, Hashimoto & Oross) or petite mutants of yeast (Gingold), which replicate their chloroplast or mitochondrial DNA in the absence of protein synthesis within the organelles. Tilney–Bassett has used genetic analysis to demonstrate a nuclear gene (P_r) which controls plastid inheritance in *Pelargonium*, and whose effects could be interpreted in terms of a nuclear coded plastid DNA polymerase. However, as Bryant makes clear, the characteristic feature of organelle DNA polymerases is how little we know about them.

In the search for other components involved in the replication of organelle DNA, possible candidates include a topoisomerase (tentatively identified by Mills & Baumgartner), and one or more proteins responsible for the attachment of organelle DNA to membranes (Rose). We can expect to see this list grow rapidly over the next few years.

Although research into organelle division and segregation has produced large amounts of information about morphological aspects of these processes, particularly in relation to chloroplasts, it will be clear to the reader that there are huge gaps in our understanding at the molecular level. It is hoped that this volume will encourage even more researchers to attempt to work out exactly how cells manage the remarkable feat of ensuring the maintenance of their organelles throughout growth and cellular division.

S.A.B. & D.L.

ACKNOWLEDGEMENTS

We are most grateful for substantial financial support provided by Amersham International, ICI Plant Protection Division, Roche Products and The Wellcome Foundation, which made it possible to include several colleagues from outside Britain amongst our list of speakers. The encouragement, advice and co-operation of the staff of Cambridge University Press has been greatly appreciated.

J. V. POSSINGHAM, H. HASHIMOTO AND
J. OROSS

Factors that influence plastid division in higher plants

Origin of plastids

The concept of plastid continuity was advanced in the nineteenth century by Schimper (1885); it proposed that plastids do not arise *de novo* but form by division of pre-existing plastids that are passed from parent cell to daughter cell during both vegetative and sexual reproduction. The currently available evidence supports this view that the division of pre-existing plastids is the main pathway by which these organelles are elaborated, and by analogy with cell division it may be appropriate now to refer to the actual event of plastid division as plastokinesis.

In the 1960s it was suggested that plastids might arise from small submicroscopic vesicles known as proplastid precursors (20–200 nm diameter) which were thought to be present in leaf buds and leaf meristems of a number of higher plants (Muhlethaler & Frey–Wyssling, 1959). The suggestion was also made that, in species such as ferns, vesicles budded from the nucleus could develop into plastids (Bell, Frey–Wyssling & Muhlethaler, 1966). Since that time little additional evidence has come forward in support of these alternative theories of plastid formation and they have largely been abandoned.

Mechanism of plastid division

It is now generally accepted that plastids arise by constriction division rather than by septation or *de novo* synthesis, and that the different plastid types such as proplastids, etioplasts, chloroplasts and even amyloplasts are capable of division (Chaley, Possingham & Thomson, 1980; Whatley, 1983; Lawrence & Possingham, 1984). Only limited information is available concerning the precise mechanism whereby plastids, especially the chloroplasts, of higher plants divide. In the case of lower plants consisting of a single cell layer, such as the green alga *Nitella axillaris* and the moss *Mnium cuspidatum*, cinematographic evidence showing the division of chloroplasts has been obtained (Green, 1964; Ueda, Tominga & Tanuma, 1970). Even in lower plants the mechanism whereby a single plastid undergoes fission into two daughter organelles is poorly understood with respect to the forces involved and the ways in which complex membranes break and reform.

Some understanding of the process of constriction division of plastids comes from the fluid mechanical models proposed by Greenspan (1977, 1978) for cytokinesis of cells. A possible sequence of events that might occur during plastid division based on this theory has been provided by Possingham & Lawrence (1983). Evidence supporting such a model is only circumstantial, and experiments are needed to determine if microdomains of lipids with variable affinities to solutes and water exist in the plastid envelope.

A number of specific structures are associated with dividing plastids, such as fuzzy plaques of electron-opaque material and electron-dense annuli in the neck region of constricting plastids (Chaly & Possingham, 1981; Leech, Thompson & Platt–Aloia, 1981) and microtubule-like structures at the ends of dividing chloroplasts (Lawrence & Possingham, 1984). Recent observation of the electron-dense rings in *Avena* indicate that there are two separate structures either side of the plastid envelope (Hashimoto, 1986). Additionally Hashimoto observed that these annuli are mainly present in plastids where the neck region has been constricted to a diameter of about 100 nm. This occurs during the final stages of plastid division.

Patterns of plastid division in leaves

The patterns of plastid division in leaves are inextricably linked with those for leaf formation and growth. These events vary between species but can be divided into two broad groups, the patterns in the leaves of dicotyledonous plants and those in monocotyledonous plants. Spinach leaves have been used in studies of plastid formation in dicotyledons (reviewed by Possingham & Lawrence, 1983), while the patterns in monocotyledons have been studied in wheat and more recently in oats (Boffey *et al.*, 1979; Boffey & Leech, 1982; Hashimoto, 1986). The sequence of events was described first for the expanding cells of leaves as both the leaf cells and plastids of exposed leaves are amenable to viewing by light microscopy. Some observations are also available concerning plastid formation in dividing leaf cells mainly from dicotyledons where cell division can continue for an extended period in leaf bases.

A complex series of changes occurs in both plastid growth and division during the development of leaves. It is against this background that the effects of environmental factors that modify plastid division are to be considered.

Plastid division in expanding leaves

There is a relatively large change in plastid number per cell between the cells of young leaves and those in fully expanded leaves. In spinach the change can be 10 fold from 10–15 plastids per cell in young (10–20 mm) leaves to 150–200 plastids in fully expanded (100 mm) leaves (Fig. 1) (Possingham & Saurer, 1969). In wheat

Fig. 1. Changes in DNA amount per chloroplast (g × 10⁻¹⁵), chloroplasts per cell, and chloroplast DNA (cpDNA) amounts per cell in mesophyll cells during spinach leaf development. DNA amounts per chloroplast are the mean of 120 measurements, and were converted to absolute terms using *Pediococcus damnosus* as a biological standard. Chloroplast numbers are the mean of 20 measurements. Plastome copy numbers were derived from DNA amounts assuming a plastome size of 1.6 × 10⁻¹⁶ g. From Lawrence & Possingham, 1986*b*.

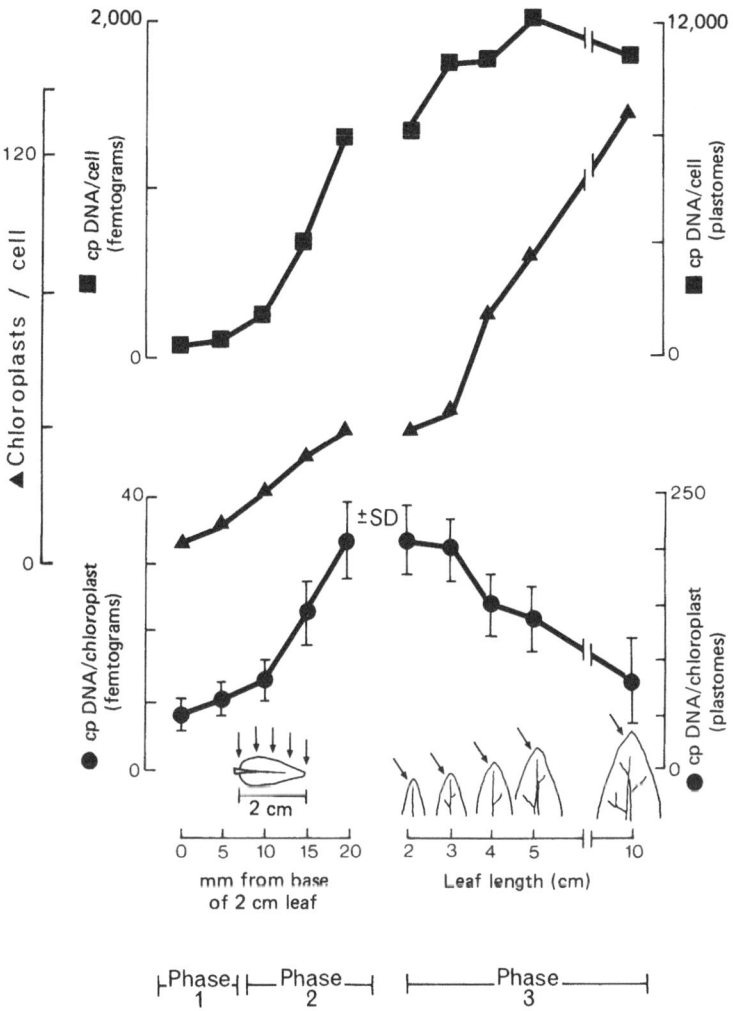

and oats the change is from 45–60 plastids per cell in young leaves with dividing cells to approximately 160 per cell in expanded leaves (Boffey *et al.*, 1979; Hashimoto, 1986). Data for oats are given in Fig. 2.

In expanding leaf cells the number of chloroplasts formed under normal growing conditions is very large. For example it takes approximately 5–6 days for a spinach leaf to grow from 20 mm to 100 mm. Over this period leaf cell number (mesophyll and palisade) changes from about 2×10^6 to 20×10^6 with chloroplast number per leaf changing from 50×10^6 to 250×10^6 over the same period. This represents an

Fig. 2. Changes in plastid number per cell (●), frequency of the dumb-bell-shaped plastids (○), cpDNA per plastid (■) and mitotic index (▲) against cellular age in the light-grown 4-day-old first leaves of *Avena*. Plastid number per cell was obtained from 150–200 protoplasts. Frequency of dumb-bell-shaped plastids was determined with 150–200 plastids from tissue homogenate. Mitotic index was determined with about 200 protoplasts. Plastid DNA amounts (g × 10^{15}) were measured by microspectrofluorometry after DAPI staining and converted to absolute values using *Pediococcus damnosus* as an internal standard. Adapted from Hashimoto, 1986.

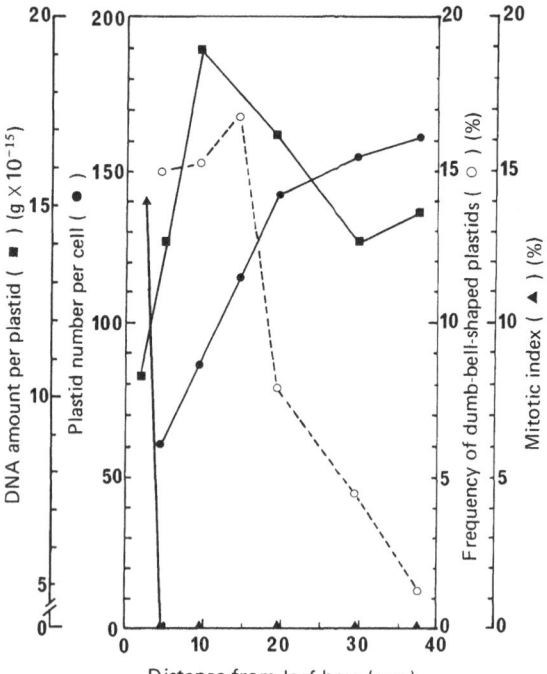

average rate of chloroplast formation of about 50×10^6 per day per leaf. Data showing changes of similar magnitude have been calculated for tobacco (Possingham, 1980).

The even distribution of [^{3}H]–thymidine between all chloroplasts in pulse chase experiments in expanding leaf cells of spinach is evidence that all the chloroplasts divide until the full complement of chloroplasts has been generated (Rose, Cran & Possingham, 1974), This suggests that during leaf growth a doubling of chloroplast numbers occurs at the time of the last division cycle. There is however a marked gradient in cell maturity between the tip and base of leaves so that chloroplast formation continues to occur at the leaf base at a time when it has ceased at the leaf tip (Saurer & Possingham, 1970; Possingham & Smith, 1972; Scott & Possingham, 1983).

In both wheat and oats there is about a fourfold increase in plastid numbers per cell between young leaves (45–60) and the cells of fully expanded leaves (150–160). Although some of this increase results from the division of proplastids, Boffey *et al.* (1979) reported that the majority of dividing plastids contain chlorophyll, have a minimal thylakoid membrane system and can be described as young chloroplasts. Based on the observation of division profiles in chloroplasts of increasing size, they also suggested that the majority of chloroplasts divide rather than a small subpopulation, a result in agreement with data for spinach.

Many if not all of the factors that influence cell expansion and final cell size probably exert some influence on final chloroplast numbers. There appear to be wide differences between species in the proportion of leaf cell surfaces that are actually covered by chloroplasts (Honda *et al.*, 1971). Free-hand cut sections of leaves of a range of species, be they annual, perennial, deciduous or evergreen, monocot or dicot, show wide differences in the proportion of cellular surfaces occupied by chloroplasts. Some cells appear full of chloroplasts while others are literally empty. Thus the feedback controls between cell size and chloroplast number appear to have a series of different end-points in different species and different cells.

Plastid division in dividing leaf cells

A knowledge of how leaves grow in cellular terms is required to understand the extent and timing of plastid division in leaves. The older simplistic concept of leaves as a two-phase system, one dominated by division and the other by expansion, is no longer tenable for most dicotyledons (Sunderland, 1960; Dale, 1976). Rather the young leaf can be viewed as a continuum of cells, varying from those in division, through to those actively expanding, to those that have completed this stage of development. In spinach new cells can be formed until the leaf is approximately half its final area (Saurer & Possingham, 1970). In many dicotyledons the majority of leaf cells are formed from divisions which take place in emerged leaves and only a small proportion come from cells formed in the apex. Thus both dividing and expanding leaf cells of dicotyledons the increases in chloroplasts are mainly from the division of chlorophyll-containing plastids (Possingham, 1980).

We have counted the numbers of chloroplasts present in cells with mitotic figures in young spinach leaves. A range of chloroplast numbers from 10 to 20 per cell was found in cells undergoing mitosis. It was not possible to conclude that chloroplast numbers double before leaf cells divide, but by the time the mitotic cycle was complete the new cells had accumulated the resting stage number of about 10 per cell. We concluded that in this plant, chloroplast division occurs both prior to and during mitosis so that the minimal level of chloroplast numbers per cell is maintained (Possingham & Lawrence, 1983; Lawrence & Possingham, 1986a,b).

A number of studies of both algae and higher plants indicate that chloroplasts tend to partition evenly to either side of the cell at mitosis (Hennis & Birky, 1984). No evidence is available concerning the mechanism involved in this partitioning in higher plants, but it appears to be highly efficient, as aplastidic cells are rarely seen in leaves. Contributing factors to this efficiency are the central position of the nucleus of meristematic cells which occupies a significant portion of the cell volume, and the formation of the metaphase plate which fills the central position in the cell thus restricting the plastids to the two ends.

Plastid division and nuclear DNA

The relationship between the levels of cell ploidy, nuclear DNA content and plastid number per cell has been well documented by Butterfass (1979). This relationship has been found to apply over a wide range of experimental knowledge indicating that the growth, development and division of plastids is nuclear controlled. It is now clear that many of the proteins involved in photosynthesis are totally or partially coded for in the nucleus, synthesised on 80S cytoplasmic ribosomes and equipped with leader sequences that enable them to be transported into plastids. There are estimates suggesting that over 70 per cent of the polypeptides of plastids are formed in this way, with a much smaller percentage being synthesised within plastids (Douce & Joyard, 1984).

In a contrary way there is good evidence to support the view that the simple proplastids which lack 70S ribosomes and are found in mutant plants such as those of 'Albostrians' barley, 'Iojap' maize and 'Lady Parker' pelargonium are formed from components that are all nuclear coded. The exception, and the only plastid-formed component, is chloroplast DNA which is replicated by an imported polymerase of nuclear origin and which passes from plastid to daughter plastid at division (Scott, Cain & Possingham, 1982). In experiments with 'Albostrians' barley it was found that both chloroplast DNA (cpDNA) levels and plastid numbers per cell in the white leaves of mutants were similar to the values measured in green areas of mutant leaves or in the leaves of control plants.

The *E.coli* chromosome is known to contain several genes which are involved in cell division (Lutkenhaus, Wolf–Watz & Donachie, 1980). These genes have been identified by temperature-sensitive mutations that cause inhibition of septum formation at high temperature with the result that long non-septate filaments form (van de Putte, Van Dillewijn & Rörsch, 1964). Yeast also are known to contain cell division genes (Hindley & Phear, 1984). It is highly probable that a series of genes also control the division of plastids. However the evidence referred to above suggests that such genes are almost certainly located in the nucleus of plant cells. Relevant here are the observations that inhibitors of 80S ribosome synthesis have a more marked effect on plastid division than inhibitors of plastid protein synthesis (Leonard & Rose, 1979).

Plastid division and chloroplast DNA

The relationship between cpDNA levels and plastid division was first investigated in expanding leaves, as methods capable of giving reliable data involved either measurement of isolated plastid preparations or the estimation of plastid and nuclear DNA levels in tissue in which cell and chloroplast number could be measured. Data from measurements of expanding leaves gave the unexpected result that plastid division can take place in the absence of cpDNA synthesis. Essentially similar trends were observed in pea, spinach, beet and wheat using a range of methodologies (Lamppa, Elliot & Bendich, 1980; Scott & Possingham, 1980, 1983; Boffey & Leech, 1982; Tymms, Scott & Possingham, 1983).

This limitation to whole-tissue samples has been removed by the development of a quantitative microspectrofluorometric method to measure plastid DNA levels of individual chloroplasts in either dividing or expanding cells (Lawrence & Possingham, 1986*a*,*b*). Fig. 3 provides a summary of observations of series of spinach leaves of differing age, covering the range from dividing through to fully expanded cells. These data together with those from the earlier studies indicate that, developmentally, spinach chloroplasts traverse a cycle with respect to their level of cpDNA (Scott & Possingham, 1983) (see Fig. 1). The chloroplasts of dividing leaf cells in young leaves and at the base of 20–mm leaves have a relatively low level of cpDNA. This level is maintained, in the dividing chloroplasts of dividing cells, suggesting there is active cpDNA synthesis prior to and during cell division. There was no evidence of a separate subpopulation of chloroplasts with higher (doubled) level of cpDNA than the average, although there were populations of cells with 2C and 4C nuclei (Fig.4). The data support the earlier finding that all the plastids of young leaves synthesise cpDNA and divide (Rose *et al.*, 1974). The individual cpDNA values for chloroplasts from each tissue fit a normal distribution curve and can be regarded as a single population. The absolute values are a consequence of increases due to cpDNA synthesis and reductions due to chloroplast divisions. It is

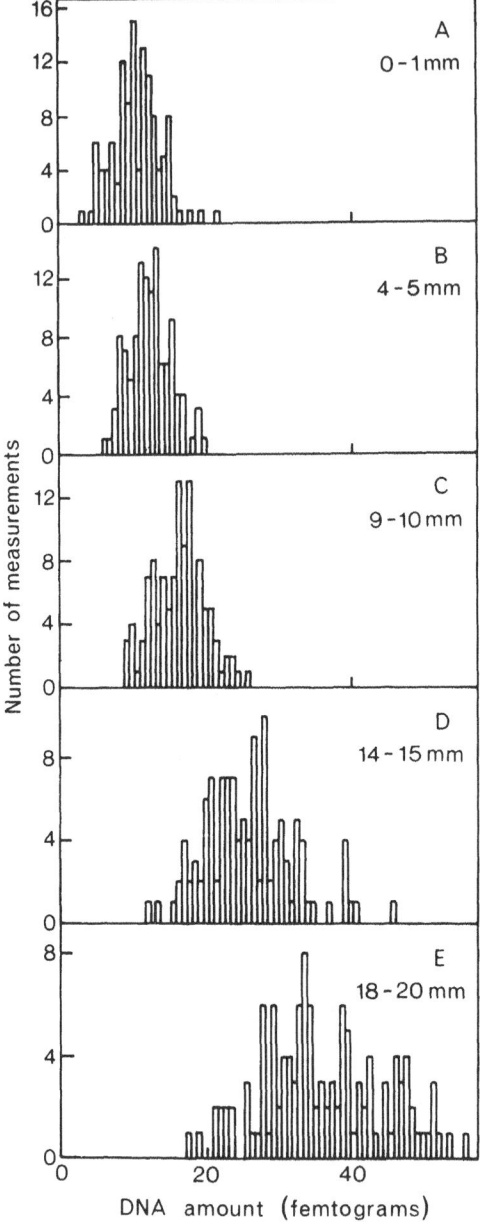

envisaged that chloroplast divisions occur at random both within the population and with respect to time.

There is a second phase of cpDNA synthesis in 20–mm leaves during the early stages of cell expansion when cpDNA levels per chloroplast double. This is followed by the final phase when chloroplast division takes place in the absence of cpDNA synthesis (Lawrence & Possingham, 1986*a,b*). A similar pattern has recently been recorded for oats in the levels of cpDNA per plastid. The level was found to be low in dividing cells, increased in the chloroplasts of partially expanded cells, and was reduced in the chloroplasts of fully expanded cells following chloroplast division (H. Hashimoto & J.V. Possingham, unpublished data).

These data could be interpreted to suggest that the level of cpDNA limits division in meristematic cells. However, even in dividing cells, the plastids are highly polyploid with respect to their number of copies of cpDNA (over 50 per plastid) and it is difficult to envisage a genome copy number of this order limiting plastid division. A similar level of cpDNA is found in the chloroplasts of mature spinach leaves as in

Fig. 3. (*Left*)Distributions of DNA amounts of 120 chloroplasts from mesophyll cells sampled at each of five positions along a spinach leaf 2 cm long. Relative DNA amounts were measured by microspectrofluorometry after DAPI staining, and were converted to absolute values using *Pediococcus damnosus* as an internal standard. The position of each sample, cut as 1–mm-wide strip, is indicated on the histograms. From Lawrence & Possingham, 1986*b*.

Fig. 4. Distributions of DNA amounts of 90 nuclei from mesophyll cells in the basal 1 mm of a spinach leaf 2 cm long. Relative DNA amounts were measured by Feulgen microdensitometry, and were converted to absolute values using *Pisum sativum* root tip nuclei as an internal standard. From Lawrence & Possingham, 1986*b*.

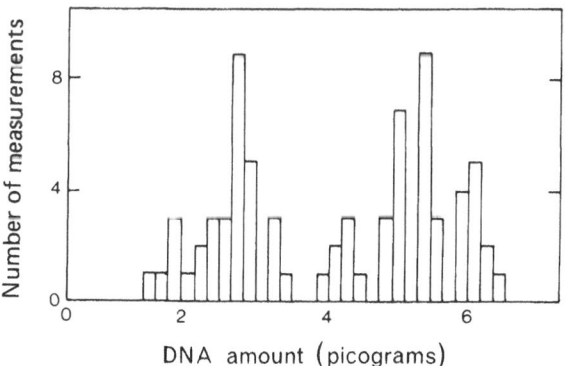

dividing cells. Currently there is no unequivocal explanation for the high levels of cpDNA in plastids. It has been suggested that the multiple copies represent a form of gene amplification which enables plastid ribosomal RNA molecules to be rapidly synthesised (Scott & Possingham, 1983).

The level of cpDNA in the plastids of epidermal cells in spinach does not undergo the same cycle as that of mesophyll chloroplasts and remains constant at the meristematic level. In fact each of the different tissues of spinach, beet and potato plants contains its own characteristic level of cpDNA. It has been suggested that the ratio of plastid to nuclear DNA may be involved in events associated with cellular differentiation (Possingham et al., 1984).

Homologies between nuclear and plastid DNA

It is now well established that in spinach the chloroplast DNA genome is repeated as short (2–3kbp) sequences within nuclear DNA (nDNA), and it has been estimated that there are about 4–5 copies of cpDNA per 2C nucleus (Scott & Timmis, 1984). Further work on the nature of these homologies by N.S. Scott & W.Y. Cheung (personal communication) has shown them to be highly complex as individual clones of nuclear DNA can have homology with up to three separated areas of the cpDNA genome. Furthermore a single chloroplast gene such as that which codes for the large subunit of Rubisco is subdivided into three separate regions, during movement to the nucleus, as portions of this gene have been found to hybridise to three different nuclear clones.

These data provide clear evidence that there has been movement of cpDNA to the nucleus of spinach during evolution. This may largely be due to nuclei having the capacity to take up and integrate 'foreign' DNA. Currently work is in progress to see if there are repeating border fragments at the junction of cpDNA and nDNA within nuclei which may provide a basis for 'recent' movement. These homologies may contribute towards the interactions that exist between the chloroplasts and nucleus of higher plants.

Light and plastid division

Light is known to induce the movement of chloroplasts within the cells of many plants (Haupt, 1982). As well, light exerts a strong influence on the growth and development of chloroplasts, particularly in those species that form proplastids or etioplasts during leaf development (Bradbeer et al., 1974; Lichtenthaler & Buschmann, 1978). However, both proplastids and etioplasts can divide in roots, shoot apices and buried cotyledons in conditions where they receive little or no light. For example bean seedlings grown for 14 days in darkness show an 11-fold increase in cell number and a 26-fold increase in plastid numbers (Bradbeer et al., 1974).

Light is known to stimulate the division of plastids in both spinach and tobacco leaf discs which have been pregrown in darkness (Boasson & Laetsch, 1969); Possingham & Smith, 1972). High light intensity (300 μmol m^{-2} s^{-1}) photosynthetically active radiation (PAR) is required to achieve maximal rates of plastid division even in situations where carbohydrates are supplied as an energy source to leaf discs. It is possible that some of the high light effects on plastid division are indirect and relate to their effects on leaf growth (Dale, 1976; Dale & Murray, 1986).

The division of chloroplasts of spinach leaf discs can be partially synchronised by transferring them to high intensity white light after growing them for 4–6 days either in darkness or in low intensity (70 μmol m^{-2} s^{-1} 480–580 nm or PAR) green or white light (Possingham *et al.*, 1975). In either low intensity light or in darkness up to 50 per cent of the plastids can be in dumb-bell configurations and about half of these separate within 12–24 hours of exposure to high intensity white light (Chaly *et al.*, 1980; Possingham *et al.*, 1975). Light is not seen as a direct trigger for chloroplast division even in this situation, and the electron transport inhibitor 3–(3,4–dichlorophenyl)–1, 1–dimethyl urea (DCMU) prevents the light-stimulated division of the large chloroplasts of spinach leaf discs grown in green light (Possingham, 1976). It seems likely that the division of chloroplasts, in contrast to that of proplastids, may require a supply of high energy compounds (ATP) from either *in situ* photosynthesis or from mitochondrial respiration.

There are a number of reports of light of different wavelengths enhancing plastid division (Possingham, 1980). For example the chloroplasts of cultured spinach discs divide equally well when illuminated either with high intensity light from lasers of 488 and 633 nm or with (PAR) white light (Possingham, 1973). As well, the division of greening bean plastids can be enhanced by red light and inhibited by subsequent far red light treatment, implicating the phytochrome system (Bradbeer *et al.*, 1974). In these experiments the red light effect on plastid division may have been coupled with the change in cell size.

Mohr (1984) has proposed that in higher plants both phytochrome and protochlorophyll play a role in plastidogenesis and, by inference, plastokinesis. Protochlorophyll does not act as a sensor pigment, rather its light dependent reduction leads to the formation of chlorophyll, an essential constituent of thylakoid membranes. On the other hand, phytochrome is regarded as a sensor pigment capable of mediating the expression of nuclear genes such as those responsible for the synthesis of the small sub unit of Rubisco and the light-harvesting chlorophyll-a/b-binding protein of photosystem II (LHCP). In experiments with mustard (*Sinapsis alba*), phytochrome has been shown to be necessary for the appearance of translatable mRNA for the nuclear-encoded small sub-unit (SSU) of Rubisco (Oelmüller & Mohr, 1986; Oelmüller *et al.*, 1986).

Diurnal effects on plastid division

It might be expected that plants grown on normal light (14 h)–dark (10 h) cycles would have a diurnal pattern of growth, and that this may influence plastid division. In preliminary experiments we have scored isolated chloroplast preparations from the central area of 20–mm leaves for the relative frequency of spherical, ovoid and dumb-bell-shaped or constricted chloroplasts at approximately 3–hour intervals throughout both the light and dark period (J. Oross & J.V. Possingham, unpublished data). At each harvest we found about 80–90 per cent of spherical chloroplasts, about 6 per cent of ovoid types and 5 per cent constricting chloroplasts. However, the percentage of constricted chloroplasts doubled in discs harvested during the early dark period. This may be due to a slowing down of the separation phase of constriction in the dark. Collectively the data suggest that chloroplast division occurs at a relatively uniform rate throughout the 24–hour period and that there was no clear evidence of a marked diurnal rhythm. A further interpretation of these data based on the frequencies of each chloroplast type could be that chloroplasts of these leaves spend about 21 h in the spherical form, and 1.4 and 1.2 h in the ovoid and constriction phases respectively. These interpretations assume a chloroplast doubling time of between 24 and 30 h (Possingham, 1976).

Temperature effects

Limited data are available concerning the effects of temperature on chloroplast division. When spinach leaf discs were cultured at temperatures ranging from 12 to 35 °C, chloroplast number per cell and chloroplast size varied inversely. At low temperatures a reduced number of large chloroplasts were formed, while high temperatures led to the formation of larger numbers of small chloroplasts relative to values in control discs grown at optimal temperatures (Possingham & Smith, 1972)

The activity of a number of chloroplast enzymes and the formation of plastid ribosomal RNA are both known to be temperature sensitive (Feierabend, 1976; Smillie, 1976). In the case of rye, Feierabend has shown that bleached proplastid-like organelles are formed when plants are grown at high temperatures. His results indicate that although high temperature inhibits plastid ribosome formation it does not inhibit plastid division or cpDNA synthesis.

Mineral nutrition effects

Leaf yellowing and chlorosis are features of plants grown in media that are deficient in virtually any of the essential nutrient elements. A number of observations have been made of the ultrastructure of the chloroplasts of yellow mineral-deficient leaves (Vesk, Possingham & Mercer, 1966; Whatley, 1971; Fido et al., 1977). A deficiency of manganese, iron and nitrogen, nutrient elements known to be chloroplast constituents, markedly reduced the numbers of chloroplasts per cell and produced yellowing of the leaves (Possingham, 1970). Recently further observations have been made of the chloroplasts of yellow nitrogen-deficient spinach leaves (Scott & Possingham, 1983). The chloroplasts of these were reduced in size and number per cell compared with control leaves which had larger cells and higher total levels of DNA per cell. The percentage of cpDNA was similar in both groups of plants. Almost by definition the essential nutrient elements influence the overall growth process of plants including both cell division and cell expansion. Accordingly it is

Fig. 5. Characteristics of leaf discs and of discs cultured in liquid nutrient media with gentle agitation for 6 days in a growth cabinet with 14–h day of 300 μmol m^{-2} s^{-1} PAR and 27 °C, and 10-h night of 25 °C. Nuclear DNA per disc (ng) obtained by measuring total DNA using Burton reaction and subtracting cpDNA measured by dot blotting. Cells, for measurement of cell number per disc × 10^3, cell area μm^2 × 10^{-2}, plastid number per cell, separated in EDTA. Symbols at day 0; leaf base (○) and leaf tip (●), at day 6: control leaf base (▲) and leaf tip (■), calcium-deficient leaf base (△) and leaf tip (□).

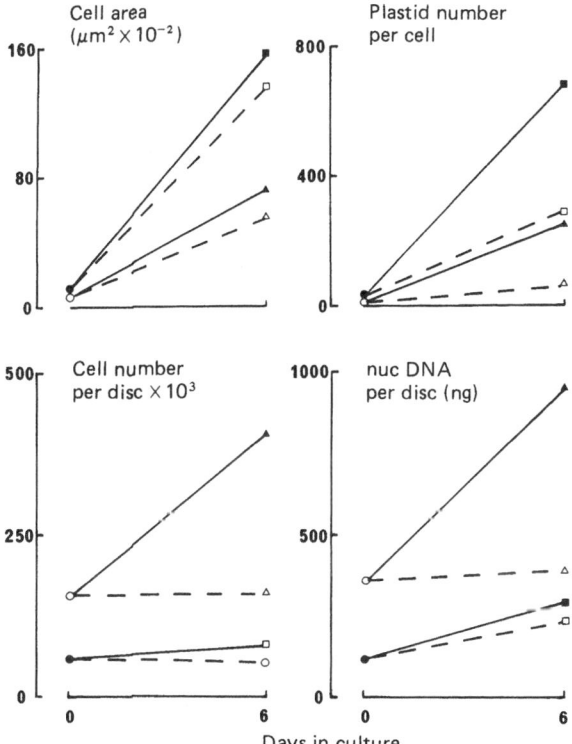

difficult to establish whether the essential mineral nutrients have direct effects on plastid division.

Cultured leaf discs have been used to study the influences of potassium and sodium on the growth of spinach and sugar-beet (Marschner & Possingham, 1975). Increasing the level of either of these nutrients increased cell size and chloroplast number per cell. The effect was more pronounced in beet than in spinach.

We have begun to examine the role of calcium in the nutrition of spinach leaf discs, cultured in shaking liquid media as agar media contain high levels of calcium. Some preliminary results are given in Figs. 5 and 6 which indicate that when leaf discs from the basal end of 20–mm leaves are grown in liquid media for a period of 6 days they enlarge as cell division continues for an extended period compared to discs cultivated on agar. There is a rise in nuclear DNA level in these discs corresponding to the increase in cell number. Chloroplast numbers per cell are similar to the values for leaf cells and to the levels in cells of discs cultured on agar (Tymms, Scott & Possingham, 1982). Where discs from the basal area of leaves were cultured on a liquid medium lacking calcium there was no increase in either cell number or DNA level once the discs were cut from the leaves. Chloroplast formation was reduced in cells of these discs.

On the other hand the cells of discs from the distal end of 20–mm leaves underwent considerable enlargement without forming any additional cells when in liquid culture media. These cells contained high numbers of chloroplasts (up to 700 per cell) even when compared to fully matured leaf cells, and they doubled their nuclear DNA levels. They were on average tetraploid but the values for individual mesophyll cells may have been higher. Distal leaf discs when cultured in media lacking calcium also increased in size by cell enlargement but formed no new cells or nuclear DNA. The cells of these discs formed additional chloroplasts in proportion to their increase in volume. The continued growth of these discs was probably based on the re-utilisation of calcium stored within and between the disc cells.

The use of liquid media led to massive increases in chloroplast number per cell in discs cultured in control media, a result that suggests the possibility that the growth of cells of discs cultured on agar may be limited by nutrient supply. It seems that chloroplast division is restrained in intact leaves compared with the levels in discs cultured in liquid media, probably because nuclear ploidy levels are lower in leaves.

Again it is difficult to separate direct from indirect effects, but it seems that calcium is required for both the division and expansion of leaf cells and for nuclear DNA synthesis; calcium may also be required for chloroplast division.

Conclusions

There are a number of endogenous and external factors which appear to affect the division of plastids. However, as so little is known about the precise

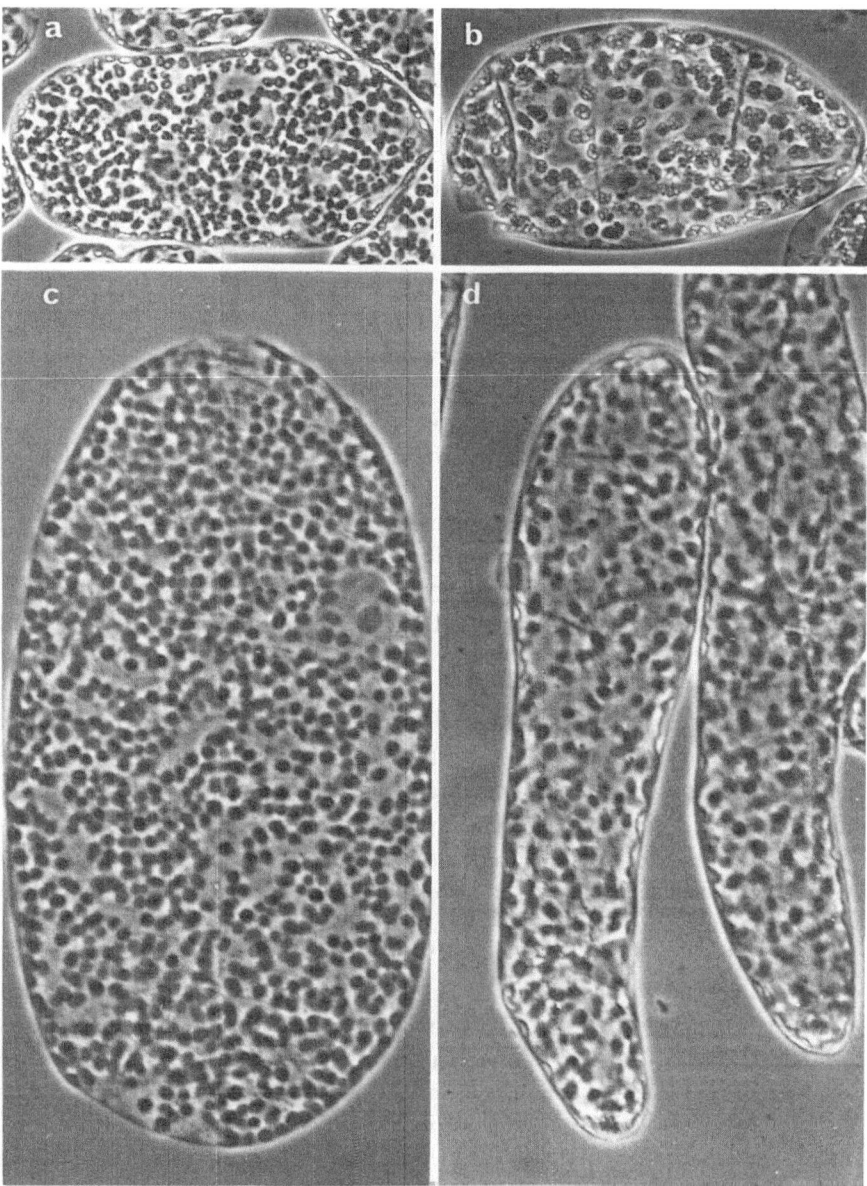

Fig. 6. Features of separated cells from discs cultured in control and calcium-deficient liquid media for 6 days in a growth cabinet. (a) Cell from leaf base disc cultured in control media; (b) cell from leaf base disc cultured in calcium-deficient media; (c) cell from disc from leaf tip cultured in control media; (d) cell from disc from leaf tip cultured in calcium-deficient media. All cells × 475.

features and mechanics of plastid division it is difficult to decide whether a number of the environmental factors which appear to exert effects on plastid division are influencing this process indirectly. Both the intensity and the quality of light affects plastid division. As with many plant processes, light has various effects: on the one hand, light appears to be required for the division of chloroplasts, but it is not essential for the division of proplastids in roots or for the division of etioplasts in the leaves of monocotyledons. It is difficult to construct an hypothesis to accommodate these widely divergent observations. Light is of course required to drive photosynthesis and carbon assimilation and so is necessary for the synthesis of chloroplast components. However it is not absolutely clear whether it is an essential trigger for the division of chloroplasts.

There is now evidence that the phytochrome system is able to modify nuclear transcription rates and so influence the synthesis rate of chloroplast constituents that are nuclear coded. Recent work suggests that a signal from plastids is also essential for the expression of nuclear genes involved in plastidogenesis (Oelmüller et al., 1986). Phytochrome has also been implicated in the marked changes that occur in chloroplast ultrastructure and physiology when some species are grown either in direct sunlight or in shaded conditions (Lichtenthaler & Buschmann, 1978). Of the external factors, light, both its intensity and quality, has the largest influence on plastid formation.

Further work is required to establish whether plastid division has either a diurnal or a circadian rhythm. It seems clear that the chloroplasts of young leaves divide at random with respect both to time and to each other so that rates of division are similar throughout the 24-hour cycle. Some synchronisation of plastid division can be brought about in the expanding cells of spinach leaf discs using low and high light intensity treatments in sequence. It is possible that these treatments in combination with low temperature may enable a closer study of the separate stages of the plastid division cycle.

There are a range of endogenous factors which operate to maintain plastid division on a fixed course and prevent it from being easily altered by external factors. Plastid DNA level is one such factor as there is a fixed and reproducible level of cpDNA per plastid in each tissue type and in each species. Although plastids are always highly polyploid and have multiple copies of their genome it is probable that there is a cpDNA level for each plastid type as a prerequisite for division. There is evidence that initially cpDNA is continuously synthesised and then halved by division so that individual values in the chloroplast population follow a normal distribution.

Nuclear DNA level almost certainly has a controlling influence on both final cell size and plastid division. Of interest here are preliminary results we have obtained for leaf discs cultured in liquid media. Particularly high rates of plastid division have been recorded under these conditions, possibly due to improved nutrition at the cell level. In these experiments calcium deficiency exerted a dramatic effect on both cell division and DNA synthesis, and in turn reduced plastid division.

Some of the present difficulty in thinking about the controls to plastid division comes from the notion that plastids have a high degree of autonomy. When it became clear that plastids have their own unique DNA and RNA, the term 'semi-autonomous' organelle was adopted. This line of thinking was reinforced by the advent of the endosymbiont theory to explain the origin of plastids and by the success of plant biochemists at inducing chloroplasts to carry out parts of the photosynthetic process *in vitro*. These findings led to the dream that chloroplasts might be cultured outside the cell, induced to divide and coaxed into performing photosynthesis *in vitro*. Because it is now clear that many components of plastids are nuclear coded, synthesised in the cytoplasm and transported into plastids, it is suggested that chloroplasts, and more so other plastid forms, should be thought of as integral parts of the cell in which they reside rather than as autonomous or semi-autonomous organelles.

Finally it is highly likely that the plant nucleus codes for plastid division genes. Such division genes are known for *E. coli*, *B. subtilis* and yeast, but as yet have not been isolated from plants. How such genes exert their effect is not known, but it is tempting to ascribe to them a direct role in the process of constriction division which is central to plastid formation.

References

Bell, P.R., Frey–Wyssling, A. & Muhlethaler, K. (1966). Evidence for the discontinuity of plastids in the sexual reproduction of a plant. *Journal of Ultrastructure Research*, **15**, 108–21.

Boasson, R. & Laetsch, W.M. (1969). Chloroplast replication and growth in tobacco. *Science*, **166**, 749–51.

Boffey, S.A., Ellis, J.R., Sellden, G. & Leech, R.M. (1979). Chloroplast division and DNA synthesis in light-grown wheat leaves. *Plant Physiology*, **64**, 502–5.

Boffey, S.A. & Leech, R.M. (1982). Chloroplast DNA levels and the control of chloroplast division in light-grown wheat leaves. *Plant Physiology*, **69**, 1387–91.

Bradbeer, J.W., Gyldenholm, A.O., Smith, J.W., Rest, J. & Edge, H.J.W (1974). Plastid development in the primary leaves of *Phaseolus vulgaris*. IX. The effect of short light treatments on plastid development. *New Phytologist*, **73**, 281–90.

Butterfass, Th. (1979). *Patterns of Chloroplast Reproduction*. Cell Biology Monographs VI. London & New York: Springer Verlag.

Chaly, N. & Possingham, J.V. (1981). Structure of constricted proplastids in meristematic plant tissues. *Biologie Cellulaire*, 203–10.

Chaly, N., Possingham, J.V. & Thomson, W.W. (1980). Chloroplast division in spinach leaves examined by scanning electron microscopy and freeze-etching. *Journal of Cell Science*, **46**, 87–96.

Dale, J.E. (1976). Cell division in leaves. In *Cell Division in Higher Plants*, ed.M.M.Yeoman, pp.315–45 London, Academic Press.

Dale, J.E. & Murray, D. (1968). Photomorphogenesis, photosynthesis, and early growth of primary leaves of *Phaseolus vulgaris. Annals of Botany*, **32**, 767–80.

Douce, R. & Joyard, J. (1984). The regulatory role of the plastid envelope during development. In *Chloroplast Biogenesis*, ed. N.R.Baker & J. Barber, pp.71–132. Amsterdam : Elsevier.

Feierabend, J. (1976). Temperature–sensitivity of chloroplast ribosome formation in higher plants. In *Genetics and Biogenesis of Chloroplasts and Mitochondria*, ed. Th. Bücher, W.Neupert, W.Sebald & S.Werner (eds)., pp. 103–10. Amsterdam : Elsevier, North Holland.

Fido, R.J., Gundry, C.S., Hewitt, E.J. & Notton, B.A. (1977). Ultrastructural features of molybdenum deficiency and whiptail of cauliflower leaves: effects of nitrogen source and tungsten substitution for molybdenum. *Australian Journal of Plant Physiology*, **4**, 675–89.

Green, P.B. (1964). Cinematic observations on the growth of chloroplasts in *Nitella. American Journal of Botany*, **51**, 334–42.

Greenspan, H.P. (1977). On the dynamics of cell cleavage. *Journal of Theoretical Biology*, **65**,79–99.

Greenspan, H.P. (1978). On fluid mechanical simulation of cell division and movement. *Journal of Theoretical Biology*, **70**, 125–34.

Haupt, W. (1982). Light–mediated movement of chloroplasts. *Annual Review of Plant Physiology*, **33**, 205–33.

Hennis, A.S. & Birky, C.W., Jr. (1984). Stochastic partitioning of chloroplasts at cell division in the alga *Olisthodiscus*, and compensating control of chloroplast replication. *Journal of Cell Science*, **70**, 1–15.

Hindley, J. & Phear, G.A. (1984). Sequence of the cell division gene CDC–2 from Schizosaccharomyces–Pombe patterns of splicing and homology to protein kinases. *Gene*, **31**, 129–34.

Honda, S.I., Hongladarom–Honda, T., Kwanyuen, P. & Wildman, S.G. (1971). Interpretations on chloroplast reproduction derived from correlations between cells and chloroplasts. *Plants*, **97**, 1–15.

Lamppa, G.K., Elliot, L.V. & Bendich, A.J. (1980). Changes in chloroplast number during pea leaf development. *Plant*, **148**, 437–41.

Lawrence, M.E. & Possingham, J.V. (1984). Observations of microtubule-like structures within spinach plastids. *Biologie Cellulaire*, **52**, 77–82.

Lawrence, M.E. & Possingham, J.V. (1986a). Direct measurement of femtogram amounts of DNA in cells and chloroplasts by quantitative microspectrofluorometry. *Journal of Histochemistry and Cytochemistry*, **34**, 761–8.

Lawrence, M.E. & Possingham, J.V. (1986b). Microspectrofluorometric measurement of chloroplast DNA in dividing and expanding leaf cells of *Spinacia oleracea. Plant Physiology*, **81**, 708–10.

Leech, R.M., Thomson, W.W. & Platt–Aloia, K.A. (1981). Observations on the mechanism of chloroplast division in higher plants. *New Phytologist*, **87**, 1–9.

Leonard, J.M. & Rose, R.J. (1979). Sensitivity of chloroplast division cycle to chloramphenicol and cycloheximide in cultured spinach leaves. *Plant Science Letters*, **14**, 159–67.

Lichtenthaler, H. & Buschmann, C. (1978). Control of chloroplast development by red light, blue light and phytohormones. In *Chloroplast Development*, ed. G. Akoyunaglau & J.H. Argyroudi–Akoyunoglou (eds)., pp. 801–15. Amsterdam : Elsevier/North Holland

Lutkenhaus, J.F., Wolf–Watz, H. & Donachie, W.D. (1980). Organisation of genes in the ftsA–envA region of the *Escherichia coli* genetic map and identification of a new fts.locus : fts 2. *Journal of Bacteriology*, **142**, 615–20.

Marschner, H. & Possingham, J.V. (1975). Effect of K$^+$ and Na$^+$ on growth of leaf discs of sugar beet and spinach. *Zeitschrift für Pflanzenphysiologie*, **75**, 6–16.

Mohr, H. (1984). Phytochrome and chloroplast development. In *Chloroplast Biogenesis*, ed. N.R. Barber, pp,305–347. Amsterdam : Elsevier.

Muhlethaler, K. & Frey–Wyssling, A. (1959). Entwicklung and Struktur der proplastiden. *Journal of Biophysical and Biochemical Cytology*, **6**, 507–12.

Oelmüller, R., Levitan, I., Bergfield, R., Rajasekhar, V.K. & Mohr, H. (1986). Expression of nuclear genes as affected by treatments acting on the plastids. *Planta*, **168**, 482–92.

Oelmüller, R. & Mohr, H. (1986). Photooxidative destruction of chloroplasts and its consequences for expression of nuclear genes. *Planta*, **167**, 106–3.

Possingham, J.V. (1970). Some effects of mineral nutrient deficiencies on the chloroplasts of higher plants. Presented at Sixth International Colloquium on Plant Analysis and Fertilizer Problems, (ISHS), Tel Aviv, Israel, pp. 155–65.

Possingham, J.V. (1973). Effect of light quality on chloroplast replication in spinach. *Journal of Experimental Botany*, **24**, 1247–60.

Possingham, J.V. (1976). Controls to chloroplast division in higher plants. *Journal de Microscopie et de Biologie Cellulaire*, **25**, 283–8.

Possingham, J.V. (1980). Plastid replication and development in the life cycle of higher plants. *Annual Review of Plant Physiology*, **31**, 113–29.

Possingham, J.V., Cran, D.G., Rose, R.J. & Loveys, B.R. (1975). Effects of green light on the chloroplasts of spinach leaf discs. *Journal of Experimental Botany*, **26**, 33–42.

Possingham, J.V. & Lawrence, M.E. (1983). Controls to plastid division. *International Review of Cytology*, **84**, 1–56.

Possingham, J.V. & Saurer, W. (1969). Changes in chloroplast number per cell during leaf development in spinach. *Planta*, **86**, 186–94.

Possingham, J.V., Scott, N.S., Tymms, M. & Lawrence, M. (1984). Quantitative relationships between nuclear and chloroplast DNA. In *Molecular Form and Function of the Plant Genome*, Abstracts, NATO Advanced Studies Institute/FEBS Advanced Course, Renesse, The Netherlands, p. B11.

Possingham, J.V. & Smith, J.W. (1972). Factors affecting chloroplast replication in spinach. *Journal of Experimental Botany*, **23**, 1050–9.

Rose, R.J., Cran, D.G. & Possingham, J.V. (1974). Distribution of DNA in dividing spinach chloroplasts. *Nature*, **251**, 641–2.

Saurer, W. & Possingham, J.V. (1970). Studies on the growth of spinach leaves (*Spinacea oleracea*). *Journal of Experimental Botany*, **21**, 151–8.

Schimper, A.F.W. (1885). Untersuchungen über die Chlorophyllkorper und die ihnen Homologen Gebilde. *Jahrbuecher für Wissenschaftliche Botanik*, **16**, 1–247.

Scott, N.S., Cain, P. & Possingham, J.V. (1982). Plastid DNA levels in albino and green leaves of the 'Albostrians' mutant of *Hordeum vulgare*. *Zeitschrift für Pflanzenphyiologie*, **108**, 187–91.

Scott, N.S. & Possingham, J.V. (1980). Chloroplast DNA in expanding spinach leaves. *Journal of Experimental Botany*, **31**, 1081–92.

Scott, N.S. & Possingham, J.V. (1983). Changes in chloroplast DNA levels during growth of spinach leaves. *Journal of Experimental Botany*, **34**, 1756–67.

Scott, N.S. & Timmis, J.N. (1984). Homologies between nuclear and plastid DNA in spinach. *Theoretical and Applied Genetics*, **67**, 279–88.

Smillie, R.M. (1976). Temperature control of chloroplast development. In *Genetics and Biogenetics of Chloroplasts and Mitochondria*, ed. Th. Bücher et al., pp. 103–10. Amsterdam : Elsevier/North Holland.

Sunderland, N. (1960). Cell division and expansion in the growth of the leaf. *Journal of Experimental Botany*, **11**, 68–80.

Tymms, M.J., Scott , N.S. & Possingham, J.V. (1982). Chloroplast and nuclear DNA content of cultured spinach leaf discs. *Journal of Experimental Botany*, **33**, 831–7.

Tymms, M.J., Scott, N.S. & Possingham, J.V. (1983). DNA content of *Beta vulgaris* chloroplasts during leaf cell expansion. *Plant Physiology*, **71**, 785–8.

Ueda, R., Tominga, S. & Tanuma, T. (1970). Cinematographic observations on the chloroplast division in *Mnium* leaf cells. *Science Report of the Tokyo Daigaku, Section B*, **13**, 129–37.

van de Putte, P.J., van Dillewijn, J. & Rörsch, A. (1964). The selection of mutants of *Escherichia coli* with impaired cell division at elevated temperature. *Mutation Research*, **1**, 121–8.

Vesk, M., Possingham, J.V. & Mercer, F.V. (1966). The effect of mineral nutrient deficiencies on the structure of the leaf cells of tomato, spinach and maize. *Australian Journal of Botany*, **14**, 1–18.

Whatley, J.M. (1971). Ultrastructural changes in chloroplasts of *Phaseolus vulgaris* during development under conditions of nutritional deficiency. *New Phytologist*, **70**, 725–42.

Whatley, J.M. (1983). The ultrastructure of plastids in roots. *International Review of Cytology*, **85**, 175–220.

TH. BUTTERFASS

Nuclear control of plastid division

If we look at a landscape, green may be the dominating colour. It depends on the continuity of plastids. Life on earth depends on plastid divisions occurring in step with cell divisions. The result of plastid division, namely, the number of plastids per cell, is not only interesting in itself but also provides the most convenient character for the study of the cytonuclear ratio (see monograph by Butterfass, 1979).

The course of events from induction to division of plastids is not established. One approach might be to study various correlations and then discuss the most probable direction of these processes. I shall adopt this approach although it is a dangerous one: what appears correlated need not be connected causally, and if, as often is the case, distributions are not normal along both coordinates, coefficients may be misleading.

In the present paper, I shall generally restrict myself to those effects which may be the results of the nuclear DNA (nDNA) amount. Genetic effects are also present, of course; they will be looked at only in passing.

Dimensions of study

Several dimensions of study are obvious.

(i) In meristems, continuity is required and ensured. In daughter cells, stochastic variations of plastid numbers call for readjustment. Higher plants lack a mechanism present in lower plants for orderly partitioning of plastids to daughter cells.

(ii) Maturing cells with increasing cell specificity may or may not, according to their function, build up larger populations of chloroplasts. This increase is a feature of development and is adaptive to function, not to preservation. The cytonuclear ratio need not, and in fact does not, remain constant while cells specialise. If chloroplast replication is not accompanied by an increase of nDNA amount (Butterfass, 1965; Cattolico, 1978; Boffey et al., 1979; Boffey & Leech, 1982; Ellis, Jellings & Leech, 1983) it poses further questions and is not considered here, although most chloroplasts arise in such differentiating cells (Possingham, 1980; Whatley, 1980).

(iii) Many cells of many taxa may become endopolyploid. Though differentation would proceed normally without any cell becoming endopolyploid, endopolyploidisation if it occurs may be adaptive. Quantitative responses are to be expected from the cell.

(iv) Related individuals or taxa may differ in ploidy. Again, quantitative responses are to be expected. The members of the group for comparison are not now related ontogenetically.

(v) Diploid taxa may differ in post-mitotic (2C) nDNA contents and for this reason also in plastid numbers.

(vi) Genetic differences, as opposed to quantitative differences in the ploidy level, may also affect plastid numbers.

In our attempts to distinguish the phenomena we must not forget that the cell is a cybernetic system showing feedback reactions. If a developing cell grows and produces more chloroplasts, these will trap more energy for cell growth; as a result new chloroplasts might be formed, and so on, until the entire system reaches an endpoint (see below). A question may be: how important in this system is the nDNA content or the cell size? Almost certainly both are involved, at least in maturing cells. The answer for meristematic cells may be different.

Proplastids are difficult to count. The scarce data show that there may be about 5–20 proplastids in a meristematic cell (Cran & Possingham, 1972; Butterfass, 1979; Whatley, 1980; Scott & Possingham, 1983; Lawrence & Possingham, 1986), i.e. pre-replication numbers may be about 5–10. This is just the range of chloroplast numbers in mature guard cells of stomata as found in about 80 per cent of all diploid species (Butterfass, 1979). As in meristematic cells, in guard cells the low number may be favoured by stabilising selection from both ends of the distribution (Mayr, 1982; see below). Like meristematic cells, guard cells do not become endopolyploid. They have arisen from meristemoids by the last mitoses occurring in a leaf. The numbers of chloroplasts in guard cells may reflect, therefore, the numbers of proplastids in post-mitotic meristematic cells and may be taken, with due caution, as a model for them. A correlation, however, has not yet been demonstrated.

Basic evidence
More nuclear DNA results in more plastids

Figs. 1 and 2 show the formal relationship of the chloroplast numbers of guard cells between diploid and induced tetraploid strains. (The terms 'diploid', 2x, and 'tetraploid', 4x, are used loosely in this paper, meaning any two levels of ploidy in which the second one is marked by having twice the number of chromosomes as the first one.) Fig. 3 compares the chloroplast numbers of diploid and of tetraploid species of the same genus from nature. Figs. 1–3 show that increases of ploidy result in increases of chloroplast numbers per guard cell. Whereas the average increases were the same in induced autotetraploids and allopolyploids, the slope of the regression line for allopolyploidy was significantly less steep. This finding has not yet been studied in detail. The difference of average changes shown by induced versus naturally polyploids will be discussed below (Table 1). Endopolyploidy is

Fig. 1. The correlation between chloroplast numbers in guard cells of induced autotetraploids (other than the colchicinated (C_0) generation) and their diploid ancestors. 115 entries. Own data and data from literature. Coefficient of correlation (r) = 0.95.

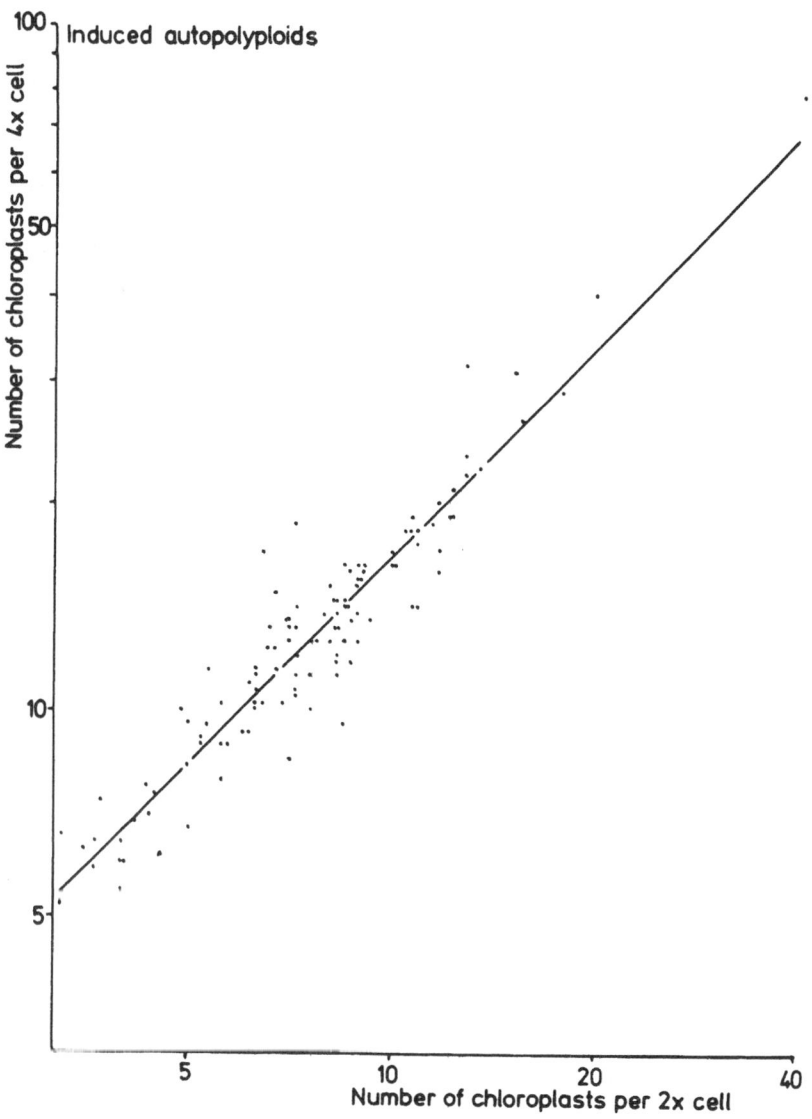

'young' autopolyploidy, and it increases the chloroplast numbers to the same extent as induced polyploidy (Butterfass, 1979; Scott & Possingham, 1983). Likewise, decreases of ploidy result in decreases of chloroplast numbers. This occurs, e.g. in haploids (Butterfass, 1979). There is also evidence that cells in G_2 (pre-mitotic) phase can entertain more chloroplasts than can cells in G_1 (post-mitotic) phase (Butterfass, 1979). A taxon with a higher nDNA content in diploid G_1 nuclei is also, on average, accompanied by a higher number of chloroplasts in guard cells (Butterfass, 1983).

Most of these observations hold if all other things are equal – genotype, growing conditions, cell type, cell age. But they may even hold in other comparisons, e.g. of gametophytes with sporophytes. A case in point is the hornwort, *Anthoceros*. It contains one chloroplast in each gametophytic cell, but two chloroplasts or one deeply lobed chloroplast in sporophytic cells (Paton, 1979). As a rule, however, the relationship is restricted to more similar cells.

Among diploid taxa, there are also positive regressions of chloroplast numbers in guard cells against the absolute amounts of nDNA. They are significantly different from zero for all taxa investigated (Butterfass, 1983). A role of the nucleotype

Fig. 2. Like Fig. 1, but induced allopolyploids of all generations. 31 entries. $r = 0.92$.

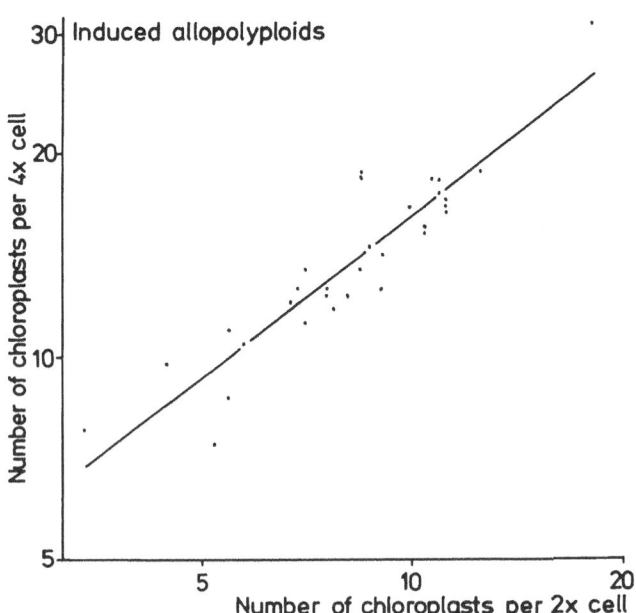

(Bennett, 1972) is indicated because non-repetitive and repetitive nDNA alike appear to be involved in chloroplast division (Butterfass, 1983). Other explanations cannot so far be excluded.

More plastids may result in various ways

If all other things are equal, the duplication of the nDNA amount is always accompanied by an increase in the number of plastids. The reverse, however, is not true (Butterfass, 1965, 1979). There are several prepatterns of chloroplast numbers per cell among the tissues of, for example, a leaf; only one of them is based on endopolyploidy if present, the others are not (Butterfass, 1979). If, however, the numbers of chloroplasts vary to an unusual extent among the cells of one tissue it may be promising to look for endopolyploidy (Butterfass, 1979).

Fig. 3. Like Fig. 1, but species pairs of different ploidies within one genus each, taken from nature. 825 available pairs pertaining to 100 genera were reduced by random selection to one entry per genus. $r = 0.84$.

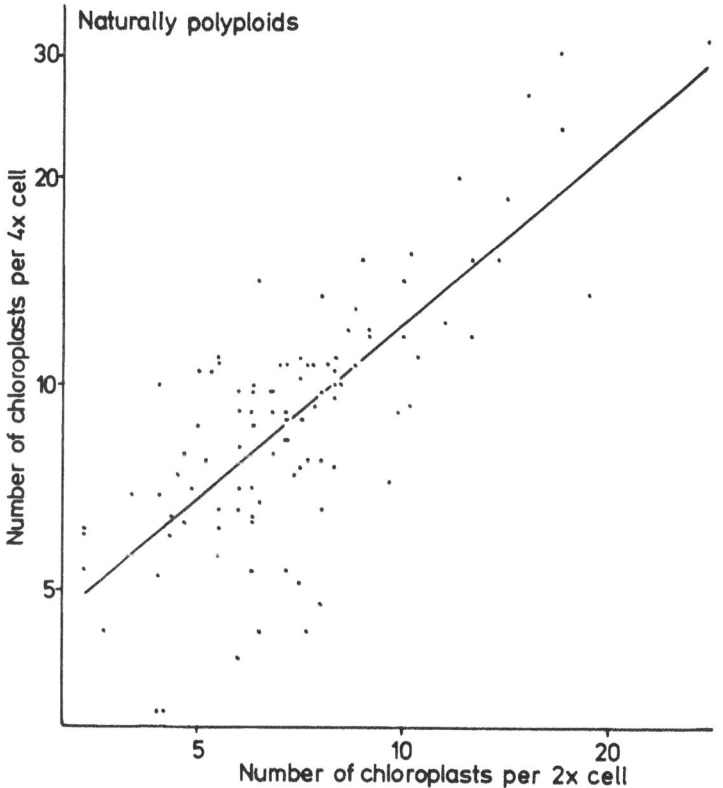

In diploid guard cells of stomata, chloroplasts occur in taxon-specific numbers even if the nDNA amounts are the same; this can be seen in Fig. 1 in Butterfass (1983). Different extra chromosomes introduced into diploid sugar-beet plants produce different changes of chloroplast numbers either by changing the level of endopolyploidy or by other mechanisms (Butterfass, 1979). These observations show that there are also genetic influences working on the division of plastids.

Sometimes the number of plastids is modified by changes of plastid size (Butterfass, 1979; Possingham, 1980). The chloroplasts are more numerous, but smaller, in sun leaf cells than in shade leaf cells, the nDNA content remaining the same. The mesophyll cells of sugar-beet trisomics-IV (type 5 of the new counting; Romagosa et al., 1986) showed a level of endopolyploidy twice that of the control; however, the number of chloroplasts remained the same, but their volume was almost doubled (Butterfass, 1979). Small cells of the diatom *Eunotia pectinalis* var. polyplastidica may contain four or eight plastids. When there are only four plastids in a cell, they are twice as large (Geitler, 1973). There are many more examples (Butterfass, 1979). During sporogenesis of mosses, but not of liverworts, the chloroplasts behave in a peculiar manner. A monoplastidic stage is invariably found, although many moss species are polyploid in nature (review in Butterfass, 1979). This case may be similar to that of naturally hexaploid *Selaginella martensii* showing only one large chloroplast in young cells, as all *Selaginella* species do. Presumably the chloroplast size increases instead of the chloroplast number. It is not known whether the single chloroplast of a young tetraploid spore mother cell of a polyploid moss is larger than that of a diploid one. The volume of all plastids taken together may be more important than the number. We understand next to nothing of the control of plastid size as part of the story. The chloroplast DNA amount may be involved. The genetic influence is obvious in spongy parenchyma cells of trisomic sugar-beet where at least seven of the nine chromosomes, if present as single extra chromosomes, result in a change in size of the chloroplasts (Butterfass, 1979).

The central hypothesis
The continuity of plastids

In mature cells, the duplication of the amount of nDNA always results in a marked increase in the number of chloroplasts. In meristematic cells, the fluctuations of nDNA between the single (2C) and the double (4C) amount may govern the replication of proplastids in the same way. Among the various controls of plastid division, so the hypothesis goes, control by the nDNA amount is predominant in meristems, leading there, on average, to an exact duplication of proplastid numbers, thus ensuring the continuity of plastids (Butterfass, 1963, 1979).

Is this a reasonable assumption? Meristematic cells are different from other cells in many characters. The proplastids are colourless and therefore of restricted use, whatever their other functions may be. They are not distinctly accumulated at the cell surface. The cells are small and lack a central vacuole. Intracellular spaces are absent, so the oxygen supply must be small and the carbon dioxide accumulation high. The nucleus occupies a large proportion of the cell volume. Because it loses its membrane periodically, the flow of materials within the cell is made easier, and the cytonuclear ratio may be more evident.

On the other hand, this mechanism would have been tailored exclusively for meristematic cells. What happens after ploidy has changed would be an inevitable consequence of a mechanism which is fundamental to dividing cells. It is not switched off in other cells of higher plants because there is no selective pressure for that. This is the reason, the hypothesis implies, why polyploid cells contain more chloroplasts than do diploid cells. Granted the principle, the plant may further benefit from it, increasing the chloroplast numbers per cell in some species and tissues by means of endopolyploidy. The nuclear control is relaxed in this situation because exact duplication of plastid numbers is no longer favoured; it would seem improbable that a mechanism selected for in meristems should work with the same efficiency in various different environments. In fact, chloroplast numbers in tetraploid guard cells are increased by only 70 per cent, on average, whereas the cell volume duplicates (Table 1). It is certainly more expensive for a cell to multiply its chloroplasts than to become larger.

A mechanism such as that outlined, which ensured proplastid division in step with nDNA replication, would have been favoured to such an extent that it would have become dominant long ago. I cannot imagine a better way of ensuring the continuity of plastids, and know of no data which would contradict the idea.

Deductions
If the principle outlined above is real, there are some interesting corollaries.

(i)   The replication of proplastids must be completed not necessarily by the time of mitosis but only by the time of the next S phase. This would explain why plastids may divide before or after mitosis.

(ii) Mitochondria are expected to behave similarly to plastids. This has been found, as far as the scarce data show. A short review can be found in Butterfass (1979).

(iii) Intracellular symbiont division may be a suitable model for plastid division (Butterfass, 1979; Whatley, 1983). The *Chlorella* cells within the cells of *Hydra viridis* (McAuley, 1981) do not divide when the host cells do not; presumably the host cell nuclei remain in the G_1 phase. But after a regenerating stimulus (perhaps inducing host nuclei to pass into G_2 phase) the symbionts begin to divide. 'Symbiont and host cell mitosis become temporarily uncoupled during regeneration' (McAuley, 1981); but finally 'algal cell division comes into harmony with host cell growth – neither outstrips the other – and the association persists' (Muscatine & Pool, 1979)

– this is just the behaviour observed in chloroplasts. Vinblastine sulphate at concentrations inhibiting animal mitosis but not nDNA synthesis led to an increase in the number of symbiotic *Chlorella* in *Hydra* cells, possibly as a result of the *Hydra* cell's passing into G_2 phase. If supernumerary *Chlorella* were injected into green *Hydra* cells, the previous number per host cell was restored within a day (Muscatine & Neckelmann, 1981). If *Chlorella* is grown in isolation it reproduces 20 times as fast as if it is growing within *Hydra* cells (McAuley, 1981). Obviously the reproduction of the *Chlorella* cells is inhibited by the host. By physiological means it is easier to increase than to decrease the division of chloroplasts (Schlayer, 1971; Butterfass, 1979). Poor growth conditions may stimulate ploidy-independent plastid division (Butterfass, 1979), hence active inhibition may be one form of nuclear influence on plastids as it is on symbionts.

(iv)   During fertilisation of higher plants, the nDNA content is duplicated, but the sperm cells of many species do not contribute plastids (Sears, 1980; Schroeder & Hagemann, 1986). In the resulting zygote, therefore, the replication of organelles might begin at once, although the nucleus is still in G_1 phase. After the nDNA has also replicated, the number of plastids might, as judged from nDNA content alone, even rise to 3–4 times the content of the unfertilised egg cell. However, I know of no evidence for this.

Minimum numbers of plastids

The numbers of chloroplasts in guard cells of diploid angiosperms vary to a surprisingly small extent; there are usually between 5 and 10 chloroplasts per cell. These numbers may be interpreted as approximating the minimum needed for avoiding apoplastidy; environmental and stochastic variations would require, on average, numbers well above 1. On the other hand, high numbers of (heterotrophic) proplastids would involve great energetic expense. Stabilising selection toward both ends of the distribution can be assumed. On the other hand, *Fritillaria imperialis*, for example, contains 50–100 chloroplasts in its guard cells and presumably a high number of proplastids in meristematic cells as well. Are so many chloroplasts really needed in this case? It seems more likely that the number is an inevitable repercussion of the excessively high 2C nDNA content present. Thus, 'minimum' needs qualification.

If large meristematic cells (and polyploid cells are large cells) were to contain as few plastids as do smaller cells, the plastids would probably no longer be arranged in such a regular configuration. A regular arrangement, however, is assumed to be the main prerequisite for the non-random partitioning of proplastids to daughter cells (Butterfass, 1969; Birky & Skavaril, 1984). Larger cells would, therefore, require more plastids. Sizes and plastid numbers of meristematic cells cannot be separated completely, therefore larger cells have larger minimum numbers. This would also hold, of course, for large-celled *Fritillaria* (see above).

Haploid clones of *Pelargonium zonale* (Daker, 1966) and *Thuja gigantea* (Pohlheim, 1968) have been kept for many years. This shows that the security allowance of the proplastid numbers of the diploid parents was great enough to allow haploids to develop normally, at least under horticultural conditions, in spite of their reduced plastid numbers. On the other hand, haploid *Trifolium hybridum* containing many apoplastidic cells (Butterfass, 1979) could, for some reason, only be kept for a few weeks.

Experience shows (Butterfass, 1979) that plastids may divide before or after the cell does; in *Euglena* (Baker, 1926) the time varies with the growing conditions (reviews: Butterfass, 1979; Possingham & Lawrence, 1983). Hence, cell division cannot be decisive for plastid division. Rather, the synthesis of nDNA sets the stage for plastid division. The plastids must have been reproduced at the latest before the next synthesis of nDNA occurred as proposed above. Plastids might be expected to reproduce faster than a cell because they are less complex. Premitotic plastid division might even have been favoured because then the number of plastids can be smaller without the risk of apoplastidic cells arising. In fact, most chloroplasts of guard cells in most species are formed already within guard-cell mother cells. Even eukaryotic *Chlorella* cells growing symbiotically within *Hydra viridis* cells divide before the host cell does (McAuley, 1981). There are some low-plastidic species, however, such as *Trifolium pratense* and *T. hybridum*, which, contrary to expectation, apparently always undergo post-mitotic plastid division in guard cells. Obviously even the pre-replication proplastid number is high enough for preventing apoplastidy in these cells.

Are cell size and chloroplast size involved?

If the nDNA amount is decisive for plastid division as presumed above, the question arises as to how the influence is exerted. A direct way would be the dependence of plastid reproduction on cytoplasmic protein synthesis (Paolillo & Kass, 1977). Unequivocal evidence for this appears to be lacking. More indirect mechanisms would depend on cell size, or on the plastids themselves dividing when their own size arrives at a measure specific for the conditions prevailing. This measure itself might be under joint control of nucleus and environment.

Fig. 4 shows the narrow correlation between chloroplast numbers and guard cell volumes among diploid species. In Fig. 5, the correlation is plotted of 4x/2x ratios between chloroplast numbers and guard cell volumes of ploidy pairs from nature. In Fig. 6, ratios between induced tetraploids and their diploid ancestors are plotted; surprisingly, no correlation could be detected (see below).

In meristematic cells it would appear logical that cell volumes and proplastid numbers duplicate exactly. If in these cells the plastid division were mediated by cell size, cell volume rather than cell surface area would be the likely determinant. The

Fig. 4. The correlation between the number of chloroplasts and the volume of guard cells of diploid dicotyledonous species. Own measurements. 43 entries from 12 genera. $r = 0.78$. Relative volumes were approximated by calculating length × width^2 (Butterfass, 1987).

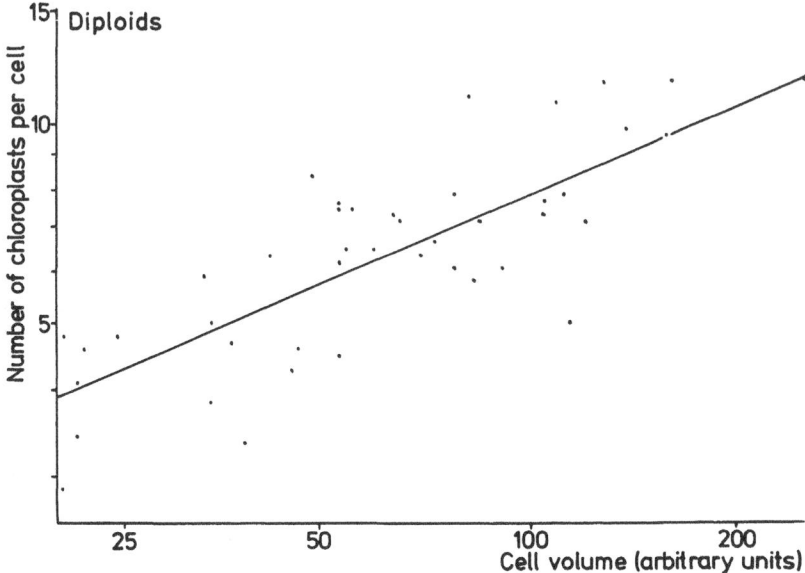

Fig. 5. The correlation between the 4x2x ratios of the numbers of chloroplasts and the volumes of guard cells of 33 species pairs from nature. Own measurements. $r = 0.81$. The ratios were calculated from relative volumes as explained for Fig. 4, all constants cancelling out (Butterfass, 1987).

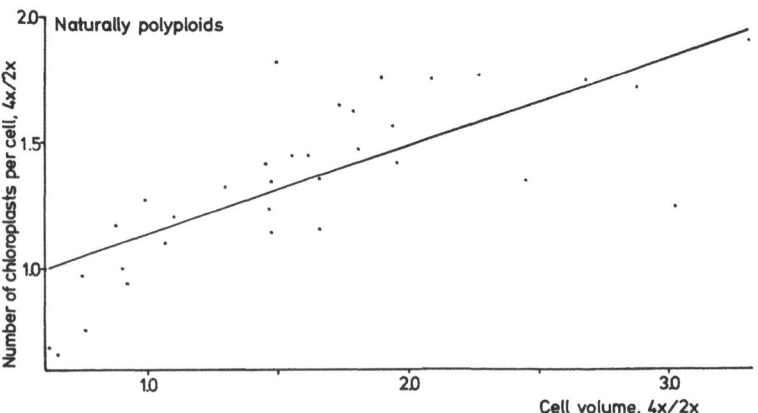

Table 1. *Relative effects of tetraploidy on guard cells:*
increases in per cent as compared with diploids

	Cell volume	Cell surface	Chloroplast number	nDNA amount
Induced	110[a]	65[b]	69	100
From nature	70	41	31	70[c]

[a]Medians, from Butterfass (1987).
[b]Calculated on the assumption of geometrical similarity.
[c]Median, calculated from nDNA data given by Bennett & Smith (1976) and Bennett, Smith & Heslop–Harrison (1982) for diploid and for tetraploid species of the same genera. $n = 39$; arithmetic mean, 67; mode, 90.

volume of guard cells of induced tetraploids was greater than that of diploids by 110 per cent (Table 1). Supposing geometrical similarity, this increase results in an increase of the surface area by 65 per cent. The 69 per cent increase found for chloroplast numbers in guard cells is not far from this, though it is just significantly higher ($n = 120$, $l.s.d._{0.05} = 4$ per cent). A volume increase by 70 per cent, as found in tetraploids from nature, corresponds to a surface increase of 41 per cent, yet the chloroplast numbers increase significantly less, namely, by 31 per cent only. Thus, surface area is not involved alone, if at all. As to cell volumes, evolutionary ageing of tetraploids reduced the increase of chloroplast numbers from 69 to 31 per cent, i. e. to

Fig. 6. Like Fig. 5, but using induced tetraploids (other than C_0 generation). 18 entries. $r = 0.14$; $r_{0.05,16}$ d.f. $= 0.47$: no correlation found.

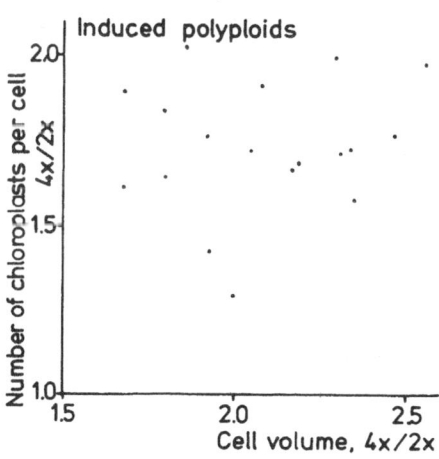

less than one half, and that of cell volumes from 110 to 70 per cent, i.e. to two-thirds. Water stress does not change the numbers of chloroplasts in guard cells to a large extent (Butterfass, 1965; Klimakhin & Firsov, 1968), but it may decrease cell sizes considerably (Zalenskij, 1904; Klimakhin & Firsov, 1968). The cell size is extremely susceptible to conditions of growth, height of leaf insertion, and to after-effects of colchicine treatment. Chloroplast numbers per cell are, by contrast, far more constant. Obviously, chloroplast number and cell size respond somewhat differently.

Among diploid species with different nDNA amounts, differences in chloroplast numbers of guard cells do not appear to be mediated mainly by cell size, either. The relative variabilities (i.e. the standard deviations divided by the arithmetic means) of cell volumes, nDNA contents and chloroplast numbers in mature diploid guard cells are significantly different, decreasing in the order given. Because different species were compared, the results might indicate that a stabilising selection for chloroplast numbers, presumably for minimum numbers, dominates the nDNA effect and restricts variation. Even the forms of variation are different, distributions of chloroplast numbers deviating less from normality than those of cell volumes and nDNA contents.

Cell size might well influence the course of events, but less so in meristematic cells and in guard cells than in mesophyll cells. It takes much less time for a meristematic cell or a guard cell to develop than for a mesophyll cell. In a mesophyll cell, as a rule, the chloroplasts divide further, whereas in a guard cell they do not, and in a meristematic cell the proplastids divide only in step with nDNA synthesis. Chloroplast number and cell surface co-develop in mesophyll cells, as found by Ellis & Leech (1985), but not in guard cells. Within sugar-beet populations of the same ploidy, chloroplast numbers and lengths of guard cells showed r (the coefficient of correlation) = 0.17 to 0.39 (Butterfass, 1979); accordingly, r^2 (the coefficient of determination) = 0.03 to 0.15, i.e. only 3–15 per cent of the total variance was common variance. Meristematic cells may behave similarly.

I see no reason for assuming that the size of a meristematic cell should be a main determinant of the number of proplastids; I would rather imagine that cell size and proplastid number are brought about independently by nDNA amount (cf. also Paolillo & Kass, 1977). Their subsequent interaction is inevitable because a character of any importance does not evolve in isolation. If one character (say, ploidy) is changed, the new 'target' for natural evolution may remain elusive for a long time because it shifts; greatly enlarged cells, as present in the beginning, will favour long-term responses of the entire plant that are different from responses to less enlarged cells as present many generations later. Cell sizes and plastid numbers cannot but develop and evolve together; the extent of their common change, however, may vary to a point where they are almost independent.

Evolutionary shifts

The narrow correlation between 4x/2x ratios of cell volumes and chloroplast numbers ($r = 0.81$) found in species pairs from nature (Fig. 5) has not been found with induced tetraploids (Fig. 6). This is all the more unexpected since an induced tetraploid is much more closely related to its diploid ancestor and reference strain than is a naturally tetraploid species to some diploid species of the same genus. The correlation as calculated ($r = 0.14$) is far from significant ($r_{0.05, 16 d.f.} = 0.47$) and is significantly smaller than that found for species pairs from nature (Fig. 5). Hence the difference seems to be real in spite of the scarcity of data. One explanation might be that both characters are co-adapting secondarily. Although both cell volume and chloroplast number increase with ploidy, they do not do so proportionately from the beginning. During the first generations after polyploidisation cell sizes may be excessively increased (Butterfass, 1987) while chloroplast numbers do not show such an effect. Chloroplast number and cell size appear to be independent in principle but become increasingly associated.

Table 1 summarises some results on evolutionary changes in these relationships.

(i) Both chloroplast numbers and cell sizes appear to have been reduced during evolution.

(ii) Cell volume and cell surface were less reduced in nature than was chloroplast number; thus, there was a trend toward a reduced density of chloroplasts in guard cells with evolutionary transformation of tetraploids. (Chloroplast areas, however, have not been measured.) The heritability of chloroplast numbers is higher than that of cell size; hence the selection for lower chloroplast numbers may have been more effective.

(iii) If the nDNA contents of tetraploids from nature are correct as published, nDNA and cell size changed to a similar extent during both tetraploidisation and subsequent evolution, but the chloroplast numbers did not.

I have assumed so far that the differences between induced tetraploids and tetraploids from nature are the result of evolution. This need not be so. In becoming tetraploid in nature, diploid species with a lower content of nDNA might have been favoured (Kostoff, Gorbatscheva & Dimitroff, 1943). If so, the differences within columns of Table 1 would not mark true changes, at least not to the extent given. Some arguments in favour of an evolutionary effect however, can be put forward, but none of them is stringent.

(i) An induced diploid moss, *Bryum corrensii*, reduced its cell volume but kept its ploidy during 11 years in culture (Wettstein, 1937; Wettstein & Straub, 1942; Butterfass, 1987).

(ii) In many species, diploid and tetraploid races exist side by side in nature (Lewis, 1979).

(iii) Induced tetraploids of *Arenaria serpyllifolia* had larger cells than had naturally tetraploids of the same species (Woess, 1941).

   (iv)   Cell size can be changed by artificial selection (e.g. Wilson & Cooper, 1970; Butterfass, 1972; Hanson, 1973; Bormotov & Smirnova, 1981).

   (v)   In diploid sugar-beet, three generations of sharp divergent selection for chloroplast numbers in guard cells were enough to produce strains with 5 or 10 instead of 7–8 chloroplasts (Butterfass, 1979). From reasons given above, a selective pressure is expected to work in nature toward minimum plastid numbers in meristematic cells and probably also in guard cells. Tetraploids also experience this pressure, and certainly they can yield to it.

The findings show that cell sizes and chloroplast numbers can be reduced. They also suggest that this actually happens in nature. Selective polyploidisation may also occur.

Evolutionary change need not be brought about by genetic means alone. There is no genuinely primary role of nDNA or of any other character. For genetic effects the nucleus must be stimulated by something, and even a nucleotypic effect would pose the question of why the nDNA content or the nuclear size is as it is. In a cell or a plant, there is a complex feedback system tending, as mentioned above, toward a steady state. The transitory increase of cell size in many induced tetraploids of the first generations (Butterfass, 1987) may just be an overshoot, i.e. a temporary excessive movement toward the final state (Bertalanffy, Beier & Laue, 1977). If so, haploids should also show an overshoot, but in the opposite direction. A change in this direction has really been found (Th. Butterfass, unpublished). The results show that at least the excessive increase of cell size in evolutionary young tetraploids need not be the unspecific result of an upsetting of the genetic equilibrium. Such a result might be directed equally in haploids as in tetraploids. An upset genetic equilibrium would be counteracted, if this is adaptive, by selection. An overshoot, in contrast, would normalise itself to some extent. Other interpretations, however, would be possible.

If the nDNA amount is suddenly changed, the complicated network of open systems as present in the plant is perturbed. The auto-adjustment of the integrated system needs much time. Nobody knows how much is 'much', but there are indications that dauermodifications, i.e. modifications sometimes surviving even generative reproduction but disappearing gradually (Jollos, 1939), may belong to this group of phenomena. If so, a time-scale measured in years would be appropriate. I feel that the times needed by a plant to cope with sudden far-reaching changes (such as the doubling of nDNA amount) in its highly complex open system have been underestimated so far in physiology, the more so as a cell may develop faster into its final condition than it can auto-adjust in response to the impulse. For an account of some of the difficulties encountered when dealing with processes on different time-scales, see Goodwin (1963).

Conclusion

This review of the relationships between nDNA amount and plastid division enables one to distinguish between several dimensions of study, namely, of meristematic cells, of differentiating and of mature cells, and the comparison of taxa with different nDNA amounts. Among vascular plants, more nDNA always involves more plastids per cell or, rarely, larger plastids. But an increase of plastid number need not indicate an increase of nDNA amount. The central hypothesis proposed postulates that the significance of the effect of nDNA content lies in meristems, ensuring there the continuity of plastids, and that this universal mechanism inevitably leads to similar side-effects in other cells and in polyploid plants. In meristematic cells and in guard cells, the primary course of events is not thought to include cell size. The evolution of tetraploid plants reduces cell size and chloroplast number to different extents and produces a new relationship between both characters. The stabilising selection toward both ends of the distribution of chloroplast numbers in guard cells tends to produce numbers as small as possible. This selection is thought to reflect the primary selection for continuity of proplastids in meristematic cells, and this continuity is what primarily counts.

Acknowledgements

My thanks are due to my colleagues, J. G. Th. Hermsen, Wageningen, Netherlands; W. Horn, Freising–Weihenstephan; A. R. Kranz, Frankfurt; R. Reimann–Philipp, Ahrensburg; G. Reuther, Geisenheim; G. Roebbelen, Goettingen; to Sueddeutsche Saatzucht– und Saatbaugenossenschaft, Waldbrunn–Oberdielbach, Nunhems Zaden, Haelen, Netherlands, and Mr H. Becela and Mr H. Grasmueck, Frankfurt a. M., for supplying seed or for other help. The reliable assistance of Mrs Ute Lehmann is gratefully acknowledged.

References

Baker, W.B. (1926). Studies in the life history of *Euglena*. I. *The Biological Bulletin*, **51**, 321–62.

Bennett, M.D. (1972). Nuclear DNA content and minimum generation time in herbaceous plants. *Proceedings of the Royal Society of London, Series B*, **181**, 109–35.

Bennett, M.D. & Smith, J.B. (1976). Nuclear DNA amounts in angiosperms. *Philosophical Transactions of the Royal Society of London, Series B*, **274**, 227–74.

Bennett, M.D., Smith, J.B., & Heslop–Harrison, J.S., (1982). Nuclear DNA amounts in angiosperms. *Proceedings of the Royal Society of London, Series B*, **216**, 179–99.

Bertalanffy, L.v., Beier, W. & Laue, R. (1977). *Biophysik des Fliessgleichgewichts*, 2nd edn. Braunschweig: Vieweg.

Birky, C.W., & Skavaril, R.V. (1984). Random partitioning of cytoplasmic organelles at cell division: the effect of organelle and cell volume. *Journal of Theoretical Biology*, **106**, 441–7.

Boffey, S.A., Ellis J.R., Selldén, G. & Leech, R.M. (1979). Chloroplast division and DNA synthesis in light-grown leaves. *Plant Physiology*, 64, 502–5.

Boffey, S.A., & Leech, R.M. (1982). Chloroplast DNA levels and the control of chloroplast division in light-grown leaves. *Plant Physiology*, 69, 1387–91.

Bormotov, V.E. & Smirnova, T.I. (1981). (The genetical nature of cell size in sugar beet.) In Russian. *Doklady Adademii Nauk Belorusskoj SSR*, 25, 446–8.

Butterfass, Th. (1963). Die Abhaengigkeit der Plastidenvermehrung von der Reproduktion der Erbsubstanz im Kern. *Berichte der Deutschen Botanischen Gesellschaft*, 76, 123–34.

Butterfass, Th. (1965). Verschiedenartige Ursachen der Plastidenvermehrung in verschiedenen Zellen. *Berichte der Deutschen Botanischen Gesellschaft*, 78, (105)–(110).

Butterfass, Th. (1969). Die Plastidenverteilung bei der Mitose der Schliesszellenmutterzellen von haploidem Schwedenklee (*Trifolium hybridum* L). *Planta*, 84, 230–4.

Butterfass, Th. (1972). Endopolyploidie und Ertrag bei diploiden und tetraploiden Zuckerrueben. III. *Theoretical and Applied Genetics*, 42, 41–3.

Butterfass, Th. (1979). *Patterns of Chloroplast Reproduction*. Cell Biology Monographs vol. 6. Vienna & New York: Springer.

Butterfass, Th. (1983). A nucleotypic control of chloroplast reproduction. *Protoplasma*, 118, 71–4.

Butterfass, Th. (1987). Cell volume ratios of natural and of induced tetraploid and diploid flowering plants. *Cytologia*, 52, 309–16.

Cattolico, R.A. (1978). Variation in plastid number. *Plant Physiology*, 62, 558–62.

Cran, D.G. & Possingham, J.V. (1972). Variation of plastid types in spinach. *Protoplasma*, 74, 345–56.

Daker, M.G. (1966). 'Kleine Liebling' a haploid cultivar of *Pelargonium*. *Nature* (London), 211, 549–50.

Ellis, J.R., Jellings, A.J. & Leech, R.M. (1983). Nuclear DNA content and the control of chloroplast replication in wheat leaves. *Planta*, 157, 376–80.

Ellis, J.R. & Leech, R.M. (1985). Cell size and chloroplast size in relation to chloroplast replication in light-grown wheat leaves. *Planta*, 165, 120–5.

Geitler, L. (1973). Bewegungs– und Teilungsverhalten der Chromatophoren von *Eunotia pectinalis* var. polyplastidica und anderer *Eunotia*-Arten bei der Zellteilung. *Oesterreichische Botanische Zeitschrift*, 122, 185–94.

Goodwin, B.C. (1963). *Temporal Organization in Cells*. London & New York: Academic Press.

Hanson, W.D. (1973). Changes in efficiencies and numbers of chloroplasts associated with divergent selections for juvenile productivity in *Zea mays* (L.). *Crop Science*, 13, 386–7.

Jollos, V. (1939). Grundbegriffe der Vererbungslehre, insbesondere Mutation, Dauermodifikation, Modifikation. *Handbuch der Vererbungslehre*, ed. E. Baur & M. Hartmann, vol. I D, III. Berlin: Borntraeger.

Klimakhin, G.I. & Firsov, I.P. (1968). (Properties of the stomatal apparatus of diploid, triploid and tetraploid beets.) In Russian. *Doklady Moskovskoj Sel'skokhozyajstvennoj Akademii Imena K. A. Timiryazeva*, 136, 55–9.

Kostoff, D., Gorbatscheva, A. & Dimitroff, P. (1943). Die Vergroesserung der Zellen in auto– und allopolyploiden Tabakpflanzen. *Zeitschrift fuer Pflanzenzuechtung*, 25, 112–16.

Lawrence, M.E. & Possingham, J.V. (1986). Microspectrofluorometric measurement of chloroplast DNA in dividing and expanding leaf cells of *Spinacia oleracea*. *Plant Physiology*, 81, 708–10.

Lewis, W.H. (1979). Polyploidy in species populations. In *Polyploidy, Biological Relevance*, ed. W.H. Lewis, pp. 103–4. New York & London: Plenum Press.

McAuley, P.J. (1981). Control of cell division of the intracellular *Chlorella* symbionts in green *Hydra. Journal of Cell Science*, **47**, 197–206.

Mayr, E. (1982). *The Growth of Biological Thought*. Cambridge, Mass., & London: Belknap Press of Harvard University Press.

Muscatine, L. & Neckelmann, N. (1981). Regulation of numbers of algae in the *Hydra–Chlorella* symbiosis. *Berichte der Deutschen Botanischen Gesellschaft*, **94**, 571–82.

Muscatine, L. & Pool, R.R. (1979). Regulation of numbers of intracellular algae. *Proceedings of the Royal Society of London, Series B*, **204**, 131–9.

Paolillo, D.J., Jr. & Kass, L.B. (1977). The relationship between cell size and chloroplast number in the spores of a moss, *Polytrichum. Journal of Experimental Botany*, **28**, 457–67.

Paton, J.A. (1979). *Anthoceros agrestis*, a new name for *A. punctatus* var. cavernosus sensu Prosk. 1958, non (Nees) Gottsche *et al. Journal of Bryology*, **10**, 257–61.

Pohlheim, F. (1968). *Thuja gigantea* gracilis Beissn. – ein Haplont unter den Gymnospermen. *Biologische Rundschau*, **6**, 84–6.

Possingham, J.V. (1980). Plastid replication and development in the life cycle of higher plants. *Annual Review of Plant Physiology*, **31** , 113–29.

Possingham, J.V. & Lawrence, M.E. (1983). Controls to plastid division. *International Review of Cytology*, **84**, 1–56.

Romagosa, I., Hecker, R.J., Tsuchiya, T. & Lasa, J.M. (1986). Primary trisomics in sugarbeet. I. *Crop Science*, **26**, 243–9.

Schlayer, G. (1971). Modifikationen des DNS-Gehalts in Zuckerruebenzellen. *Planta*, **98**, 294–9.

Schroeder, M.–B. & Hagemann, R. (1986). Ultrastructural studies on plastids of generative and vegetative cells in Liliaceae. 6. *Acta Botanica Neerlandica*, **35**, 243–8.

Scott, N.S. & Possingham, J.V. (1983). Changes in chloroplast DNA levels during growth of spinach leaves. *Journal of Experimental Botany*, **34**, 1756–67.

Sears, B.B. (1980). Elimination of plastids during spermatogenesis and fertilization in the plant kingdom. *Plasmid*, **4**, 233–55.

Wettstein, F.v. (1937). Experimentelle Untersuchungen zum Artbildungsproblem. I. *Zeitschrift fuer induktive Abstammungs- und Vererbungslehre*, **74**, 34–53.

Wettstein, F.v. & Straub, J. (1942). Experimentelle Untersuchungen zum Artbildungsproblem. III. *Zeitschrift fuer induktive Abstammungs- und Vererbungslehre*, **80**, 271 80.

Whatley, J.M. (1980). Plastid growth and division in *Phaseolus vulgaris. New Phytologist*, **86**, 1–16.

Whatley, J.M. (1983). Plastids – past, present, and future. *International Review of Cytology*, Supplement 14, pp. 329–73.

Wilson, D. & Cooper, J.P. (1970). Effect of selection for mesophyll cell size on growth and assimilation in *Lolium perenne* L. *New Phytologist*, **69**, 233–45.

Woess, F.v. (1941). Experimentelle Untersuchungen zum Artbildungsproblem an *Arenaria serpyllifolia* und *Arenaria Marschlinsii. Zeitschrift fuer induktive Abstammungs- und Vererbungslehre*, **79**, 444–72.

Zalenskij, V.R. (1904). (Contributions to the quantitative anatomy of different leaves of the same plants.) In Russian. *Izvestiya Kievskago Politekhnicheskago Instituta Imperatora Aleksandra II*, god IV, vol. 1. Otdelya estestvenno-istoricheskij i agronomicheskij. Kiev.

Woess, F.v. (1941). Experimentelle Untersuchungen zum Artbildungsproblem an *Arenaria serpyllifolia* und *Arenaria Marschlinsii*. *Zeitschrift fuer induktive Abstammungs- und Vererbungslehre*, **79**, 444–72.

Zalenskij, V.R. (1904). (Contributions to the quantitative anatomy of different leaves of the same plants.) In Russian. *Izvestiya Kievskago Politekhnicheskago Instituta Imperatora Aleksandra II*, god IV, vol. 1. Otdelya estestvenno-istoricheskij i agronomicheskij. Kiev.

R. M. LEECH AND K. A. PYKE

Chloroplast division in higher plants with particular reference to wheat

A major factor influencing the photosynthetic capacity of a leaf is the number of chloroplasts in its mesophyll cells. For a given leaf anatomy photosynthetic capacity will depend on the size, number and activity of the chloroplasts under unit leaf area (see Leech & Baker, 1983 for a review). Chloroplast number has been shown to be genetically determined and the mean chloroplast number per leaf mesophyll cell is remarkably constant within a species. Indeed, in the genus *Triticum* (wheat) the mean chloroplast number is a diagnostic characteristic for many species (Leech, 1986; Pyke & Leech, 1987). An understanding of the origins and control of the increase of chloroplasts during normal leaf development is critical to an understanding of the control of leaf function. It is now well established that up to 90 per cent of the chloroplasts in a mature leaf mesophyll cell are the products of the division of young green photosynthetically active chloroplasts and not of meristematic proplastids (see Possingham, 1980; Possingham & Lawrence, 1983 for a review). The replication of young chloroplasts in post-mitotic expanding leaf cells has been shown to occur in all species which have been appropriately examined and its great importance for the efficient functioning of the mature leaf cell is now recognised. The characteristics and control of this process of chloroplast replication which occurs universally in the post-meristematic cells of all young expanding leaves are the subjects of this chapter.

It is important to distinguish the process of chloroplast division from the quite distinct process of proplastid division which occurs in the leaf meristem (Whatley, 1979) and also the root meristem (Chaly & Possingham, 1981). In leaf meristematic cells plastid division keeps pace with, or only slightly exceeds, cell division. In contrast, in the expanding leaf cells, cell division has ceased and rapid chloroplast division makes a major contribution to the size of the mature chloroplast complement of each mesophyll cell. The young green chloroplasts which replicate in post-meristematic leaf cells are already 4.5–6.5 μm long (Boffey *et al.*, 1979) and have well developed grana (Possingham & Saurer, 1969; Leech, 1976; Chaly, Possingham & Thomson, 1980). Our own estimates for wheat show that the youngest leaf mesophyll cells, which have just ceased to divide, have between 10 and 15 plastids while the mature mesophyll cell of a typical hexaploid wheat leaf contains between 130 and 150 chloroplasts. All the cycles of chloroplast replication are completed in less than 12 hours and often all the chloroplasts of the cell divide synchronously. Millions of chloroplasts are produced within a few hours in every young expanding leaf.

Despite the apparent universality and importance of chloroplast division in expanding leaves, detailed observations of dividing chloroplasts have only been published relatively recently. Randolf (1922) provided the first clear evidence for chloroplast division by the dual demonstration of the presence of constricted chloroplasts in post-meristematic leaf cells and also showed that an increase in chloroplast number occurred in chronologically older leaf cells. It is to be regretted that Randolf's observations were not immediately confirmed, or referred to in subsequent reviews, but chloroplast division was 'rediscovered' in 1969 by Possingham & Saurer who showed the presence of constricted chloroplasts and a fivefold increase in chloroplast number in young, expanding spinach leaves. Shortly afterwards Ridley & Leech (1970) discovered that young green chloroplasts from spinach and broad-bean leaves could divide outside the cell, and similar observations with young isolated tobacco chloroplasts were made by Kameya & Takahashi (1971) and Kameya (1972) (see Leech, 1980 for a review). Several series of detailed light-microscope studies have since confirmed the correlation between the presence of dumb-bell-shaped chloroplasts and an increase in chloroplast number in expanding leaves of species of wheat (Boffey *et al.*, 1979), *Phaseolus* (Whatley, 1980) and spinach (Chaly *et al.*, 1980) and in cultured leaf discs from tobacco (Boasson & Laetsch, 1969) and spinach (Chaly *et al.*, 1980). Possingham & Lawrence (1983) have reviewed the evidence from several systems which supports the now-established conclusion that the size of the chloroplast population in a mature leaf cell is determined by the number of rounds of chloroplast division occurring during cell expansion in the developing leaf. Moreover, there is convincing evidence that all of the young chloroplasts in the developing leaf cell replicate, i.e. division is not confined to a subpopulation of plastids (Lawrence & Possingham, 1986). There is also no evidence supporting the suggestion that chloroplasts may develop from tiny initials, i.e. organelles beyond the resolving power of the light microscope: no such initials have ever been demonstrated in young leaf cells.

The wheat leaf as an experimental system for the study of chloroplast replication

Reviewers of the work published in the past twenty years on chloroplast division in higher plants have been able to draw few general conclusions about the characteristics and mechanisms of the chloroplast replication process in different plants. This is almost certainly because different groups of authors have chosen very diverse types and ages of leaf tissue for their studies and so rather few common characteristics of chloroplast replication have been recognised. However, there is no doubt that when genetically uniform plants are grown under carefully controlled and standardised conditions, both leaf and chloroplast development are remarkably stable and highly reproducible (Leech, 1986). Chloroplast development is known to be a very plastic process and responds rapidly to episodic environmental changes, and it might be expected that if a replicating chloroplast is subjected to relatively mild stress it will respond rapidly. Clearly when analysing a process such as chloroplast division in which normal control processes might be expected to be particularly sensitive and easily affected by stress conditions, it is very important that in the first instance observations of chloroplast replication should be undertaken in a leaf tissue which is not in any way stressed. In this paper we have therefore chosen to collect together the detailed observations made over several years in our laboratory in York on chloroplast replication in young naturally developing green wheat leaves. Leaf and chloroplast development in modern cultivars of wheat is very stable and highly reproducible (see Leech, 1984, 1985, 1986 for reviews). The young wheat leaf was chosen for our work for the following reasons.

(1) Chloroplast replication and cell division are separated in time; within a single leaf these two processes are separated in space.

(2) Wheat leaf tissue and its cells are amenable to detailed structural and biochemical analysis, i.e. their metabolic and structural components can be readily isolated in a relatively undamaged state and, in particular, intact highly active chloroplasts can be obtained.

(3 Wheat genetics is a very advanced field of study and many viable genetic lines are available in which the effect of changes in nuclear and/or cytoplasmic components can be studied.

(4) The wheat plant has been chosen as one of the species for intensive investigation by molecular biologists. As a result a great deal is known about the structure and function of the wheat nuclear and chloroplast genomes.

(5) The linear structure and growth characteristics of the leaf of wheat enable the development of its cells to be timed rather accurately so it is possible to calculate *in vivo* rates of synthesis and degradation relatively unambiguously for proteins and many other components.

For these reasons young leaves of wheat provide an excellent system in which to study leaf and chloroplast development. Cell division is confined to the basal intercalary meristem, and the young leaf consists of a linear gradient of sequentially older cells with the oldest cells near the tip and the youngest nearest the base of the leaf (Boffey *et al.*, 1979; Boffey, Sellden & Leech, 1980). Plastid development occurs within the developing leaf mesophyll cells so a linear gradient of plastid development is also present. Chloroplast division is confined to a discrete band of cells separated from the basal meristem by a 'band' of developing cells in which no plastid division occurs. This leaf growth pattern is also found in several genera of the Graminae including *Aegilops, Avena, Digitaria, Lolium, Hordeum, Triticale, Triticum, Zea* and several wild grass species.

Evidence that chloroplasts divide in the young leaf

Unambiguous evidence that chloroplast division occurs in leaf cells requires the demonstration of both

(1)   the presence of profiles of chloroplasts in 'the dividing state', i.e. 'dumb-bells', and

(2)   an increase in chloroplast number per cell.

(1) By careful observation of slices of young wheat leaf tissue in the light microscope it is possible to pin-point the area of the leaf in which the maximum proportion of dumb-bell-shaped chloroplasts are observed: it is in a band of tissue 1.5–2.5 cm from the leaf base. As many as 10 per cent of the chloroplasts in the cells of this region are dumb-bell-shaped and elongated in outline. In some cells all the chloroplasts are dumb-bell-shaped, indicating a high degree of intracellular synchronisation of the division process. The arrangement in a 7-day-old wheat leaf of the hexaploid *Triticum aestivum* cv. Maris Dove is shown in Fig. 1. The morphological characteristics of the dividing chloroplasts can be most clearly observed in Nomarski differential interference optics and are seen particularly clearly if micrographs of the same cell are taken at successive planes of focus through the cell. A series of three such pictures is shown in Fig. 2, and by observing all three pictures together, it is clear that almost all chloroplasts in the cell are dumb-bell-shaped in outline. The very slender dimensions of the connecting isthmus in very many of the dividing chloroplasts as the two daughter plastids separate can also be recognised. This central isthmus is sometimes less than one-tenth the diameter of the rest of the plastid. Measurements of the elongated dumb-bell-shaped chloroplasts show that they are 40–50 per cent longer than the more rounded profiles in younger and in older cells. The constriction occurs perpendicular to the long axis of the chloroplast and is usually equidistant from the two ends of the plastid.

(2) The largest increase in chloroplast number is observed for those mesophyll cells which are older than the cells in which dumb-bellshaped profiles are most numerous. These are cells in a band between 15 mm and 25 mm from the leaf base as shown in Fig. 1. As the cells develop, therefore, the peak of dumb-bell-shaped chloroplast formation is followed closely but slightly later by a two- to threefold increase in chloroplast number. This clearly demonstrates the precursor–product relationship between the dumb-bell-shaped chloroplast and an increase in chloroplast number. From now on the dumb-bellshaped chloroplasts will be referred to as 'replicating' or 'dividing' chloroplasts.

Because the time sequence of chloroplast replication is reflected as a separation in space in the young wheat leaf tissue, it is possible to establish the precise timing of the division process and to estimate from measurements of the rates of cell and leaf elongation the lifetime of a chloroplast division profile. The doubling of chloroplast number occurs as cells move from 18 mm to 21 mm above the leaf base, and this stage of development takes 4 hours. At any time of sampling during this period, only about 7 per cent of all the plastids are constricted, hence the average lifetime of a division profile is $0.07 \times 4 \times 60$ = approx. 20 minutes (Boffey *et al.*, 1979). Careful analysis of the proportion of each morphological form of chloroplast allows a calculation of the relative lengths of time of each phase of the chloroplast replication cycle. The mean lifetime of the most highly constricted form immediately prior to fission is 6 minutes, i.e. the time when the isthmus is at its narrowest: the chances of observing this most constricted form are therefore minimal. The paucity of previous observations in the literature recording chloroplasts in this phase of division is almost certainly accounted for by the short life of the highly constricted morphological form. We estimate that the complete process of chloroplast development in wheat takes approximately 6 hours, chloroplast replication from initiation to the separation of the

Fig. 1. Diagram of the first leaf of a 7-day-old hexaploid wheat plant *Triticum aestivum*. The zone of mitosis and the zone where constricted chloroplasts are found are marked. Plastid number per cell was measured using Nomarski interference optics by the method of Boffey & Leech (1982) except for the youngest plastids (45 per cell). This number was calculated from leaf sections observed by transmission electron microscopy as described in Boffey *et al.* (1979). The transverse bands record the positions of the tissue sections used for chloroplast number determinations.

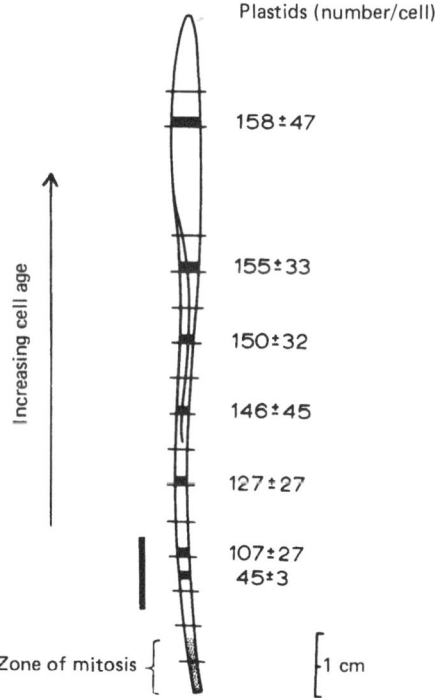

Plastids (number/cell)

158 ± 47

Increasing cell age

155 ± 33

150 ± 32

146 ± 45

127 ± 27

107 ± 27
45 ± 3

Zone of mitosis

1 cm

two daughter plastids can be completed in less than an hour, and the lifetime of the narrowest isthmus form is between 5 and 10 minutes.

Chloroplast division in vitro

Chloroplasts in young mesophyll cells are very tightly packed so it is very difficult to follow the division process *in vivo* . In order to investigate the sequence of

Fig. 2. Three micrographs of the same isolated wheat cell photographed at different focal planes using Nomarski interference optics. The cells were separated in 100 mM EDTA after fixation in 3.5 per cent glutaraldehyde (Boffey *et al.*, 1979). Almost all the chloroplasts are dumb-bell-shaped. (Micrographs by Anita Jellings.)

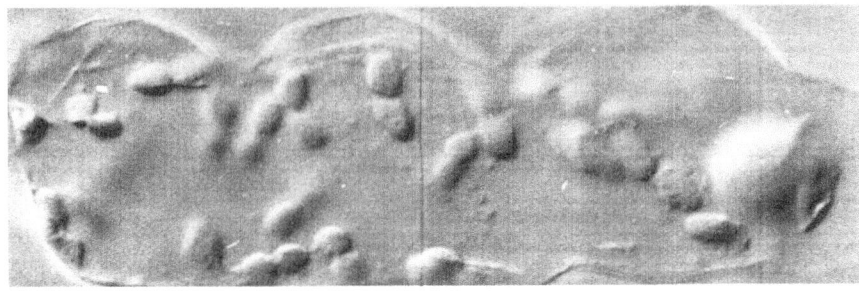

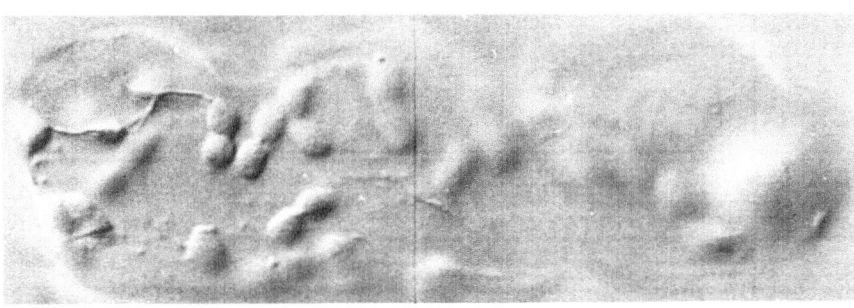

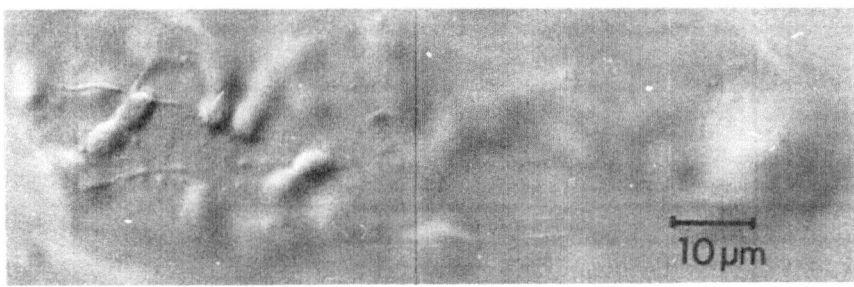

events and the changes in chloroplast shape and size which are characteristic of the replication process, the dividing chloroplasts are more easily observed after isolation from the cell. It was demonstrated by Ridley & Leech (1970) and later by Kameya & Takahashi (1971) that once the chloroplast division process has been initiated it can be completed outside the cell. Provided the osmolarity of the isolation medium is adjusted to between 300 and 500 mM the morphological characteristics of dividing chloroplasts can be preserved outside the cell and, very occasionally, the two daughter halves can be seen to separate. In a series of light microscopical observations Ray Ellis (1981), in this laboratory, established that the proportions of the different forms of replicating chloroplasts in wheat cells were maintained after isolation, and the characteristics of the chloroplasts in suspension *in vitro* truly reflected their features in the cell, both quantitatively and qualitatively. The range of form and the contours of dividing chloroplasts of wheat observed in isolated suspensions using the Nomarski interference optics is illustrated in Fig. 3. By observing the behaviour of chloroplasts in a hanging drop suspension in this way, the sequence of morphological changes occurring during chloroplast replication can be established (Leech, Thomson & Platt–Aloia, 1981). Four different types of constricted plastid shape can be recognised, and these are illustrated in Table 1; the proportion of each type of chloroplast found in suspensions from progressively older cells are recorded. The most common form of replicating chloroplast is an elongated or a 'peanut' shape and represents 84 per cent of the total profiles in the youngest and 98 per cent of the total profiles in the oldest cells respectively. Careful observation of the chloroplasts in suspended drops shows that the 'peanuts' and elongated forms are totally interconvertible. In contrast, the most highly constricted dumb-bell forms never represented more than 2–6 per cent of the total plastids observed. These are always chloroplasts which in face view appear as an elongated, but not greatly constricted, figure of eight. In a small proportion of chloroplasts the two halves of the dumb-bell are frequently observed to pivot round the central isthmus at approximate right-angles to each other. This twisting of the two halves around the central axis which was discovered in isolated suspensions was also characteristic of the peanut-shaped plastids.

The sequence of morphological changes during chloroplast replication
From the combined observations of four independent observers in this laboratory we have been able to construct a sequence of events during the replication of young chloroplasts in wheat (Fig. 3). The rounded dumb-bell appearance precedes the assumption of the 'peanut' shape and the chloroplast then becomes progressively constricted until, immediately prior to fission, only a very narrow isthmus exists. Fission is momentary and has been rarely observed. In isolated suspensions the growth of the daughter plastids has never been shown to occur, but in the cells of the wheat leaf, following fission, the daughter chloroplasts approximately double in size

Table 1. *The characteristics of young chloroplasts in the state of division in suspensions isolated from leaf tissue sections of young green wheat leaves*

Leaf tissue sections (cell age)	Percentage of dividing chloroplasts			
	24 h	28 h	32 h	36 h
Elongated or 'peanut'	84	89	87	98
Figure of eight	13	2	3	0
Constricted	2	2	6	0
L-shaped	1	7	4	2

before the next round of division commences. The second round of chloroplast division is considerably less synchronised than the first, and occurs during the following 12 hours of leaf expansion. Most frequently the two fission products resemble each other closely in shape and size, and immediately after separation the daughter plastids are about half the volume of their parents.

An important question is whether all the young chloroplasts in a wheat leaf cell divide or whether division is restricted to a sub-population of the chloroplasts responsible for the increase in chloroplast number in the cell. The observation that in some cells all the chloroplasts are simultaneously in the dumb-bell configuration lends strong support to the contention that all the chloroplasts have the potential to form dumb-bells, but it does not prove that all the chloroplasts complete the division process and actually produce two daughter plastids. More definitive proof comes from measurement of the shape and size of the chloroplasts before, during and after replication, and such measurements suggest most strongly that all the plastids of the cell population are able to divide. Each daughter plastid is an oblate spheroid and is approximately half the volume of its parent immediately after separation, so it follows that the major axis of each daughter plastid should be 0.8 of its parent and attached daughters should have an axis length of 1.6 of the pre-division length of the parent chloroplast. Since the ratio of length of constricted to non-constricted plastids remains constant at 1.5 in wheat before, during and after replication, it can be concluded that all the chloroplasts in the population are capable of division (Boffey *et al.*, 1979). In addition, we have never observed very small chloroplasts or any organelle which could be described as a small 'initial' using the light microscope or in sections of young wheat leaves viewed by transmission electron microscopy. (Possingham & Saurer (1969) and Boasson & Laetsch (1969) also showed that increases in chloroplast number could occur in leaf cells where no proplastids had ever been observed.) We can therefore conclude that the increase in cellular chloroplast number

Fig. 3. (1) Isolated dividing chloroplasts from young wheat leaves (*Triticum aestivum* cv. Maris Dove) observed using phase-contrast (left side) and Nomarski interference optics (right side). (2) Plasticine models illustrating the range of morphological form observed in the suspensions of isolated chloroplasts. (3) Proposed sequence of changes occurring during chloroplast division in wheat. (Modified from Leech *et al.*, 1981).

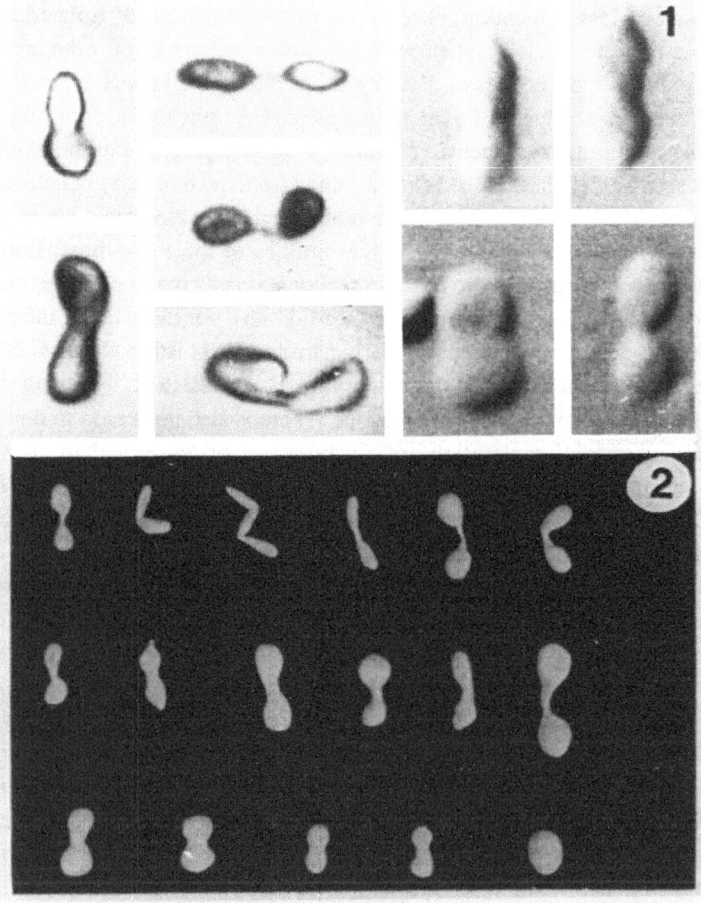

in wheat mesophyll cells can be entirely accounted for by the division of pre-existing chloroplasts.

Internal morphological changes during chloroplast replication

In addition to establishing the morphological sequence of changes occurring during chloroplast replication, observation of the behaviour of isolated replicating chloroplasts in suspension also provided us with a sound basis for the interpretation of the appearance of profiles of dividing chloroplasts in thin sections in the electron microscope. Knowledge of the three-dimensional morphological form of replicating chloroplasts in *in vitro* suspension enables us to interpret the outline of chloroplast profiles seen in the wheat cells. When sections from the thin band of cells containing replicating chloroplasts are sectioned and observed in the electron microscope, all the profiles found can be matched with the outlines of the three-dimensional shapes previously seen in suspension. In these sections of cells many elongated and dumb-bell profiles of chloroplasts were seen (see Fig. 4b), but the relative infrequency of replicating profiles showing a particularly narrow central isthmus was to be expected from the three-dimensional shape previously established by using Nomarski interference optics in the light microscope. There is also no reason to doubt that the daughter plastids of replication are always more or less equal in size, and that sometimes the two attached daughter plastids lie at right angles to each other (see examples in Leech *et al.*, 1981; Webber *et al.*, 1984). Inequalities in shape or size can be explained completely when the plane of section is taken into account.

In thin sections of wheat leaf tissue viewed in the electron microscope, the replicating chloroplasts are sectioned in many planes at random and very few planes of section fall exactly at right angles to the isthmus. Such a profile of a replicating spinach chloroplast is shown in Fig. 4a; this is a particularly beautiful electron micrograph and was taken by Denis Greenwood several years ago. Many of the grana have been sectioned absolutely at right angles to the thylakoid membranes, the isthmus is clearly visible, and complete profiles of both the daughter plastids can be observed. In the final stages of division, isolated wheat chloroplasts look even more constricted in side view and the connecting isthmus is only one-tenth the diameter of the daughter plastids, but in surface view these chloroplasts have a very fat figure-of-eight appearance and the isthmus is more than three-quarters the width of the daughter plastids. Chloroplast profiles relating to both views can be found in thin sections of the appropriate cells. An elongated constricted wheat chloroplast profile is shown in Fig. 4b.

Transmission electron micrographs also provide valuable additional views of the internal structure of the dividing chloroplast and enable changes in the internal membrane structure and in the chloroplast envelope membranes to be followed as the chloroplast divides. Observation of profiles of chloroplasts which have been cut at right angles to the isthmus can be identified as representing a variety of stages in the replication process by reference to the light-microscope observations. As chloroplast division progresses and the constriction zone narrows to an isthmus, internal membrane changes are consistently observed. As division progresses, fuzzy plaques of electron-opaque material are frequently, but not invariably, seen covering, or even displacing, envelope membranes in the region of the isthmus. We have suggested (Leech *et al.*, 1981) that these plaques probably reflect molecular changes in the arrangement of the membrane components of the isthmus region just before the separation of the daughter halves, and the plaques perhaps represent the deposit of

Fig. 4. (a) Profile of a dividing chloroplast in a young spinach (*Spinacia oleracea*) leaf. Preparation and photography by Denis Greenwood and published with his permission. (b) Profile of a dividing chloroplast in a young wheat leaf *Triticum aestivum* cv. Maris Dove). Preparation and photography by K.A. Platt–Aloia and published with her permission. (K.A. Platt–Aloia & R.M. Leech, unpublished.)

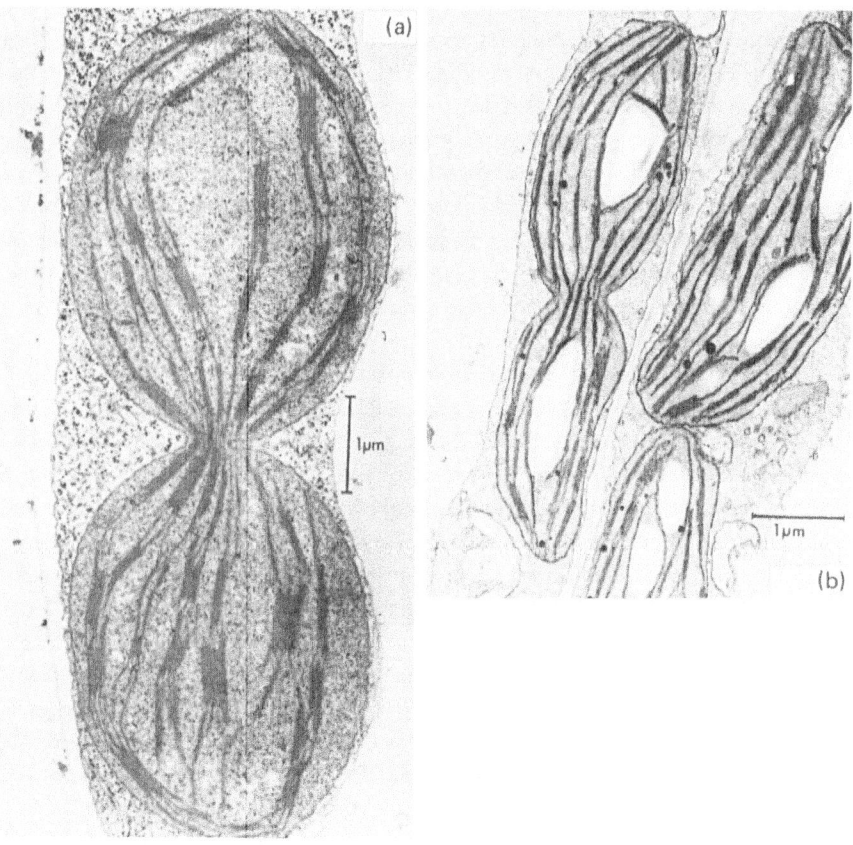

some components important in the formation of the constriction zone. Increasing accumulation, or aggregation, or changes in hydrophobicity could result in visualisation of the components in this separation zone. Sometimes the inner envelope membrane shows invaginations at this stage, but in the final stages of constriction the delineation of the envelope membranes becomes almost impossibe. However, just prior to separation, contiguous double membranes appear to be present around both daughter plastids. After separation of the two daughters, sometimes myelin-like aggregations may be seen between them (Leech *et al.*, 1981). As far as we are aware, the presence of microtubules inside or outside a dividing chloroplast has never been formally reported for wheat, although microtubule-like structures have been reported in spinach (Lawrence & Possingham, 1984) but no electron micrographs have been published.

Observation of the arrangement of the grana membranes in replicating chloroplasts, particularly their presence in the narrowing isthmus region, raises questions about the redistribution of internal membranes and other chloroplast components during division. The grana in young dividing chloroplasts have stacks with up to 10 partitions. How do these stacks redistribute during division? (In *Phaseolus* Whatley *et al.* (1982) have suggested that the granal membranes, which are part of a spiral ribbon, tear apart.) As yet no serial sections are available from wheat but we have observed that in highly constricted chloroplasts, although granal membranes are only rarely observed in the isthmus, small, apparently 'matching' granal stacks are often present at its two ends. These grana typically have only half the number of partitions found in the larger grana disposed more centrally in the daughter chloroplasts.

Chloroplast DNA and chloroplast replication

Considerable importance is also attached to the characteristics of the mode of redistribution of chloroplast DNA during the chloroplast replication process. In young wheat leaves, chloroplast DNA synthesis is probably complete before the young plastid begins to divide, and chloroplast DNA synthesis also occurs quite independently of nuclear DNA synthesis (Dean & Leech, 1982*a*; Boffey & Leech, 1982; Boffey, 1985). (A similar separation of DNA synthesis in the nucleus and in the young chloroplast occurs in young spinach leaves (Scott & Possingham, 1983).) In young wheat leaves it is possible to count the cellular number of chloroplasts and simultaneously measure their DNA content. Before plastid replication takes place in wheat, the number of plastome copies can rise to over a thousand in each plastid, and after division is completed, the mature chloroplasts contain betwen 200 and 300 copies of chloroplast DNA (Boffey & Leech, 1982). The amount of chloroplast DNA per cell remains constant, but is diluted during replication and partitioned out to the increasing number of daughter chloroplasts. In species of wheat of different ploidies, nuclear and chloroplast DNA are in a constant proportion (Bowman, 1986). Since in hexaploid wheat there are more than 40 000 plastome copies per cell, the amount of chloroplast DNA seems unlikely to be a major restrictive factor in the continuing development of the chloroplast.

The regions in the chloroplast containing DNA can be visualised relatively precisely under UV illumination following staining with the specific fluorochrome DAPI (4',6–diamidino–2–phenylindole) which specifically binds to DNA and was introduced for chloroplast work by James & Jope (1978). After staining, successive focal planes in isolated replicating chloroplasts in suspension can be brought into focus and the staining areas revealed. Mature wheat chloroplasts contain between 7 and 16 discrete fluorescent areas after staining with DAPI, and these are located in the peripheral stroma regions, often in a discrete band (Sellden & Leech, 1981). (A similar arrangement has been described in other Graminae, e.g. barley (Scott, Cain & Possingham, 1982), but other dicotyledonous species seem to possess more centrally located chloroplast DNA (Possingham & Lawrence, 1983) which is associated with the granal membranes in spinach (Rose, 1979).) During chloroplast replication in wheat no special relocalisation of DNA has been observed, but presumably an approximately equal number of DNA-containing regions are redistributed to each daughter chloroplast relatively routinely.

The control of chloroplast replication

Cellular development and chloroplast development and division are continuously interdependent in young leaf mesophyll cells of wheat. The extent of chloroplast replication and growth within the leaf cells during their development greatly influences the final characteristics of the photosynthetic machinery under leaf area in the mature leaf. How does the cell control the number of chloroplasts within it? We can consider the nature of possible controls of chloroplast replication by considering the process as three sequential events which are probably controlled to some extent independently of each other. Initiation first occurs in response to an as yet unidentified signal, the chloroplast begins to elongate and the characteristic constriction is formed centrally. In young wheat cells it seems likely that all the chloroplasts in the cell respond more or less simultaneously to the signal to replicate. In cells of an appropriate age virtually all the chloroplasts may simultaneously be in a state of division and, given the short duration of the replication process and the time-scale for the isthmus formation (5–10 minutes), it is certainly not surprising that this synchrony, although frequently searched for, has not always been observed in other plants. Synchrony of chloroplast replication indicates a cellular or nuclear initiation signal rather than one emanating from the individual chloroplasts. The nature of such a signal is entirely unknown at the present time.

The second phase during the chloroplast replication process follows initiation and is seen as a progressive elongating of the plastid and narrowing of the isthmus. As this process progresses at different rates in chloroplasts in the same cell, its control almost certainly resides in the chloroplast. All our own observations on this second phase of chloroplast division can be explained (Leech *et al.*, 1981) if a theoretical model similar to that proposed by Greenspan (1977, 1978) for cytokinesis is adopted. If this model, based on changes in the dynamic instability of surface tension forces of the chloroplast envelope, is extended to cover the formation of an isthmus in a dividing chloroplast, then chloroplast isthmus formation, once triggered, must be a spontaneous event.

The third phase of chloroplast replication, i.e. termination of the replication process, is just as fundamental to the division process as its initiation and may be considered as the last stage of replication. We have surveyed many species and cultivars of wheat and determined the chloroplast numbers in their mature leaf cells. Our findings led us to investigate further the proposition that cell size is a dominant factor in controlling chloroplast number. Several lines of evidence now very strongly support this conclusion. The most compelling evidence comes from two sources:

(1)   *mean* chloroplast numbers estimated for *populations* of mesophyll cells in leaf tissue, and

(2)   *actual* numbers of chloroplasts in the *individual* cells of the same leaf. Such individual cells show variation in cell size and also in chloroplast number.

Mean cellular chloroplast numbers and polyploidy in leaf cell populations
Mature leaf cells. When chloroplast replication ceases, the final chloroplast number in that cell is achieved and is subsequently not altered. When the mean chloroplast number for a sample of cells from the population of mesophyll cells is established, it is found to be a very stable and consistent number for similar leaves of plants of the same species grown on separate occasions. This is strikingly demonstrated in our studies of wheat in which we have examined both wild and cultivated species and also highly selected cultivars of hexaploid wheats (Jellings & Leech, 1984; Leech, 1986; Pyke & Leech, 1987). The mean chloroplast number is constant within the same leaf of plants of the same species or cultivar, but there are considerable differences between species. Indeed, the mean chloroplast number can be used as a diagnostic characteristic for each particular species or cultivar.

Particularly dramatic differences in mean chloroplast number are associated with changes in ploidy level in wheat (Dean & Leech, 1982*b*; Ellis, Jellings & Leech, 1983; Jellings & Leech, 1984; Leech, Leese & Jellings, 1985; Pyke & Leech, 1987). In one of the first studies, using three non-isogenic species of *Triticum*, one diploid, one tetraploid and one hexaploid (Dean & Leech, 1982*b*), a doubling in ploidy level was accompanied by an approximate doubling in chloroplast number. (A similar situation was shown for isogenic diploid, tetraploid, hexaploid and decaploid alfalfa (Meyers *et al.*, 1982) and in *Festuca* (Joseph, Randall & Nelson, 1981).) In wheat, artificially produced genetic hybrids provide extremely useful material for the examination of the relationship between chloroplast number increase and nuclear ploidy levels. We (Leech, 1981) examined in detail three hybrids in which a

Table 2. *Chloroplast number per cell in species of* Aegilops *and*
Triticum *and in hybrids in which hexaploid nuclei (n) have been*
introduced into alien cytoplasms (c) (15–20 cells per sample.)

Plants	Chloroplasts/cell
(1) *Triticum tauschii* (2x)	57 ± 3
Triticum aestivum cv. Chinese Spring (6x)	130 ± 7
T. aestivum (n) × *T. tauschii* (c)	167 ± 9
(2) *Aegilops ovata* (4x)	94 ± 4
Triticum aestivum cv. Chinese Spring (6x)	130 ± 7
T. aestivum (n) × *A. ovata* (c)	145 ± 8
(3) *Triticum timopheevi* (4x)	118 ± 5
Triticum aestivum cv. Maris Ranger	150 ± 8
T. aestivum (n) × *T. timopheevi* (c)	156 ± 10

Data from A.J. Jellings & R.M. Leech, unpublished.

hexaploid *Triticum aestivum* nucleus had been artificially introduced into an alien
cytoplasm of a plant of different ploidy. In two of these hybrids the cytoplasm was
diploid (*Aegilops ovata* and *Triticum tauschii*), and in one tetraploid (*Triticum
timopheevi*). The results of the examination of the mean chloroplast number in the
first leaves are shown in Table 2. It can be seen that in all these hybrid plants, the
mean chloroplast number per cell reflects the hexaploid genetic constitution of the
nucleus and not the genetic constitution of the cytoplasm. These analyses provide
further evidence for the dominant effect of the nuclear genome size on mean cellular
chloroplast number.

Although increases in mean chloroplast number have frequently been correlated
with increases in nuclear genome size (Butterfass, 1959; 1973, 1979, 1983; this
volume; Joseph *et al.*, 1981; Meyers *et al.*, 1982; Dean & Leech, 1982*b*; Ellis *et al.*,
1983; Pyke & Leech, 1987), no causal relationship has yet been identified. Indeed,
several persuasive lines of evidence suggest that the influence of nuclear genome size
on chloroplast number is *indirect* and operates via the effect of nuclear genome size
on cell size. Cell size itself appears to exert a major control on chloroplast number.
We have several lines of evidence from our studies of wheat which lead us to this
conclusion. In general, our approach has been to investigate leaf material in which the
influence of cell size on chloroplast number can be considered in the absence of
complications of changes in nuclear genome size.

*Changes in mean chloroplast number in populations of leaf cells during natural leaf
development – cell size and chloroplast number correlations.* In the leaves of many

Table 3. *Mesophyll cell plan area and chloroplast number in primary, fifth and flag leaves of two hexaploid wheat species (standard errors in parentheses)*

Type of leaf	Primary	Fifth	Flag
Triticum aestivum cv. Avalon			
Cell plan area (μm^2, n=90)	3587(97)	3005(82)	2049(53)
Chloroplasts (no./cell)	134	125	108
Triticum macha			
Cell plan area (μm^2, n=90)	4273(113)	3152(99)	1866(69)
Chloroplasts (no./cell)	194	157	120

wheat species we have examined, chloroplast number increases occur in time with the expansion of the cell, and cell expansion ceases a short time before the chloroplast number finally stabilises. In mature mesophyll cells of hexaploid wheat, 70 per cent of the cell surface area is covered with chloroplasts. This extremely regular and reproducible pattern suggests to us that chloroplast division and growth continues as long as there is sufficient space available to accommodate the fully expanded chloroplasts, and it then ceases.

A tight cell size to chloroplast number correlation can be induced experimentally if the developing wheat leaf meristem is treated with maleic-hydrazide to prevent cell plate formation. Giant cells are produced which, at maturity, may contain up to 600 chloroplasts, compared with a normal number of about 150. As can be seen from the results shown in Fig. 5a, in cells of intermediate size, chloroplast number also closely follows cell size.

Variation of cell size and chloroplast number with leaf position. In hexaploid wheat species, specifically *Triticum aestivum* cv. Avalon and *Triticum macha*, the mean mesophyll cell size differs markedly with leaf position. The mean cell size in primary leaves is larger than in the fifth leaf, in which mean cell size is larger than in the flag leaf. These differences provide an additional opportunity to study the relationship between cell size and chloroplast number. The relevant cell sizes and chloroplast numbers are recorded in Table 3. It can be seen that the decrease in mesophyll cell size which accompanies the increase in leaf number is entirely mirrored by a reduction in the number of chloroplasts per cell. Again, this evidence supports the conclusion that cell size is a major factor controlling chloroplast number in wheat.

Fig. 5. (a) Chloroplast number per cell and cell area for individual mature cells of the first leaf of 7-day-old *Triticum aestivum* cv. Maris Dove. The plants were either grown normally (Boffey & Leech, 1982) or in 1 mM maleic hydrazide. The experiments were performed by Ray Ellis and are published with his permission. (J.R. Ellis & R.M. Leech, unpublished.) (b) Mean chloroplast number per cell and mean mesophyll cell plan area for cells of third leaves of 18-day-old *Triticum aestivum* cv. Avalon. The plants were grown hydroponically in a range of nitrate nutrient solutions (Hewitt, 1966). (K.A. Pyke & R.M. Leech, unpublished.)

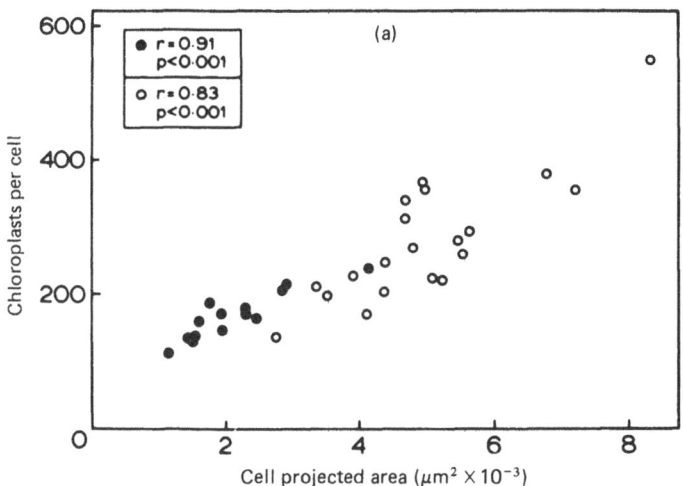

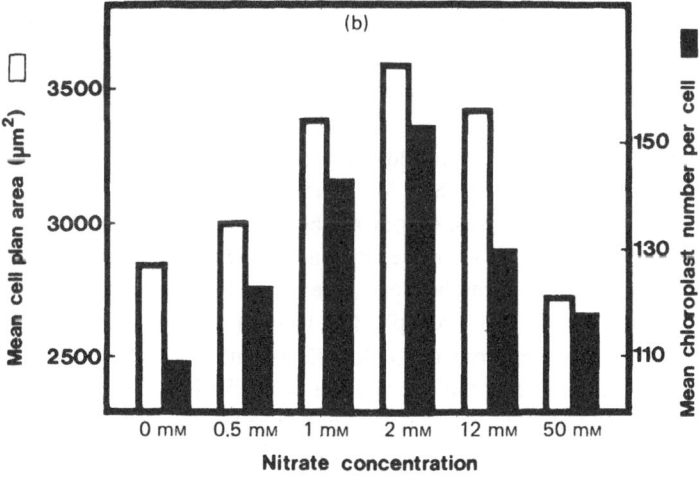

The relation between cell size and chloroplast number in extreme environmental conditions. The response to such conditions is usually in individual chloroplast size rather than in chloroplast number (Jellings, Usher & Leech, 1983a,b). In order to test the effect of nitrogen availability on mesophyll cell size and chloroplast number in leaves of *Triticum aestivum*, plants grown in hydroponic nutrient solutions containing nitrate at six levels of concentration between 0 mM and 50 mM were compared. The results are shown in Fig. 5b. It can be seen that there is a direct effect of the nitrate concentration on mean cell size, with the largest cells developing in plants grown in 2 mM nitrate, with smaller cell sizes at higher and lower molarities. However, the important relationship is again demonstrated, i.e. that larger cells indeed have larger numbers of chloroplasts and that the cell size to chloroplast number ratio is constant.

Chloroplast numbers for individual cells

Chloroplast number/cell size relationships can best be followed in detail if the individual cells of the leaf population are examined singly and chloroplast numbers and cell sizes for individual cells are considered rather than the mean values for the cell populations of many cells (Pyke & Leech, 1987). Fig. 6a is a scatter diagram of chloroplast number per cell plotted against size of cell in *Triticum aestivum*. The measurements were taken from 36 individual separated mesophyll cells, sampled from the fully expanded leaf tissue of the first leaf of 10 seedlings. It can be seen that for the individual cells of the leaf population, the number of chloroplasts per cell is closely correlated with the cell plan area. A particularly critical experiment was performed by J.R. Ellis in this laboratory, who was able to separate the effects of cell size from those of nuclear genome content by examining the relationship between chloroplast number of individual cells and cell size in isogenic diploid and tetraploid *Triticum monococcum*. The results are shown in a scatter diagram in Fig. 6b. An extremely tight correlation is demonstrated between the number of chloroplasts and the size of the cell *irrespective of ploidy level*. We have recently extended these observations to study in greater detail the underlying variation in chloroplast number between the individual mesophyll cells within a population of a single leaf (Pyke & Leech, 1987). We examined two diploid and four tetraploid species of wheat and 15 genotypes of the hexaploid *Triticum aestivum*. For each species or genotype, the individual cell plan areas were plotted against the number of chloroplasts in that cell. It can be clearly seen from the results plotted in Fig. 7 that cells of similar size, but of different ploidies, have similar numbers of chloroplasts.

We conclude that the number of chloroplasts within a cell is closely correlated ($p < 0.001$) with the size of the cell and this relationship is consistent in wheat for species of different ploidies over a wide range of cell sizes. Cells with an abnormally large or small number of chloroplasts for a given cell size are never observed.

Conclusions

We can now pose the question: How does the cell control the number of chloroplasts within it in relation to its own size? Our results have shown that in wheat the variation in nuclear DNA amount in individual cells is unlikely to be able to account for the variable chloroplast numbers found in the population of mesophyll cells (coefficient of variation greater than 20 per cent). The degree of endopolyploidisation within the leaf cell populations is also very low (Ellis *et al.*, 1983) and is insufficient to explain a causal relationship between nuclear DNA content and chloroplast number in this genus.

The implication that cell size is an important controlling factor on chloroplast number suggests that some physical limitation on further chloroplast replication may occur. Evidence suggests that this limitation is related to the internal cell surface area over which the chloroplasts are disposed. The type of chloroplast 'cover' of this area will be related to the product of the size of the individual chloroplasts and the number of them in each cell. Chloroplast 'cover' is a very consistent factor within a species, although it can vary widely between species (Honda *et al.*, 1971) and also between cell types within the same leaf. In hexaploid wheat, division and expansion of chloroplasts results in a consistent proportion of the cell surface being covered, of the order of 70 per cent (Ellis & Leech, 1985). When chloroplast 'cover' is calculated and related to cell size, in many comparisons in wheat there is a much tighter

Fig. 6. Scatter diagrams of chloroplast number plotted against cell size, in (a) cells of the first leaf of *Triticum aestivum* and in (b) diploid (open circles) and tetraploid (filled circles) cells of first leaves of *Triticum monococcum*. In (b) the measurements were taken from 20 separate mesophyll cells sampled from four seedlings of each ploidy (from Ellis & Leech, 1985).

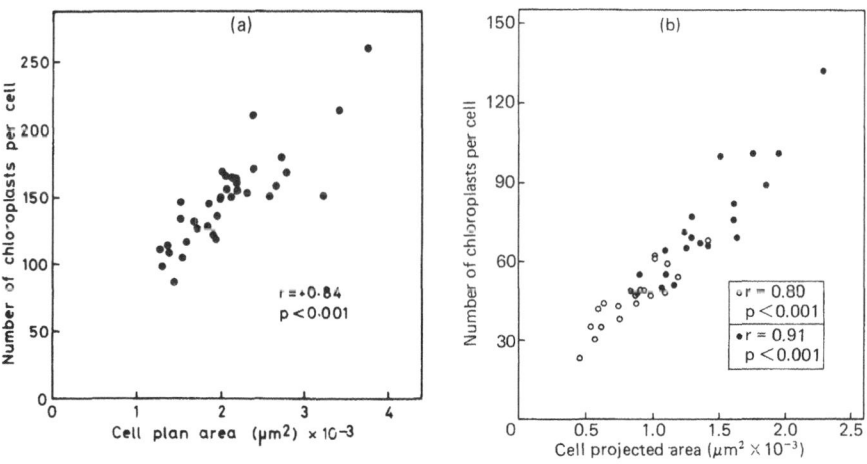

correlation between these two parameters than between cell size and chloroplast number. On current evidence it would seem that this relationship is under strict genetic control, and it clearly has very important implications in consideration of the biochemical interactions within the cell. The same amount of chloroplast 'cover' can be achieved in a variety of ways, for example by a few large chloroplasts or many smaller ones. Between these two extremes there will be a large difference in the area of chloroplast envelope available for transport into and out of the chloroplast in relation to the total chloroplast volume.

In this chapter we have emphasised the importance of the interdependence of chloroplast replication and enlargement and cellular development in young wheat leaves. Chloroplast replication is clearly an integral and inevitable part of cellular development. At present the signals which initiate and terminate the chloroplast

Fig. 7. A plot of mesophyll plan area (x) and number of chloroplasts per cell (y) for two diploid and four tetraploid species of wheat and 15 hexaploid genotypes of *Triticum aestivum*. Each point represents an individual cell. $r = 0.91$, $p < 0.001$. Diploids, (◊); tetraploids, (□); hexaploids, (○).(From Pyke & Leech, 1987.)

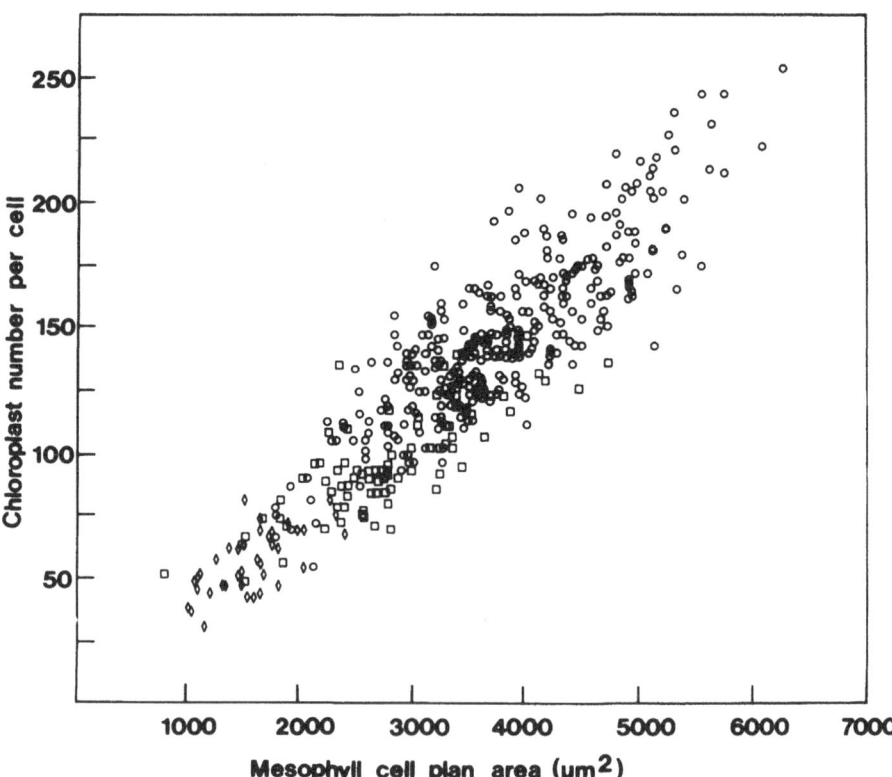

replication process are not understood and the nature of the mechanism which brings about the division of the chloroplast and its components into two halves has yet to be determined. It may be helpful in further experimentation to consider chloroplast division, its initiation and its termination as distinct phases probably subject to different controls. Our own studies incline us to the view that initiation and termination are under nuclear/cytoplasmic control and that once initiation has occurred the control of the division process is from within the chloroplast. New investigations will be needed to establish whether this interpretation is the correct one.

Acknowledgements

We are very grateful to Denis Greenwood, Raymond Ellis, Anita Jellings and Katherine Platt–Aloia for allowing us to use unpublished micrographs and experimental results. We are also most grateful to Dr C.N. Law and Dr R.B. Austin of the Plant Breeding Institute, Cambridge for gifts of wheat cultivars and genetic lines.

References

Boasson, R. & Laetsch, W.M. (1969). Chloroplast replication and growth in tobacco. *Science*, **166**, 749–51.

Boffey, S.A. (1985). The chloroplast division cycle and its relationship to the cell division cycle. In *The Cell Division Cycle in Plants*, ed. J.A. Bryant & D. Francis, pp. 233–46. Cambridge University Press.

Boffey, S.A., Ellis, J.R., Sellden, G. & Leech, R.M. (1979). Chloroplast division and DNA synthesis in light-grown wheat leaves. *Plant Physiology*, **64**, 503–5.

Boffey, S.A. & Leech, R.M. (1982). Chloroplast DNA levels and the control of chloroplast division in light-grown wheat leaves. *Plant Physiology*, **69**, 1387–91.

Boffey, S.A., Sellden, G. & Leech, R.M. (1980). The influence of cell age on chlorophyll formation in light-grown and etiolated wheat seedlings. *Plant Physiology*, **65**, 680–4.

Bowman, C.M. (1986). Copy numbers of chloroplast and nuclear genomes are proportional in mature mesophyll cells of *Triticum* and *Aegilops* species. *Planta*, **167**, 264–74.

Butterfass, T. (1959). Ploidie und Chloroplastenzahlen. *Berichte der Deutschen Botanischen Gesellschaft*, **72**, 440–51.

Butterfass, T. (1973). Control of plastid division by means of nuclear DNA amount. *Protoplasma*, **76**, 167–95.

Butterfass, T. (1979). *Patterns of Chloroplast Reproduction: A Developmental Approach to Protoplasmic Plant Anatomy*. Cell Biology Monographs vol. 6. Wien: Springer-Verlag.

Butterfass, T. (1983). A nucleotypic control of chloroplast reproduction. *Protoplasma*, **118**, 71–4.

Chaly, N. & Possingham, J.V. (1981). Structure of constricted proplastids in meristematic plant tissues. *Biologie Cellulaire*, **41**, 203–10.

Chaly, N., Possingham, J.V. & Thomson, W.W. (1980). Chloroplast division in spinach leaves examined by scanning electron microscopy and freeze etching. *Journal of Cell Science*, **46**, 87–96.

Dean, C. & Leech, R.M. (1982a). Genome expression during normal leaf development. 1. Cellular and chloroplast numbers and DNA, RNA and protein levels in tissues of different ages within a seven day old wheat leaf. *Plant Physiology*, **69**, 904–0.

Dean, C. & Leech, R.M. (1982b). Genome expression during normal leaf development. 2. The direct correlation between ribulose bisphosphate carboxylase content and nuclear ploidy in a polyploid series of wheat. *Plant Physiology*, **70**, 1605–8.

Ellis, J.R. (1981). 'Cellular development and chloroplast development in the first leaf of wheat seedlings.' D.Phil. thesis, University of York.

Ellis, J.R., Jellings, A.J. & Leech, R.M. (1983). Nuclear DNA content and the control of chloroplast replication in wheat leaves. *Planta*, **157**, 376–80.

Ellis, J.R. & Leech, R.M. (1985). Cell size and chloroplast size in relation to chloroplast replication in light-grown wheat leaves. *Planta*, **165**, 120–5.

Greenspan, H.P. (1977). On the dynamics of cell cleavage. *Journal of Theoretical Biology*, **65**, 79–99.

Greenspan, H.P. (1978). On fluid mechanical simulations of cell division and movement. *Journal of Theoretical Biology*, **70**, 125–34.

Hewitt, E.J. (1966). *Sand and Water Culture Methods Used in the Study of Plant Nutrition*. Technical Communication No. 22 of the Commonwealth bureau of horticulture and plantation crops. East Malling, Maidstone, Kent.

Honda, S.I., Hongladarum–Honda, T., Kwanyren, P. & Wildman, S.G. (1971). Interpretations of chloroplast reproduction derived from correlation between cells and chloroplasts. *Planta*, **97**, 1–15.

James, T.W. & Jope, C.A. (1978). Visualization by fluorescence of chloroplast DNA in higher plants by means of the DNA-specific probe 4'6–diamidino–2–phenylindole. *Journal of Cell Biology*, **79**, 623–30.

Jellings, A.J. & Leech, R.M. (1984). Anatomical variation in first leaves of nine *Triticum* genotypes, and its relationship to photosynthetic capacity. *New Phytologist*, **96**, 371–82.

Jellings, A.J., Usher, M.B. & Leech, R.M. (1983a). Variation in the chloroplast to cell area index in *Deschampsia antarctica* along a 16° latitudinal gradient. *British Antarctic Survey Bulletin*, **61**, 13–20.

Jellings, A.J., Usher, M.B. & Leech, R.M. (1983b). Chloroplast size in tall and short phenotypes of *Poa flabellata* on South Georgia. *British Antarctic Survey Bulletin*, **59**, 41–6.

Joseph, M.C., Randall, D.D. & Nelson, C.J. (1981). Photosynthesis in polyploid fescue. II. Photosynthesis and ribulose 1,5–bisphosphate carboxylase of polyploid tall fescue. *Plant Physiology*, **68**, 890–4.

Kameya, T. (1972). Cell elongation and division of chloroplasts. *Journal of Experimental Botany*, **23**, 62–4.

Kameya, T. & Takahashi, N. (1971). Division of chloroplasts *in vitro*. *Japanese Journal of Genetics*, **46**, 153–7.

Lawrence, M.E. & Possingham, J.V. (1984). Observations of microtubule-like structures within *Spinacia oleracea* plastids. *Biologie Cellulaire*, **52**, 77–82.

Lawrence, M.E. & Possingham, J.V. (1986). Microspectrofluorometric measurement of chloroplast DNA in dividing and expanding leaf cells of *Spinacia oleracea*. *Plant Physiology*, **81**, 708–10.

Leech, R.M. (1976). The replication of plastids in higher plants. In *Cell Division in Higher Plants*, ed. M. Yeoman, pp. 135–9. London: Academic Press.

Leech, R.M. (1980). The survival, division and differentiation of higher plant plastids outside the leaf cell. In *Chloroplasts*, ed. J. Reinert, pp. 225–35. Results and Problems in Cell Differentiation, vol. 10. Berlin: Springer–Verlag.

Leech, R.M. (1981). Chloroplast number in wheat. In *Abstracts of XIIIth International Botanical Congress*, p. 224.

Leech, R.M. (1984). Chloroplast development in angiosperms: current knowledge and future prospects. In *Chloroplast Biogenesis*, ed. N.R. Baker & J. Barber, pp. 1–21. Amsterdam: Elsevier.

Leech, R.M. (1985). The synthesis of cellular components in leaves. In *Control of Leaf Growth*, ed. N.R. Baker, W.J. Davies & C.K. Ong, pp. 93–113. Cambridge University Press.

Leech, R.M. (1986). Stability and plasticity during chloroplast development. In *Plasticity in Plants*, ed. D.H. Jennings & A. Trewavas, pp. 121–53. Cambridge University Press.

Leech, R.M. & Baker, N.R. (1983). The development of photosynthetic capacity in leaves. In *The Growth and Functioning of Leaves*, ed. J.E. Dale & F.L. Milthorpe, pp. 271–307. Cambridge University Press.

Leech, R.M., Leese, B.M. & Jellings, A.J. (1985). Variation in cellular ribulose bisphosphate carboxylase content in leaves of *Triticum* genotypes at three levels of ploidy. *Planta*, 166, 259–63.

Leech, R.M., Thomson, W.W. & Platt–Aloia, K.A. (1981). Observations on the mechanism of chloroplast division in higher plants. *New Phytologist*, 87, 1–9.

Meyers, S.P., Nichols, S.L., Baer, G.R., Molin, W.T. & Schrader, L.E. (1982). Ploidy effects in isogenic populations of alfalfa. I. Ribulose 1,5–bisphosphate carboxylase, soluble protein, chlorophyll, and DNA in leaves. *Plant Physiology*, 70, 1704–9.

Possingham, J.V. (1980). Plastid replication and development in the life cycle of higher plants. *Annual Review of Plant Physiology*, 31, 113–29.

Possingham, J.V. & Lawrence, M.E. (1983). Controls to plastid division. In *International Review of Cytology*, vol. 84, ed. G.H. Bourne & J.F. Danielli, pp. 1–56. London: Academic Press.

Possingham, J.V. & Saurer, W. (1969). Changes in chloroplast number per cell during leaf development in spinach. *Planta*, 86, 186–94.

Pyke, K.A. & Leech, R.M. (1987). The control of chloroplast number in wheat mesophyll cells. *Planta*, 170, 416–20.

Randolf, L.F. (1922). Cytology of chlorophyll types of maize. *Botanical Gazette*, 73, 337–75.

Ridley, S.M. & Leech, R.M. (1970). Division of chloroplasts in an artificial environment. *Nature*, 227, 463–5.

Rose, R.J. (1979). The association of chloroplast DNA with photosynthetic membrane vesicles from spinach chloroplasts. *Journal of Cell Science*, 36, 169–83.

Scott, N.S., Cain, P. & Possingham, J.V. (1982). Plastid DNA levels in albino and green leaves of the 'albostrians' mutant of *Hordeum vulgare*. *Zeitschrift für Pflanzenphysiologie*, 108, 187–91.

Scott, N.S. & Possingham, J.V. (1983). Changes in chloroplast DNA levels during growth of spinach leaves. *Journal of Experimental Botany*, 34, 1756–67.

Scllden, G. & Leech, R.M. (1981). Localization of DNA in mature and young wheat chloroplasts using the fluorescence probe 4',6–diamidino–2–phenylindole. *Plant Physiology*, 68, 731–4.

Webber, A.N., Baker, N.R., Platt–Aloia, K. & Thomson, W.W. (1984). Appearance of a state 1–state 2 transition during chloroplast development in the wheat leaf: Energetic and structural considerations. *Physiologia plantarum*, 60, 171–9.

Whatley, J.M. (1979). Plastid development in the primary leaf of *Phaseolus vulgaris* variation between different types of cell. *New Phytologist*, 82, 1–10.

Whatley, J.M. (1980). Plastid growth and division in *Phaseolus vulgaris*. *New Phytologist*, 80, 1–16.

Whatley, J.M., Hawes, C.R., Horne, J.C. & Kerr, J.D.A. (1982). The establishment of the plastid thylakoid system. *New Phytologist*, 90, 619–30.

Sellden, G. & Leech, R.M. (1981). Localization of DNA in mature and young wheat chloroplasts using the fluorescence probe 4',6–diamidino–2–phenylindole. *Plant Physiology*, **68**, 731–4.

Webber, A.N., Baker, N.R., Platt-Aloia, K. & Thomson, W.W. (1984). Appearance of a state 1–state 2 transition during chloroplast development in the wheat leaf: Energetic and structural considerations. *Physiologia plantarum*, **60**, 171–9.

Whatley, J.M. (1979). Plastid development in the primary leaf of *Phaseolus vulgaris* variation between different types of cell. *New Phytologist*, **82**, 1–10.

Whatley, J.M. (1980). Plastid growth and division in *Phaseolus vulgaris*. *New Phytologist*, **80**, 1–16.

Whatley, J.M., Hawes, C.R., Horne, J.C. & Kerr, J.D.A. (1982). The establishment of the plastid thylakoid system. *New Phytologist*, **90**, 619–30.

J.M. WHATLEY

Mechanisms and morphology of plastid division

It is generally accepted that plastids never arise *de novo*, but rather undergo self replication. When the number of plastids in each cell is low, the process of replication must, of necessity, keep pace with cell division, but when the plastid population is large, such an interrelationship is much less apparent. Although plastids are believed to divide by a form of binary fission, details of the mechanisms involved are but poorly understood. What, then, is known about these mechanisms? Hard evidence is in short supply, but perhaps some indication of the mechanisms involved can be obtained from systems in which plastid division has been interrupted or delayed in such a way that individual elements of the finely tuned and integrated process are separately expressed.

In the last decade, biochemical and ultrastructural information has accumulated which strongly supports the hypothesis, first put forward by Schimper (1883), that chloroplasts evolved from photosynthetic endosymbionts. The endosymbionts envisaged as ancestral to the chloroplasts of the red (Rhodophyte) and the green (Chlorophyte) algae are considered to have been prokaryotic; those ancestral to the plastids of the diverse Chromophyte algae (dinoflagellates, diatoms, browns, etc.) and of the Euglenophytes may have been eukaryotic (Gibbs, 1981; Whatley & Whatley, 1981). If the ancestors of plastids in green algae and their descendants, the land plants, were once free-living prokaryotic algae which themselves replicated by binary fission, might the division process in modern prokaryotes provide some insight into that in plastids? Might the integrated patterns of division shown by present-day hosts and their endosymbionts lead us to a better understanding of those of the cell and its plastids?

Plastids in land plants

As plastids develop from the proplastid state into mature chloroplasts, their shape changes from that of a sphere into a conspicuously oblate spheroid (Fig. 1a,c,d) – the characteristic discoid form. The transition between these two stages is marked by the plastid becoming pleomorphic (Fig. 1b). During development, the minor axis of the plastid changes little in length, but in leaf mesophyll and other actively photosynthetic cells, the major axes can increase to more than five times their original length. Throughout this time of overall plastid growth, the profiles of the plastids as seen in face view remain circular, except during those brief and intermittent periods when plastid division is in progress.

Plastids can divide at any stage of their development from the proplastid to the recently mature chloroplast and, in each cell, all plastids appear to be capable of division (Whatley, 1986). Proplastids and chloroplasts both seem to follow the same sequential steps during the division process: (1) elongation – in the direction of one major axis (Fig. 1e,f); (2) progressive annular constriction – in a plane at right angles to the major axes and usually about the midpoint (Fig. 1g); (3) assumption of a dumb-bell shape with a narrow interconnecting isthmus (Fig. 1h) and the segregation of plastid DNA and thylakoids; (4) separation of the two daughter plastids following their twisting in orientation about the connecting neck (Figs. 2–5). Although most of these steps are easily identified, questions remain about their precise order and about the mechanisms involved in the formation of each.

Fig. 1. (a–d) Three dimensional views of changes in plastid shape during development. (e–h) Face views of changes in plastid shape during plastid division. (e) Is the direction of pre-division plastid extension random or controlled? (f) Possible mechanisms involved in pre-division extension; are they unidirectional or bidirectional? (1) Longitudinal attachment of (actin) microfilaments; (2) generalised single axis plastid growth; (3) localised single axis plastid growth; (4) polar attachment of (actin) microfilaments. (g) Possible mechanisms involved in constriction: (1) contractile system in the cytoplasm; (2) envelope constriction produced by forces of surface tension; (3) contractile system in the plastid stroma. (h) The twisting and final separation of the daughter plastids.

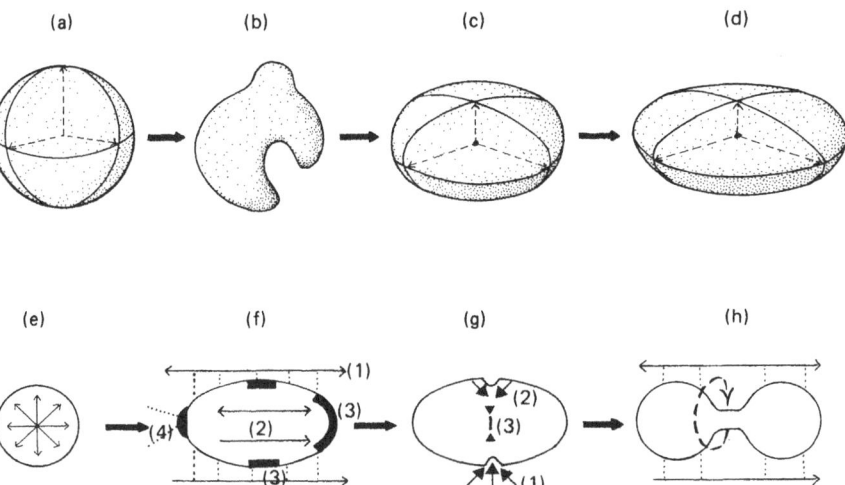

Fig. 2. Elongated and constricted chloroplasts in young leaves of *Phaseolus vulgaris*. Block courtesy of Dr C.R. Hawes; tissue fixed and stained with osmium tetroxide and zinc iodide. Bar, 1.0 μm.

Fig. 3. Chloroplast of *Spinacia oleracea* at a late stage of division. A single thylakoid strand continues to link the two putative daughter plastids. Reproduced, with permission, from Whatley (1971). Bar, 1.0 μm.

Fig. 4. The isthmus between two putative daughter plastids (Pa and Pb). A pair of electron-dense plaques (→) lie within the lumen of the plastid envelope. Bar, 0.5 μm

Fig. 5. A dividing chloroplast of *Phaseolus vulgaris* just before separation. The thylakoid systems of the future daughter plastids have rotated with respect to each other. Bar, 1.0 μm

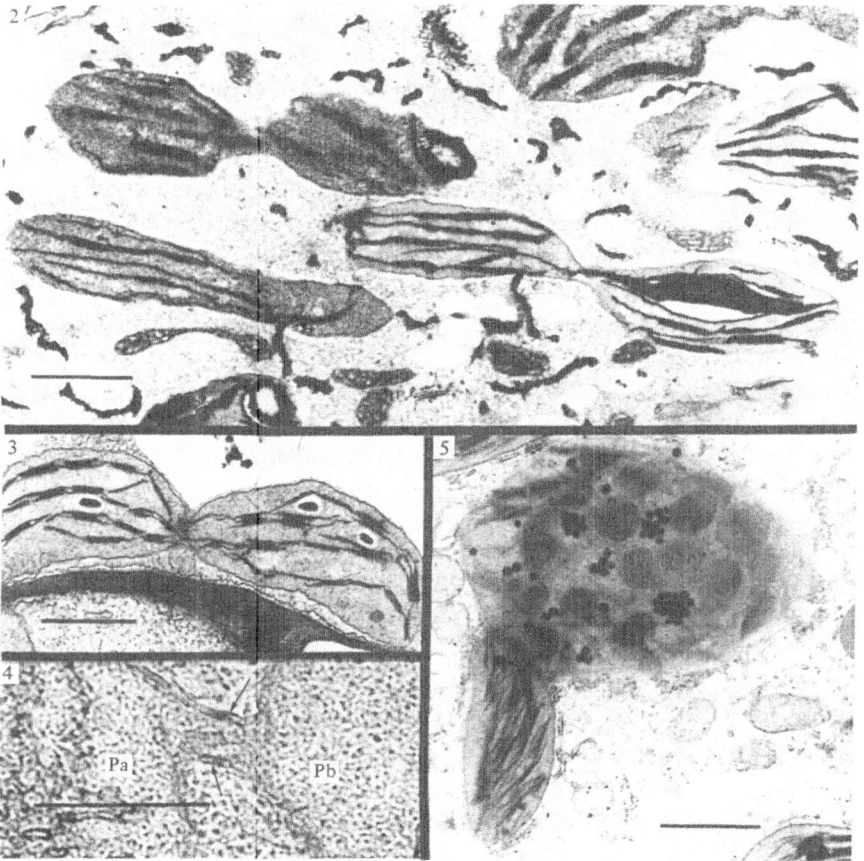

Elongation and polarity

The interrelationship between the processes of elongation and constriction is not completely clear. Leech and her associates (Leech, Thomson & Platt–Aloia, 1981) proposed a tentative time sequence of events, based on their light-microscope observations of isolated chloroplasts from wheat leaves. They suggested that plastids entering division can first be recognised as fat dumb-bells which only later become elongated. These elongated plastids are often attenuated at their distal ends (peanut-shaped) and have a central constriction which is only visible in face view. As division proceeds, the connecting isthmus becomes narrower and the ends more rounded and, later, the daughter plastids separate. In young leaf tissue of several species examined in the light microscope and in the transmission and scanning electron microscopes, elongated plastids are found which can be shown to lack constrictions. It has been suggested in earlier papers that it is this elongated stage which marks the beginning of plastid division (Whatley, 1980; Whatley *et al.*, 1982; Possingham & Lawrence, 1983). The discrepancy between these observations may be the result of differences in species, in growing conditions or in experimental treatment; it may also indicate that the processes of elongation and constriction are under separate control and that their relative timings are not fixed. It is convenient here to treat the two as separate processes and to include in the section on elongation a discussion of plastid polarity.

The elongation process brings about a noticeable increase (commonly a doubling) in the length of the plastid along one major axis. It should, however, be recognised that this pre-division extension is quite distinct from the extension growth associated with the overall increase in plastid size. The first is along one axis only and is intermittent; it is limited to the early phase of each cycle of division. The second is multidirectional, though still mainly within the plane of the major axes; it is essentially a continuous process throughout the development of many types of cell and results in a uniform increase in diameter of the circular plastid profiles. These two forms of extension growth frequently, but certainly not always, operate together.

In young primary leaves of *Phaseolus vulgaris*, the mode and the major size classes of plastids in the epidermal cells remain constant throughout the period of plastid division following germination (Whatley, 1980). This indicates an absence of overall plastid growth. However, the positive skewness of the data illustrates the presence of a subpopulation of elongated (i.e. pre-division) plastids and there is a concomitant increase in plastid number per cell file. In the epidermal cells, therefore, the plastids are approximately doubling in length prior to division and producing daughter plastids which are of about the same size as their parents. By contrast, in the palisade cells, there is a conspicuous overall increase in plastid size both during the period of plastid division and afterwards. The positive skewness of the data and the increase in plastid number again show that the plastids are dividing. In the palisade cells, therefore, overall growth and predivision extension of the plastids operate together and the daughter plastids in each successive generation are larger than their parents; but, when division has stopped, overall growth of the plastids continues to operate independently. This is the most common pattern. Much less frequently, plastids decline in size when they fail to double in length prior to division (Butterfass, 1979, 1980). However, I know of no example of plastid division in the absence of some degree of pre-division single axis extension growth.

Though pre-division extension appears to be an integral part of plastid replication, we know virtually nothing about the mechanisms involved. How is the direction of extension controlled? Is each plastid so polarised that it always elongates in the same direction or is the direction random for each separate cycle of division? Perhaps direction is determined by the cytoskeletal framework of the cell itself. Is elongation unidirectional or bidirectional (Fig. 1f)? Do the forces which produce extension result from 'pulling' by the cell – i.e. by stretching – or from 'pushing' by plastid growth? The former seems unlikely on any significant scale, as there is no evidence of the disruption, stretching or thinning out of plastid components such as membranes or granal stacks, except perhaps within the dumb-bell isthmus during the final stages of division.

Many bacteria and cyanobacteria are filamentous, a growth habit which indicates a continuing directional trend which is uninfluenced by the surrounding environment. The giant amoeba *Pelomyxa palustris* has three species of endosymbiotic bacteria of which one often forms short filaments of 2–8 cells following division. This bacterium is unusual in that it also has one end wall which is highly distinctive and can be used as a structural marker to identify cell polarity. It is clear that within each filament, the distinctive wall always forms at the same end of every daughter cell (Whatley, 1976). By contrast, membrane-labelling techniques have shown that daughter cells of the mycoplasma *Spiroplasma citri* develop reversed polarity and are thus mirror images of each other (Garnier, Clerc and Bové, 1984). These latter investigations also suggest that prior to division of *Spiroplasma* by binary fission, growth of the plasma membrane of this (wall-less) prokaryote is at first restricted to one end; later, a second zone of growth is initiated adjacent to a central constriction, though whether this second phase of growth begins before or after the onset of constriction is not clear. The authors point out that the system of division in this mycoplasma is very similar to the unit cell model for *Escherichia coli* proposed by Donachie & Begg (1970), in which the bacteria grow at the single apical site when they are less than, and at the two sites when they are greater than, a critical length.

Plastids, too, can occasionally form short filaments or pseudofilaments (Fig. 6). Do they also show polarity and have similar well-defined sites of localised envelope extension? Kolkwitz (1899) is stated by Fritsch (1935) to have shown that the chloroplasts of the green alga *Spirogyra* elongate by both apical and 'intercalary' growth, but the subject of localised organelle growth seems to have received little attention in more recent times. Might the attenuated ends of elongating (peanut-shaped) plastids described by Leech and her associates (Leech *et al.*, 1981) indicate such specific sites of localised growth (Fig. 1f), or might they represent sites at which the plastids are attached to the cytoskeletal framework of the cell?

Brown & Lemmon (1984) investigated the root meristematic cells of *Selaginella* and *Isoetes*. Each such cell contains a single plastid which divides at the same time as the cell divides. They found that the long axis of the plastid isthmus was always aligned at right angles to the preprophase band of microtubules. The authors concluded from their observations that the plastid had a definable axis and suggested that a system of microfibrils and microtrabeculae might be responsible for stabilising the plastid. They further noted that only after the plastid had taken up its new position within the cell did its poles become linked to cytoplasmic microtubules.

Generally plastids lack an obvious structural marker which can be used to determine polarity. An exception may be the chloroplasts of the coenocytic green algae *Caulerpa*, *Halimeda* and related genera. These plastids have at one tip a concentric arrangement of membranes (the concentric lamellar system or polar thylakoid body) which has been postulated to be the site of thylakoid origin (Borowitzka & Larkum, 1974; Calvert, Dawes & Borowitzka, 1976). Borowitzka & Larkum (1974) and Giles & Sarafis (1974) independently observed that in apparently dividing plastids, the two concentric lamellar systems lay at opposite ends of each plastid. The positioning of the two concentric lamellar systems at the plastid poles could indicate that the plastids in the Caulerpales, like *Spiroplasma citri*, have mirror-image polarity. Perhaps with the new histological and immunological techniques now

Fig. 6. A 'filamentous' chloroplast from *Equisetum telmateia* with three constrictions. At each constriction the thylakoid system has become almost or completely split. Bar, 1.0 μm

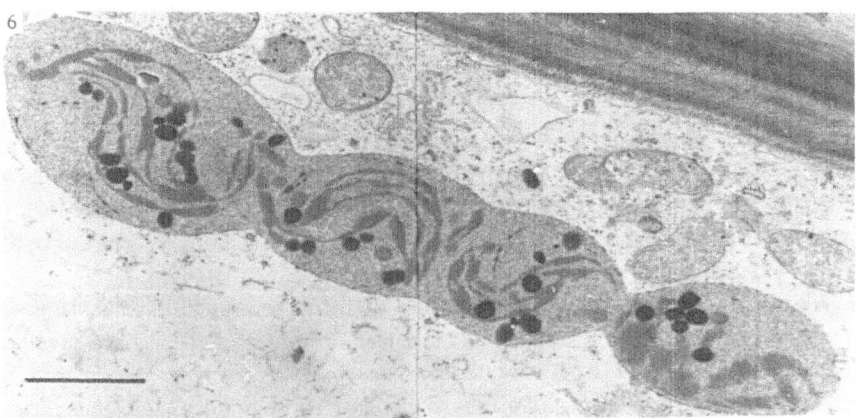

becoming available for plant cells, the time has come when we can at last find out if the plastids or angiosperms, too, can show inherent polarity or localised membrane and thylakoid growth during division.

A positive correlation between chloroplast number per cell and cell size has been recorded for several species, but other observations indicate that the two criteria are independent of each other (summarised in Butterfass, 1979). Butterfass has concluded that although chloroplast number depends on nuclear DNA amount, a direct correlation between cell size and chloroplast number cannot be completely ruled out. Is there any obvious way in which cell expansion might directly influence the process of plastid division?

Even if the plastids of higher land plants are polarised, the directional and spatial constraints imposed by the cytoskeletal framework of the expanding cell and its developing central vacuole must play a part in controlling directional extension of the plastids. In very young tissues which lack central vacuoles, cell expansion at first tends to be multidirectional. In young palisade cells of this type in primary leaves of *Phaseolus vulgaris* the direction of pre-division plastid extension is very varied (Fig. 7a). In leaves which have almost completed expansion but in which chloroplast division still continues, the plastids are confined to the thin layer of peripheral cytoplasm within which the possible directions for plastid extension must be severely limited. Between these two extremes are immature palisade cells (Fig. 7b) in which the central vacuole is not yet established and in which cell expansion is limited to a plane perpendicular to the leaf surface (Dale, 1964; Whatley, 1980, 1986). In these cells the direction of extension of almost all plastids except those adjacent to the upper and lower cell walls is approximately parallel to that of cell extension. Does this conspicuous change in the direction of extension of most plastids (*c.* 25°) during early leaf development perhaps result from a major change in pitch of their cell's cytoskeleton or cytomatrix? Note that the observations on plastid alignment shown in Fig. 7 are part of a preliminary investigation and need to be extended to include additional statistical data.

Is there any other indication that elements of the cell cytoskeleton might influence the direction of pre-division plastid extension? May the plastid envelope become so linked to the cytomatrix that cell expansion directly promotes plastid extension (Fig. 1f)? Examples of direct attachment of plastids to some cytoskeletal elements are certainly known – to microtubules in some mosses during sporogenesis (Brown & Lemmon, 1982, 1983) and to actin microfilaments in algae such as *Chara*, (Williamson, 1985), *Nitella* (Green, 1964; Palevitz & Hepler, 1975) and *Mougeotia* (Marchant, 1976; Haupt, 1982). Freeze–fracture studies of the giant internodal cells of *Chara* (McLean & Juniper, personal communication) clearly show that the actin cables are attached by short microfilaments to the envelopes of the adjacent chloroplasts. Most of the known linkages between cytoskeletal elements and plastids are associated with plastid movement, but the *Chara* plastids are static. As the actin cables and the chloroplasts have the same parallel alignment, the cables are more likely to be responsible for the anchorage and alignment of the plastids in these cells.

A similar role for actin filaments has recently been suggested in a lower land plant. An investigation of light-dependent chloroplast movement was carried out using high-voltage electron microscopy of whole critical point dried epidermal cells of *Selaginella helvetica* (Cox *et al.*, 1987). It was shown that the chloroplasts were connected by bundles of tightly packed microfilaments which in turn seemed to interconnect with, but were distinct from, the filaments of the cytomatrix. The staining properties of these bundles implied that their microfilaments were composed of actin. It was concluded that, though the cytomatrix system may play a part in chloroplast movement, the behaviour of the microfilament bundles indicated that their role was more likely one of anchorage.

It is now believed that actin strands may also have a cytoskeletal function in higher plant cells (Parthasarathy *et al.*, 1985). The possible significance of the micro-

Fig. 7. Tracings from montages of electron micrographs of palisade cells from leaves of *Phaseolus vulgaris* sampled on the second (a) and the fifth (b) day after seed germination. The tracings show the orientation of the elongated and constricted plastids within these cells. The bar graphs record the numbers of plastids in each size class of 10° intervals. Angles were measured as deviations from the plane of the leaf surface.

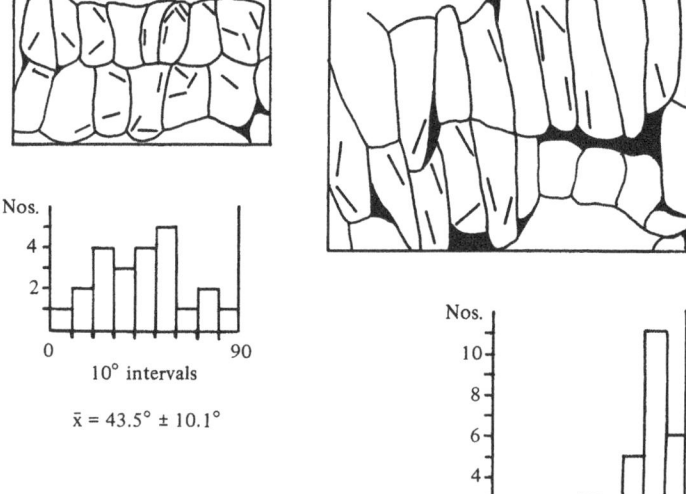

Day 2   (a)   Day 5   (b)

Nos.

$\bar{x} = 43.5° \pm 10.1°$

$\bar{x} = 68.7° \pm 7.5°$

10° intervals

filaments in either stretching or, more realistically, in maintaining extension of the plastids, has not to my knowledge been investigated. At first sight the direct involvement of cytoskeletal elements in maintaining plastid extension might seem unlikely, and certainly when chloroplasts are isolated in a carefully selected suspending medium they can retain their extended form (Leech *et al.*, 1981). Nevertheless, when cell expansion is interrupted or halted, the potentially dividing plastids can fail to form dumb-bells and instead appear to round up and to fold their thylakoid systems about the expected plane of division.

Fern croziers provide a good illustration of a tissue in which cells undergo a natural interruption and subsequent resumption of expansion. The crozier provides in each

Fig. 8. The shapes and dimensions of chloroplasts and cells in coiled and uncoiling croziers of *Pilularia globulifera*. (●–●) cell length; (○–○).chloroplast mean axial ratio.

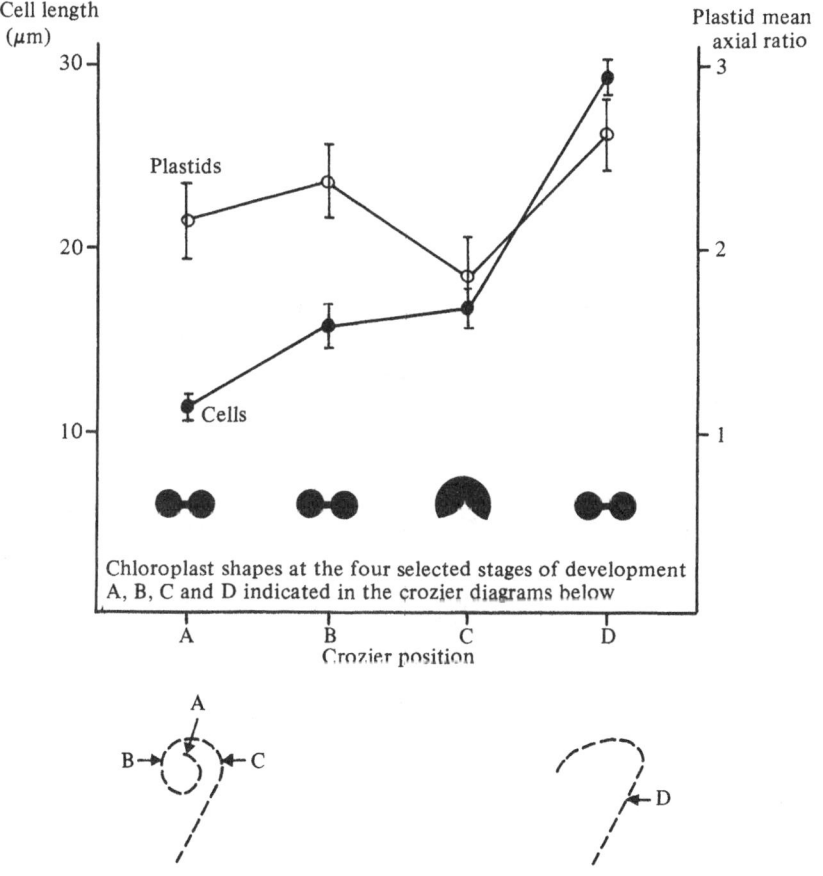

cell file a sequence of developmental stages extending from the apical cell at the tip to differentiated cells towards the base. Near the tip, successive cells show some increase in size as a result of the expansion which accompanies meristematic activity; older cells show no such change in size. It is only with the onset of crozier uncoiling that these older cells begin to expand once more. During the two separate periods of cell expansion in *Pilularia globulifera* (Whatley, 1975), elongated, constricted and dumb-bell-shaped plastids are common, but in the intervening stationary period they appear to be replaced by the 'folded' plastids described (Figs. 8–11).

The various observations described above, though certainly not definitive, seem to suggest that growth of the plastid envelope and thylakoid system must surely be necessary for the pre-division extension as well as for the overall plastid growth. Though an innate plastid polarity may play some part, elements of the cell cytoskeleton (actin?) may well help to control the direction of extension growth and may also be responsible for maintaining the plastid temporarily in the pre-division extended state.

Fig. 9. Dividing chloroplast of *Pilularia globulifera*; from Stage B in Fig. 8. Bar, 1.0 μm.

Fig. 10. Apparently dividing but rounded chloroplast of *Pilularia*; from Stage C in Fig. 8. Bar, 1.0 μm

Fig. 11. Dividing chloroplast of *Pilularia* ; from Stage D in Fig. 8. Bar, 1.0 μm.

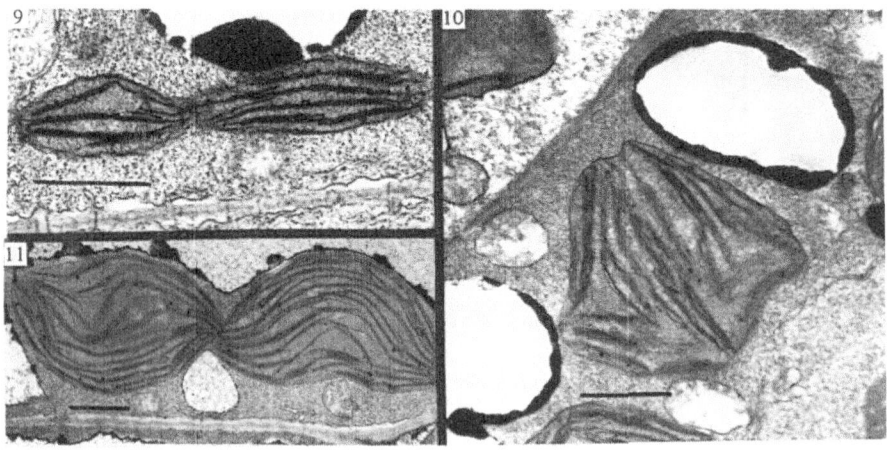

Constriction

When a soap bubble is stretched, it becomes centrally constricted, the resulting dumb-bell shape being a direct response to an attempted reduction in surface tension (Thompson, 1928). A similar model for cytokinesis based on surface tension forces was put forward more recently by Greenspan (1977; 1978) who suggested that an equatorial contractile belt of actin/myosin filaments might also be involved. Leech and her associates (Leech *et al.*, 1981) have suggested that their observations on chloroplast division could be explained by surface tension forces as in the Greenspan model. Is the dumb-bell shape assumed by the dividing plastid, then, a purely physical phenomenon, or are other contractile mechanisms involved (Fig. 1g)? Might the constricting surface tension forces so squeeze the plastid about its median plane that they alone are responsible even for pre-division plastid extension? This last seems unlikely as a sole cause, since plastids are often found which are elongated yet lack an annular constriction (Fig. 2).

The constriction of dividing plastids is usually central, but can sometimes be asymmetrical. In young primary leaves of *Phaseolus vulgaris* the rate of plastid division approximately keeps pace with that of cell division during the earliest stages of seed development on the parent plant, but later, shortly after germination of the ripe seed, the rate of plastid division increases and the number of plastids per palisade cell rises from about 10 to about 35 (Whatley, 1980, 1986). When cell division stops, the rate of plastid division declines, and some days later, plastid division also comes to an end. Dividing plastids with central constrictions are characteristic of the early and late periods of replication; plastids with asymmetrical constrictions are characteristic of the intermediate period, when the rate of plastid division is at its peak. Might this asymmetry of constriction indicate that the rate of pre-division plastid extension has temporarily become out of phase with the separate process of constriction? Must it not also imply that forces other than surface tension are involved in plastid constriction?

In land plants, as the dividing plastids become dumb-bell-shaped, the narrow neck connecting the two daughter plastids is often encircled by an annulus. This structure was first observed as a pair of electron-dense plaques (Fig. 4) positioned on either side of the isthmus and lying in the lumen between the two membranes of the plastid envelope. Only later were the plaques identified as segments of an annulus (Suzuki & Ueda, 1975; Chaly & Possingham, 1981).

It has been suggested that the separation of some dividing bacteria and of cyanobacteria is promoted by the formation of the intervening wall. In some species the wall material is preferentially deposited around the invaginating septum to form a greatly thickened annulus (Plate 13:10 in Drews & Weckesser, 1982). Cyanelles are photosynthetic structures found in a number of anomalous unicellular protists; they have been postulated to be intermediate in status between cyanobacteria and chloroplasts (Aitken & Stanier, 1979; Whatley, John & Whatley, 1979). Those in

Cyanophora paradoxa, like the chloroplasts of red and green algae and land plants, have an envelope comprising two membranes with no visible cell wall between. However, when the cyanelles divide, fibrillar electron dense material is deposited as an annulus within the lumen of the envelope at the constricting neck which forms between the daughter cyanelles (Pickett–Heaps, 1972). This fibrillar material is sensitive to lysozyme, suggesting the presence of peptidoglycan, a prokaryotic wall component, which has indeed been shown to be present (Schenk, 1970; Aitken & Stanier, 1979). Though the *Cyanophora* annulus is more massive, it nevertheless bears some resemblance to that associated with the isthmus of dividing plastids. However the possibility that the chloroplast annulus might represent the residue of an ancestral prokaryotic cell wall must be reconsidered in the light of more recent information.

Although images of dividing plastids can be found which appear to show that the electron-dense plaques lie entirely within the envelope lumen as described above (Fig. 4), Hashimoto (1986) has published other micrographs of plastids in *Avena sativa* which appear to show with equal clarity that the envelope lumen is electron lucent. In Hashimoto's micrographs, each 'single' plaque can be seen to consist of a pair of stained deposits, lying on either side and to the exterior of the membranes. Hashimoto correctly points out that such a configuration implies that the annulus does *not* in this case lie within the envelope lumen but rather comprises two rings, the first lying on the cytoplasmic side of the outer plastid envelope and the second on the stromal side of the inner envelope. Because the appearance of electron-dense images depends on both the number of 'stain atoms' in the path of an electron beam and the angle at which a structure is cut, the annular plaques most often seen in thin sections appear blurred. Only in these few plastid sections in which the membranes at the isthmus are cut in exact median transverse section can the position of the annulus be determined with reasonable accuracy, and even then the image is not completely sharp. An obvious discrepancy therefore remains to be resolved and must await the development of a treatment which can stain or otherwise delineate the annulus or paired annuli with much greater precision than is at present possible.

The paired annuli described by Hashimoto perhaps represent the more probable configuration. Some support for the existence of an independent cytoplasmic annulus comes from the work of Mita *et al.* (1986) who described an electron-dense annulus associated with the constriction in dividing chloroplasts of the unusual red alga *Cyanidium caldarum*. In this species the annulus is a single ring which is much more massive than that associated with angiosperm chloroplasts. It is clearly confined to the cytoplasm and positioned some distance away from the chloroplast envelope. This annulus could be identified not just at the dumb-bell stage of plastid division, but even during the very earliest stages of constriction. No annulus was observed inside the chloroplasts.

As yet we know virtually nothing about the structure or function of these annuli. May they, perhaps, be (actin/myosin) contractile rings which, as in the Greenspan model for cytokinesis, help to constrict the plastids during division? The cytoplasmic 'space' created by constricting chloroplasts frequently becomes occupied by other cell organelles. Some of these organelles can become conspicuously deformed in shape. Significantly, the organelles in question are all associated with the endomembrane system. They include microbodies (Figs. 15, 20, 21 in Pickett–Heaps, 1972), vacuoles, endoplasmic reticulum and even segments of nuclear envelope, but not mitochondria which appear to be less deformable (Figs 3, 11, 12 and 13). The shapes assumed by these organelles, particularly the microbodies illustrated by Pickett–Heaps, suggest the existence of a force of considerable strength exerted in a direction perpendicular to the chloroplast constriction. There is no evidence to indicate whether this seemingly strong force might represent the 'sucking' of organelles into a cytoplasmic 'vacuum' created by the constricting plastid, or the 'pulling' of them, perhaps – and it is here that association with the endomembrane system may be relevant – by a contractile system of cytomatrix microfilaments which

Fig. 12. A dividing chloroplast of *Phaseolus vulgaris*. Note that part of the nucleus appears deformed and occupies the cytoplasm within the chloroplast constriction (→). Bar, 1.0 μm.

Fig. 13. A dividing plastid of *Equisetum telmateia*. Note that the sheath of endoplasmic reticulum (→) continues to run parallel to the envelope as the plastid becomes constricted. Bar, 1.0 μm.

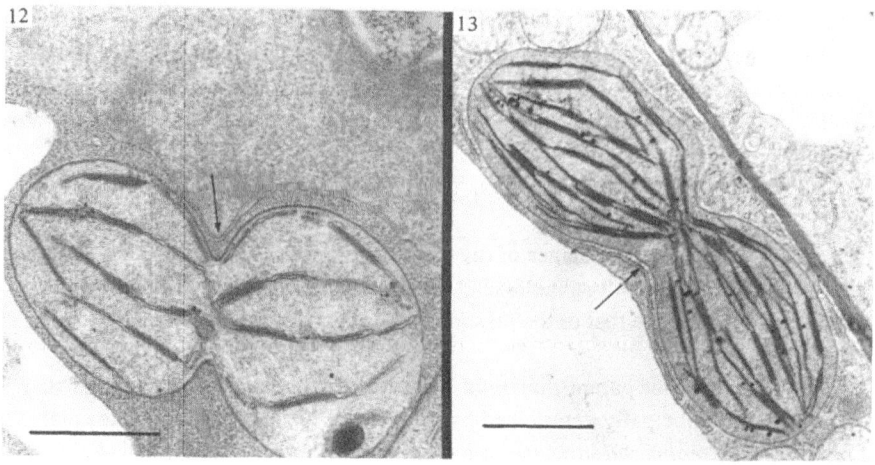

firmly but temporarily link a short segment of the organelle's membrane or envelope to that of the chloroplast isthmus during constriction.

A range of structures superficially resembling microfilaments and microtubules have been identified in several genera of filamentous cyanobacteria. Some of these, arrayed with plate-like bodies, have been seen in close parallel alignment with the cyanobacterial cross-walls (Jensen, 1985). Actin-like proteins and 4–nm filaments or fibrils have been discovered in other prokaryotes, for example in wall-less species which are capable of changing their shape (Searcy, Stein & Searcy, 1981; Townsend & Archer, 1983; Garnier et al., 1984; Williamson, Brink & Zieve, 1984). Plastids, too, frequently undergo minor changes in shape, and a range of filamentous and tubular structures have frequently been identified in the plastid stroma (Chaly & Possingham, 1981; Lawrence & Possingham, 1984). The presence in plastids of an actin-like component has thus always been a possibility, though it has usually been judged to be a remote one. It is therefore particularly exciting that McCurdy & Williamson (1987) have discovered a pea chloroplast protein which, by several chemical and immunological criteria, resembles vertebrate and algal actins. The authors established that this actin-related protein ($41\ 000$–M_r) resided within the outer envelope membrane, and was apparently associated with the stromal rather than with the thylakoid fraction. A second actin-like protein ($58\ 000$–M_r) was also identified, but was less thoroughly investigated.

The discovery of an actin-like protein inside chloroplasts raises the interesting possibility that a myosin-like protein might also be present. If this were so, might angiosperm plastids have their own contractile structure – the inner annulus described by Hashimoto – capable of working in conjunction with an outer contractile ring situated in the cytoplasm? Such a contractile ring within the stroma might help to constrict the thylakoid system as well as the plastid and, in addition, help to maintain the dumb-bell shape of dividing plastids following their isolation. Improving techniques may soon make it possible to solve even the problem of these enigmatic annuli.

Separation

During the final stages of division, the plastid contents must be distributed between the daughter organelles. Only segregation of the thylakoids will be considered here, since that of the plastid DNA is described by Dr Rose in Chapter 10 in this volume.

The proportion of parent thylakoid material which enters each daughter plastid appears to be principally determined by the site at which constriction takes place. In leaves of *Phaseolus vulgaris* the thylakoid system forms a broad spiral ribbon of about two coils (Whatley et al., 1982). When the developing chloroplasts divide, the thylakoids of each coil appear to be severed in turn, usually from the exterior towards the interior. The severed and possibly split ends of the thylakoids fuse together seemingly at random and thus produce daughter spirals which become more complex with each division cycle. The period of thylakoid division can be both variable and prolonged. In some plastids, the thylakoid system appears already to have divided when envelope constriction has scarcely begun. In other plastids, the two thylakoid systems remain linked even during the final dumb-bell stage of division. The thylakoid link across the isthmus is usually a single stroma lamella which appears porous or vesicular, suggesting that severing of the spiral ribbon may involve thylakoid dedifferentiation (Whatley, 1980). These various observations suggest that separation of the thylakoid system may involve a mechanism which operates under a control which is different from that which separates the daughter plastids.

Shortly before separation, the two segments of the plastid dumb-bell and/or their thylakoid contents (Fig. 5) rotate with respect to each other (Whatley, 1980; Leech *et al.*, 1981; Whatley *et al.*, 1982). Might rotation of the plastids perhaps result from torsion set up by components of an attached cytomatrix? Alternatively, if the thylakoid system alone rotates, might this rotation be simply a readjustment with respect to each other of the two now independent spirals? In his simulation of cytokinesis, Greenspan (1978) found that the forces of surface tension were insufficient to separate constricted oil droplets. For dividing plastids we do not know what forces are necessary for separation or what role, if any, rotation plays in the process.

Following separation of the daughter plastids, the envelope membranes of each must fuse together. Leech and her associates have observed that as division nears completion, envelope membranes at the isthmus become locally disordered and myelin-like material may be formed. They suggest this may represent a transitory stage in the necessary membrane reorganisation. Plastid separation is very rapid (Possingham & Lawrence, 1983) and thus information about the final stages of division is particularly difficult to obtain.

The separate elements of plastid division

Many separate processes are needed to bring about organelle division, but it is difficult to find conditions under which the individual events involved can be distinguished. It is not unreasonable to consider that the various processes which combine to produce plastid division might all end at different times, though the order in which they might do so would be expected to vary in different species or under different growth conditions. It is, for example, established that during the later stages of plastid division in wheat and spinach, replication of plastid DNA stops well in advance of chloroplast division (Boffey & Leech, 1982; Scott & Possingham, 1983).

In leaves of *Phaseolus vulgaris*, during the week immediately after plastid division has stopped, and only then, some plastids develop aberrantly. They double or more in length and their thylakoid systems can become constricted, rotate or divide, even in the absence of envelope constriction (Figs 14, 15). In equivalent post-division cells of *Equisetum telmateia* and in the non-expanding cells of the *Pilularia* crozier, thylakoid constriction and separation can even take place in the absence of plastid elongation (Whatley, 1971, 1975). Do these examples represent plastids in which some of the separate division mechanisms have continued to operate when others have stopped? Aberrant plastids similar to those in *Phaseolus* have previously been observed in *Mimosa pudica* and have been interpreted as resulting from plastid fusion (Esau, 1972). However, the timing and the available cell space observed in *Phaseolus* would seem to favour the first alternative.

Occasionally in leaves of *Phaseolus* that are unusually slow to expand, and more often in older cells of *Equisetum*, one finds populations of apparently dividing plastids in which the inner plastid envelope has become conspicuously invaginated, either unilaterally or, much less often bilaterally, at the expected site of constriction (Fig. 16). These median invaginations are segments of small sheets which, in thin section, appear longer, narrower and straighter than the many small (usually tubular) invaginations that can be found at any position around the envelope of many plastids. A variety of causes and interpretations have been suggested for this well-recorded configuration, but it still remains unexplained (Gifford & Stewart, 1967; Cran & Possingham, 1972; Dyer, 1976). Nevertheless the formation of such invaginations could well indicate that, during plastid division, extension of the inner and the outer envelope membranes may be independently controlled and indeed may even reflect the independent origins of the two membranes from a host vacuolar and an endosymbiotic cell membrane respectively (Whatley, 1976). If these various anomalous plastids do indeed illustrate some of the separate mechanisms of plastid division, further investigation of plants grown under conditions which modify the rate of growth, for example, might well provide new information about the division process.

Conclusions

This survey of the mechanisms and morphology of plastid division has illustrated the scarcity of substantiated information on the subject and has, indeed, raised more questions than it has answered. The process of plastid division has been followed using the light microscope and there is no reason to doubt that it is accomplished by binary fission, following the establishment of a more or less central constriction; several separate mechanisms are clearly involved. During division the plastid becomes elongated to a greater or lesser extent. This pre-division elongation involves extension of the thylakoid system and of both the inner and outer envelope membranes. We do not know how the direction of extension is controlled or whether this control is exerted by the cell (perhaps by means of the cytomatrix), by the plastid (perhaps by its innate polarity) or by both. It is highly probable that extension is the result of membrane growth, but we do not know if this is localised (perhaps initially apically, as in some prokaryotes) or general. The occasional atypical median invaginations of the inner plastid envelope could point to: (a) separate mechanisms for elongation of the inner and outer envelope membrane and (b) centrally localised extension of the inner envelope which is comparable to the secondary extension of the prokaryotic plasma membrane at the central constriction. Both the thylakoid system and the plastid envelope can become constricted, but we do not know if the two processes are interdependent, and, perhaps, produced in response to a common contractile system. There is, however, some evidence that division of the thylakoid system is at least partially independent of plastid division and that it may involve

Fig. 14. A chloroplast in a 3-week-old leaf of *Phaseolus vulgaris*. There is no detectable increase in chloroplast number between weeks 2 and 3. However, in this elongated plastid, the thylakoid system has apparently divided and twisted. Bar, 1.0 μm.

Fig. 15. A chloroplast from a leaf of the same age as in Fig. 14. In this elongated plastid the thylakoid system has a conspicuous median constriction (▲) as well as additional misorientation (→) suggesting even further constriction. Bar, 1.0 μm.

Fig. 16. An elongated plastid from *Phaseolus vulgaris* which has an extensive median invagination of the inner plastid envelope (→). Bar, 1.0 μm.

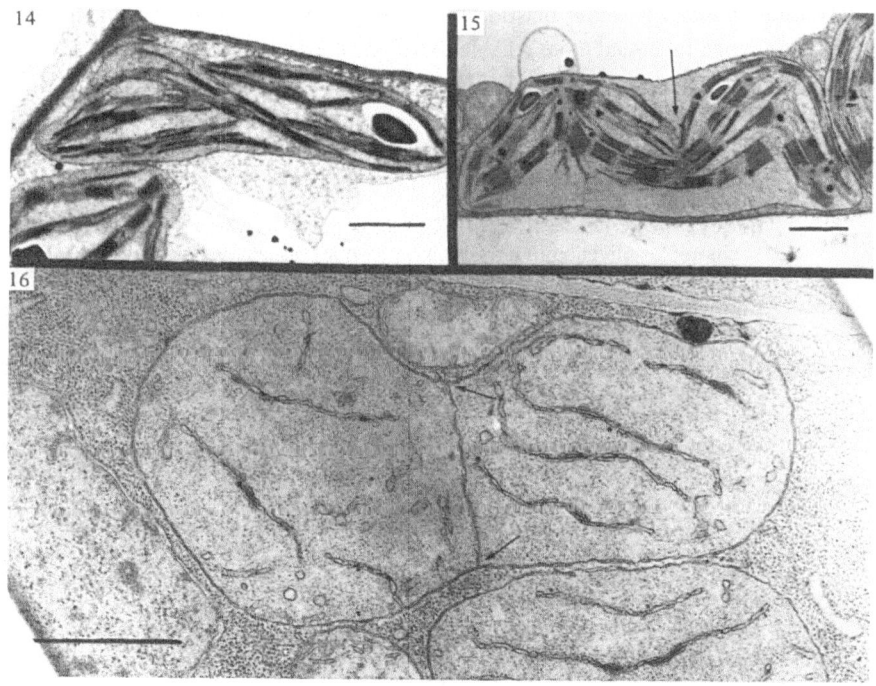

dedifferentiation or vesicularisation of segments of the thylakoid ribbon in each coil. It has been suggested that envelope constriction results from forces of surface tension, but recent observations seem to point to the presence of a (supplementary?) contractile system. If there is such a system, we have yet to determine if it is located in the cytoplasm, in the plastid stroma or even in both. Shortly before separation, the thylakoid systems of the daughter plastids appear to rotate with respect to each other. We do not know whether this change in thylakoid orientation results from the twisting of the plastids or just of the thylakoids. Nor do we know the cause of this rotation. We are equally ignorant of both the cause and the method of final plastid separation.

Although during the past decade our knowledge of the mechanisms and morphology of plastid division has not increased as much as one might, perhaps, have hoped, we may now have arrived at a stage when further advances can realistically be expected. Investigations of the roles of actin both in the cytoplasm and in the plastids will surely be extended, and, in the plastids, myosin and similar proteins must be sought. Using histochemical and immunological techniques similar to those already carried out successfully with *Spiroplasma* and other prokaryotes, it should now be possible to determine the polarity and patterns of membrane growth of the plastids. When these and other studies are put together with the information now becoming available – about plastid DNA, for example – it should be possible to produce a much fuller and more reliable description of how plastids divide.

References
Aitken, A. & Stanier, R.Y. (1979). Characterization of peptidoglycan from the cyanelles of *Cyanophora paradoxa*. *Journal of General Microbiology*, **112**, 219–23.

Boffey, S.A. & Leech, R.M. (1982). Chloroplast DNA levels and the control of chloroplast division in light-grown wheat leaves. *Plant Physiology*, **69**, 1387–91.

Borowitzka, M.A. & Larkum, A.W.D. (1974). Chloroplast development in the Caulerpalean alga *Halimeda*. *Protoplasma*, **81**, 131–44.

Brown, R.C. & Lemmon, B.E. (1982). Ultrastructure of sporogenesis in the moss *Amblystegium riparium*. I. Meiosis and cytokinesis. *American Journal of Botany*, **69**, 1096–107.

Brown, R.C. & Lemmon, B.E. (1983). Microtubule organization and morphogenesis in young spores of the moss *Tetraphis pellucida* Hedw. *Protoplasma*, **116**, 115–24.

Brown, R.C. & Lemmon, B.E. (1984). Plastid apportionment and preprophase microtubule bands in monoplastidic root meristem cells of *Isoetes* and *Selaginella*. *Protoplasma*, **123**, 95–103.

Butterfass, Th. (1979). Patterns of chloroplast reproduction. *Cell Biology Monographs*, vol. 1. Wien and New York: Springer–Verlag.

Butterfass, Th. (1980). The continuity of plastids and the differentiation of plastid populations. In *Chloroplasts*, ed J. Reinert, pp. 29–44. Berlin: Springer–Verlag.

Calvert, H.E., Dawes, C.H. & Borowitzka, M.A. (1976). Phylogenetic relationships of *Caulerpa* (Chlorophyta) based on comparative chloroplast ultrastructure. *Journal of Phycology*, **12**, 149–62.

Chaly, N. & Possingham, J.V. (1981). Structure of constricted proplastids in meristematic plant tissues. *Biologie Cellulaire*, **41**, 203–10.

Cox, G., Hawes, C., van der Lubbe, L. & Juniper, B.E. (1987). High-voltage electron microscopy of whole, critical point dried plant cells. II. Cytoskeletal structures and plastid motility in *Selaginella*. *Protoplasma*, **140**, 173–86.

Cran, D.G. & Possingham, J.V. (1972). Two forms of division profile in spinach chloroplasts. *Nature*, **235**, 142.

Dale, J.E. (1964). Leaf growth in *Phaseolus vulgaris*. I. Growth of the first pair of leaves under constant conditions. *Annals of Botany, N.S*, **28**, 579–89.

Donachie, W.D. & Begg, K.J. (1970). Growth of the bacterial cell. *Nature*, **227**, 1220–4.

Drews, G. & Weckesser, J. (1982). Function, structure and composition of cell walls and external layers. In *The Biology of Cyanobacteria*, Botanical Monographs, vol. 19, pp. 333–57. Oxford: Blackwell Scientific Publications.

Dyer, A.F. (1976). The visible events of mitotic cell division. In *Cell Division in Higher Plants*, ed. M.M. Yeoman, pp. 50–110. London: Academic Press.

Esau, K. (1972). Apparent temporary chloroplast fusion in leaf cells of *Mimosa pudica*. *Zeitschrift für Pflanzenphysiologie*, **67**, 244–54.

Fritsch, F.E. (1935). *The Structure and Reproduction of the Algae*, vol. 1. Cambridge University Press.

Garnier, M., Clerc, M. & Bové, J.M. (1984). Growth and division of *Spiroplasma citri*: elongation of elementary helices. *Journal of Bacteriology*, **158**, 23–8.

Gibbs, S.P. (1981). The chloroplasts of some algal groups may have evolved from endosymbiotic eukaryotic algae. *Annals of the New York Academy of Sciences*, **361**, 193–207.

Gifford, E.M. & Stewart, K.D. (1967). Ultrastructure of the shoot apex of *Chenopodium album* and certain other seed plants. *Journal of Cell Biology*, **33**, 131–42.

Giles, K.L. & Sarafis, V. (1974). Implications of rigescent integuments as a new structural feature of some algal chloroplasts. *Nature*, **248**, 512–3.

Green, P.B. (1964). Cinematic observations on the growth and division of chloroplasts in *Nitella*. *American Journal of Botany*, **51**, 334–42.

Greenspan, H.P. (1977). On the dynamics of cell cleavage. *Journal of Theoretical Biology*, **65**, 79–99.

Greenspan, H.P. (1978). On fluid mechanical stimulations of cell division and movement. *Journal of Theoretical Biology*, **70**, 125–34.

Hashimoto, H. (1986). Double ring structure around the constricting neck of dividing plastids of *Avena sativa*. *Protoplasma*, **135**, 166–72.

Haupt, W. (1982). Light-mediated movement of chloroplasts. *Annual Reviews of Plant Physiology*, **33**, 205–33.

Jensen, T.E. (1985). Cell inclusions in the Cyanobacteria. *Archiv für Hydrobiologie*, Suppl. 71, 33–73.

Kolkwitz, R. (1899). Die Wachstumsgeschichte der Chlorophyllbänder von *Spirogyra*. *Festschrift für Schwendener*, pp. 271–87.

Lawrence, M.E. & Possingham, J.V. (1984). Observations of mictotubule-like structures within spinach plastids. *Biologie Cellulaire*, **52**, 77–82.

82   J.M. WHATLEY

Leech, R.M., Thomson, W.W. & Platt-Aloia, K.A. (1981). Observations on the mechanism of chloroplast division in higher plants. *The New Phytologist*, **87**, 1–9.

McCurdy, D.W. & Williamson, R.E. (1987). An actin-related protein inside pea chloroplasts. *Journal of Cell Science*, **87**, 449–56.

Marchant, J. (1976). Actin in the green algae *Coleochaete* and *Mougeotia*. *Planta*, **131**, 119–20.

Mita, T., Kanbe, T., Tanaka, K. & Kuroiwa, T. (1986). A ring structure around the dividing plane of the *Cyanidium caldarum* chloroplast. *Protoplasma*, **130**, 211–13.

Palevitz, B.A. & Hepler, P.K. (1975). Identification of actin *in situ* at the ectoplasm-endoplasm interface. *Journal of Cell Biology*, **65**, 29–38.

Parthasarathy, M.V., Perdue, T.D., Witztum, A. & Alvernaz, J. (1985). Actin network as a normal component of the cytoskeleton in many vascular plant cells. *American Journal of Botany*, **72**, 1318–23.

Pickett–Heaps, J.D. (1972). Cell division in *Cyanophora paradoxa*. *The New Phytologist*, **71**, 561–7.

Possingham, J.V. & Lawrence, M.E. (1983). Controls to plastid division. *International Review of Cytology*, **84**, 1–56.

Schenk, H.A.E. (1970). Nachweis einer lysozymempfindlichen Stützmembran der Endocyanellen von *Cyanophora paradoxa* Korschikoff. *Zeitschrift für Naturforschung*, **25b**, 656.

Schimper, A.F.W. (1883). Uber die Entwicklung der Chlorophyllkörner und Farbkörper. *Botanische Zeitung*, **41**, 105–14.

Scott, N.S. & Possingham, J.V. (1983). Changes in chloroplast DNA levels during growth of spinach leaves. *Journal of Experimental Botany*, **34**, 1756–67.

Searcy, D.G., Stein, D.B. & Searcy, K.B. (1981). A mycoplasma-like archaebacterium possibly related to the nucleus and cytoplasm of eukaryotic cells. *Annals of the New York Academy of Sciences*, **361**, 312–23.

Suzuki, K. & Ueda, R. (1975). Electron microscope observations on plastid division in root meristematic cells of *Pisum sativum* (L). *Japanese Botanical Magazine*, **88**, 319–21.

Thompson, D.A.W. (1928). *On Growth and Form*. Cambridge University Press.

Townsend, R. & Archer, D.B. (1983). A fibril protein antigen specific to *Spiroplasma*. *Journal of General Microbiology*, **129**, 199–206.

Whatley, J.M. (1971). The chloroplasts of *Equisetum telmateia* Erhr: a possible developmental sequence. *The New Phytologist*, **70**, 1095–102.

Whatley, J.M. (1975). Chloroplast structure in coiled and uncoiling croziers of *Pilularia globulifera*. *The New Phytologist*, **74**, 413–20.

Whatley, J.M. (1976). Bacteria and nuclei in *Pelomyxa palustris*: comments on the theory of serial endosymbiosis. *The New Phytologist*, **76**, 111–20.

Whatley, J.M. (1980). Plastid growth and division in *Phaseolus vulgaris*. *The New Phytologist*, **86**, 1–16.

Whatley, J.M. (1986). Patterns of plastid replication during plant development. In *The Chondriome – Chloroplast and Mitochondrial Genomes*, ed. S.H. Mantell, G.P. Chapman & P.F.S. Street, pp. 3–36. New York: John Wiley & Sons.

Whatley, J.M., Hawes, C.R., Horne, J.C. & Kerr, J.D.A. (1982). The establishment of the plastid thylakoid system. *The New Phytologist*, **90**, 619–29.

Whatley, J.M., John, P. & Whatley, F.R. (1979). From extracellular to intracellular: the establishment of mitochondria and chloroplasts. *Proceedings of the Royal Society of London* (series B), **204**, 165–87.

Whatley, J.M. & Whatley, F.R. (1981). Chloroplast evolution. *The New Phytologist*, **87**, 233–47.

Williamson, D.L., Brink, P.R.& Zieve, G.W. (1984). Spiroplasma fibrils. *Israel Journal of Medical Sciences*, **20**, 829–35.

Williamson, R. (1985). Immobilisation of organelles and actin bundles in the cytoplasm of the alga *Chara corralina* Klein ex. Wild. *Planta*, **163**, 1–8.

W.D. DONACHIE

The replication of a free-living prokaryote: *Escherichia coli*

It is generally accepted that the mitochondria and chloroplasts found in present-day eukaryotic cells share prokaryotic ancestors with the free-living bacteria and their relatives (Cohen, 1970). Although that divergence took place about 10^9 years ago, these modern organelles retain their cellular integrity to a remarkable extent. They not only possess remnants of an independent genome and protein-synthesising system but also, despite extreme specialisation and loss of all capacity to grow and multiply outside the eukaryotic cell, continue to propagate *as* cells within cells. Chloroplasts and mitochondria, like their free-living ancestors, grow and divide in an organised and regular manner, and it may therefore be useful to outline what we know about the regulation and mechanism of the morphogenetic cycle in a free-living prokaryote. By far the best-understood prokaryote cell (or, indeed cell of any kind) is the common enteric bacterium *Escherichia coli*. Many of the main features of its growth and division are now known, in large part due to the enormous number of mutants which are available and to the excellence of its genetics and methods of gene manipulation. What follows is a brief outline of the ways in which *E. coli* cells maintain their integrity, shape and composition during growth and multiplication.

E. coli cells are small rods with rounded ends. They grow by elongation and multiply by binary fission. Their average proportions (length to width) remain constant in different culture media, but their absolute dimensions depend on growth rate (see below). Their minimum dimensions (a new-born cell in a very slowly-growing population) are about 1.5×0.5 μm and their maximum (a fast growing cell just before division) about 6.0×1.0 μm (see Donachie & Robinson, 1987). These cells contain between 1 and 4 copies of the genome, which consists of a covalently closed circular DNA molecule about 1.4 μm in circumference and containing about 4×10^6 base pairs (m.wt. about 2.3×10^9). The potential coding capacity of the genome is about 2000–3000 proteins of average size, and about 1200 individual genes have been identified (see Bachmann, 1983). In the cell the DNA is found in the form of nucleoids, defined bodies consisting of tightly packed DNA strands associated with RNA polymerase, nascent RNAs, ribosomes and a histone-like protein (HU). *E. coli* belongs to the gram-negative subdivision of the eubacteria and therefore has a complex cell envelope consisting of two lipoprotein membranes with a layer of peptidoglycan in the 'periplasmic space' between them. The outer surface of the outer membrane contains lipopolysaccharides, and there are also a number of different envelope 'organelles' such as flagella and pilae which are important in motility and in anchoring cells to surfaces (as well as in certain kinds of gene transfer). For a review of the composition and function of the cell envelope see Di Rienzo, Nakamura & Inouye, 1978.

Maintenance of cell shape

Free-living prokaryotes have high internal osmotic pressures, estimated at 4–5 atmospheres for *E. coli* and up to 20 atmospheres for gram-positive bacteria, which have thicker walls (Mitchell & Moyle, 1956). The peptidoglycan layer of the cell envelope provides the mechanical strength required to prevent such cells from bursting in low osmotic strength media. Peptidoglycan is unique to prokaryotes and, in gram-negative bacteria such as *E. coli*, consists of a monomolecular net of glycan chains cross-linked by peptide bridges. This single covalently closed giant molecule encloses the entire cell membrane. It not only provides mechanical stability but also gives the cell its characteristic shape. Purified 'sacculi' consisting only of this peptidoglycan layer retain the shape of the bacterial cell from which they have been prepared. The specialised precursors for this macromolecule (which include D–amino acids, N–acetyl glucosamine and N–acetyl muramic acid) are synthesised in the cytoplasm, but the glycan chains are polymerised and cross-linked to the existing peptidoglycan net by enzymes in the cell membrane. Seven such enzymes have been found in *E. coli*. Of these, five are present in high numbers of molecules per cell and appear to be responsible for the overall synthesis of the peptidoglycan layer. However the two remaining enzymes, which are present at only about 20 and 50 molecules per cell respectively, appear to be responsible for shaping the sacculus. Mutants (*ftsI*) which lack one of these proteins (called 'penicillin-binding protein 3', or PBP3) grow as ever-elongating cylinders ('filaments') but cannot divide, while mutants (*pbpA*) which lack the other (PBP2) can divide but grow as spheres (Spratt, 1975). Double mutants which lack both morphogenetic proteins grow as ever enlarging spheres (Begg & Donachie, 1985). (For those unfamiliar with bacterial genetic techniques it should be pointed out that since such morphogenetic defects are lethal to the cell, 'conditional' mutants are used in which the mutant defect is expressed only under certain conditions, such as a high growth temperature.) Specific β–lactam drugs are available which target one or the other of these two proteins and produce the characteristic shape changes in normal cells.

Mutants (*rodA*) in a third, low abundance membrane protein also grow and divide as spheres. No specific enzymatic activity has yet been identified for the RodA protein but clearly it also is involved in shaping the peptidoglycan. Interestingly, there is now evidence that the RodA protein interacts directly with both the PBP2 and PBP3 proteins in the membrane (Begg, Spratt & Donachie, 1986). These three proteins may therefore form a 'morphogenetic complex' in the cell membrane which modifies the growing peptidoglycan layer during the alternating elongation and septation phases of the cell cycle.

Cell division

As we have seen, cell division in *E. coli* involves the formation of an ingrowing peptidoglycan septum which requires the presence of the PBP3 protein in the morphogenetic complex. However a number of other proteins are also specifically required for cell division. The genes coding for four of these (*ftsQ*, *ftsA*, *ftsZ* and *envA*) are adjacent to one another within a cluster of genes, all of which code for proteins required for peptidoglycan synthesis. These include *ftsI*. (Two other genes lie at one end of the cluster. One of these is of unknown function and the other is required for insertion of proteins into, or their transport across, membranes.) Unlike PBP3 (the *ftsI* gene product) the other four division proteins are not found in the cell membrane but they are equally essential for cell division. Indeed one of them, the FtsZ protein, appears to be required for the initiation of septation (Begg & Donachie, 1985) and to be rate limiting for division (Ward & Lutkenhaus, 1985). Thus, in contrast to PBP3 protein which can be produced in excess without any effect on the morphogenesis of the cell, over-production of FtsZ induces the production of extra divisions. These extra divisions are seen in the production of 'minicells' which are budded off from the cell poles when FtsZ is overproduced (Ward & Lutkenhaus, 1985). The proteins PBP3, FtsQ and FtsA are required for subsequent stages in septation, with FtsA required only for the last stage. The EnvA protein is required for the final splitting apart of the double peptidoglycan layer of the septum, allowing invagination of the outer membrane to form the two new poles of the sister cells (Normark, Bowman & Bloom, 1971; Wolf–Watz & Normark, 1976).

The product of yet another gene (*minB*) appears to be required to prevent the further use of cell poles as sites for septum formation (Teather, Collins & Donachie, 1974). This block can be overcome by excess FtsZ protein, as we have seen. The products of other genes may also be specifically required for division but this is not yet known. For a review of the genes involved in *E. coli* cell division, see Donachie, Begg & Sullivan, 1984.

Genome replication and segregation

Chromosomal DNA replication is closely regulated by cell growth. Initiation of replication occurs at a unique site (*oriC*), and DNA synthesis proceeds bidirectionally from *oriC* until the replication forks meet at a point (*terC*) which is on the opposite side of the circular chromosome. The initiation step is a distinct process from subsequent replicative synthesis of DNA (see Baker *et al.*, 1986), and it is this step which is regulated in relation to cell growth.

Initiation takes place in cells only when their mass reaches certain critical values. At these values, the ratio of mass to the number of copies of chromosomal origins is always constant (M_i, the 'initiation mass': Donachie, 1968). (Fast-growing cells have more copies of the origin than slow-growing cells but the same rule always holds.) The molecular basis for this growth-dependent regulation of initiation is not yet known, but there appears to be only one protein (DnaA protein) which is required for initiating but not for subsequent DNA replication. Unfortunately it is not yet known how the activity or synthesis of this protein is regulated in relation to cell growth and mass. For a description of the molecular events involved in the initiation step, see Baker *et al.*, 1986.

Once initiated, DNA replication goes on at a rate which is largely independent of the actual growth rate of the cells (which in *E. coli* can vary from as fast as one doubling every 20 minutes to as slow as one doubling in several days, depending on the nutrient supply). Under most laboratory culture conditions it takes about 40 minutes for the replication forks to reach the terminus. (If the doubling time of the cells is less than this, new replication forks are initiated at the duplicated chromosome origins before the preceding forks have reached the terminus, to give the characteristic 'overlapping rounds' of chromosome replication of fast-growing bacteria: see, Donachie, Jones & Teather, 1973 for a fuller description of this process.)

All these processes take place in the cell when the DNA is in the form of tightly folded nucleoids, and the completion of replication is followed by the separation of two sister nucleoids. The separation of replicated genomes into nucleoids and the positioning of these nucleoids in the centres of the daughter cells require the action of other proteins. DNA gyrase is required throughout the replication process but it is also required specifically for the decatenation of the replicated sister chromosomes before they can physically separate (Steck & Drlica, 1984). Partition of the separate nucleoids and their correct positioning in the cell requires DNA gyrase and perhaps other 'Par' proteins. Without their action, DNA is replicated but remains in one or a few enlarging masses per cell (Hirota, Ryter & Jacob, 1968; Hussain *et al.*, 1987).

Integration of cell growth, DNA replication and cell division

Cell division must be regulated in respect to the number and location of nucleoids. The fact that it is so regulated is shown by the very rare occurrence of cells lacking DNA in normal cultures. Not more than one in a thousand cells are found to have no nucleoid (N^- cells). However the situation is quite different in *par* mutants. When nucleoid segregation is blocked, cell division continues at a normal rate (i.e. one per unit mass doubling) but septa form between the ends of the single enlarged nucleoid and the cell pole, so as to cut off N^- cells. The N^- cells appear to be of random sizes, and this, together with the very accurate placement of septa exactly midway between nucleoids in normal cells, has given rise to the hypothesis that septa cannot form within a certain minimum distance of a nucleoid (Hussain *et al.*, 1987). Partition of sister nucleoids by the Par proteins is therefore required to provide an inhibition-free zone in the cell centre in normal cells. In the absence of partition, inhibition-free zones will eventually form at the DNA-free cell ends and it is here that septa will form.

Such a mechanism not only will ensure that septa will be laid down normally in the right place but also will prevent their formation for a period of time if, for some reason, completion of chromosome replication and segregation has been delayed. Such conditions occur when cells are shifted from a culture medium which gives slow growth to one which allows more rapid growth. Cells respond by growing more rapidly within minutes of such a 'shift-up'. However, as we have seen, the rate of movement of replication forks around the chromosome is not affected by such a change in growth rate. The frequency of initiation is increased to keep pace with the more rapid rate of increase in cell mass, but the frequency of termination cannot increase to the new rate until 40 minutes later. If the frequency of cell division were to increase immediately after a shift-up, then the consequence could only be the formation of N^- cells. In fact this never happens and the frequency of septa formation does not increase until the frequency of termination increases, 40 minutes after the shift-up. This 'rate maintenance' effect on cell division can be understood if septa can form only after nucleoid partition has taken place.

This model predicts that if DNA replication were blocked, then cell division would be blocked for a period but would eventually resume to produce N^- cells. This is indeed what happens if DNA synthesis is inhibited, but only under certain conditions. This is because blocks to DNA synthesis in normal cells usually result in the induction of specific inhibitor proteins which prevent cell division. These inhibitors will be discussed below, and here it is sufficient to say that if the formation of these inhibitors is prevented, inhibition of DNA replication results in only an initial inhibition of cell division. The blocked cells continue to grow and elongate and eventually start to produce N^- cells from their ends.

Such a model, invoking 'zones of inhibition', is a purely formal one at the moment, and its reinterpretation in terms of molecular action will have to wait for a characterisation of the mode of action of the Par proteins themselves.

Endogenous inhibitors of division

In contrast to the special situation just discussed, inhibition of DNA synthesis normally leads to an almost immediate inhibition of cell division also. This is because most inhibitors of DNA synthesis cause the induction of a set of genes, the 'SOS regulon', which include one (*sfiA*) which codes for a specific inhibitor of cell division. The way in which this comes about is quite well understood and an outline will be given here, but for a full description of the SOS system, specialised reviews should be consulted (Little & Mount, 1982; D'Ari, 1985). Inhibition of DNA synthesis and certain kinds of damage lead to the exposure of single-stranded sections of DNA. These bind the RecA protein, and this causes an allosteric change in this protein such that it is able to bind in turn to a second protein, LexA, and cause it to undergo a spontaneous cleavage. The LexA protein, which is inactivated in this way, is the common repressor for all of the genes which form the SOS regulon. Inactivation of LexA therefore leads to increased levels of transcription of all of these genes and the production of various proteins which have roles in DNA repair. The SfiA protein is produced only under these conditions and it binds specifically to FtsZ protein and therefore blocks the initiation of new septa. Division is inhibited until the DNA block is repaired and the single-stranded regions removed. RecA protein then returns to its original state, LexA production resumes, repression of the SOS genes is re-established, and further production of SfiA protein stops. Existing SfiA molecules are rapidly destroyed by certain proteases, and cell division resumes.

The *sfiA* gene is present in all strains of *E. coli* (and indeed there is evidence that it is highly conserved during evolution, as is the *ftsZ* gene: Corton, Ward & Lutkenhaus, 1987) but certain strains carry genes for other division inhibitors. One of these (the SfiC inhibitor) is coded for by an exogenous element which appears to be a remnant of a phage genome which has inserted into the chromosome (D'Ari & Huisman, 1983). Expression of this inhibitor gene is normally repressed but it is also induced by activation of RecA protein by single-stranded DNA. In this it resembles the *kil* gene of phage λ. When λ integrates into the host chromosome, most of its genes, including *kil*, become repressed by a repressor, CI, which is produced by the λ itself. The CI repressor molecule also is cleaved by combination with activated RecA protein, causing the induction of λ replication in cells with damaged DNA. The *kil* function blocks division in these cells during λ replication. Recently another division-inhibitor gene (*dicA*) and its repressor gene (*dicB*) have been found in the chromosome. In this case the inhibitor seems never to be activated under normal circumstances because of the action of the *dicB* gene. Perhaps the inhibitor is produced under special circumstances, but DNA sequence analysis of the *dicA dicB* region has shown strong homologies to phage genomes (J.-P. Boucher, personal communication) and it therefore seems most likely that this also is a relic of a phage genome which now has no function.

It seems likely that the function of the various endogenous inhibitors of cell division is to prevent the wasteful production of N⁻ cells during periods when DNA replication is blocked for one reason or another. This would be advantageous to the host cell and also to any resident viruses which would be able to make full use of the energy supplies of a possibly doomed cell.

Counting divisions

Cells normally divide once after each nucleoid doubling. We are so accustomed to such behaviour that it seldom occurs to us to wonder about the 'once'. How is it decided that only a single septum will form at the time when division is allowed? That this is a real problem becomes clear when we examine the division of various mutants. We have already mentioned that Par mutants replicate their DNA but do not partition it into separate nucleoids, and we have also said that such cells grow long and divide off N⁻ cells from their ends. Septa in these cells form in random positions within the DNA-free zone, but careful counting shows that the number of septa formed is no different from the number formed in Part cells, in which one septum forms between each pair of nucleoids. The number of septa which can be formed anywhere in the cell is therefore fixed and doesn't depend on the number of separate nucleoids. That the number of septa which can be formed per unit cell mass is strictly limited is shown even more dramatically in another mutant, *minB*. In this mutant the cell poles (which may be considered to be old septal sites) continue to be available for further division. When a division occurs at a pole, a small N⁻ minicell is budded off but this is not an 'extra' division. For every division which produces a minicell, a 'normal' division is lost. In this mutant, polar division sites compete on equal terms with normal sites between nucleoids, but the total number of divisions remains unchanged at one per nucleoid pair. Cells behave as if they produce some 'division factor' in packets, one per mass doubling, and that each such quantum is used up entirely in the formation of a single septum (Teather *et al.*, 1974).

No one knows the nature of this strange factor but it is exciting to find that over-production of a single protein species (FtsZ) can apparently by itself increase the number of divisions which take place in each cell cycle (Ward & Lutkenhaus, 1985).

Bacterial morphogenesis and organelle morphogenesis

The morphogenesis (maintenance of cell shape during growth and division) of *E. coli* and other bacteria depends largely on the regulation of the shape and growth of the peptidoglycan sacculus. Mitochondria and chloroplasts have no such rigid layer and therefore few of the morphogenetic processes described here can be expected to operate in these organelles. However, some bacterial cells can still grow and divide as so-called 'L-forms' which lack peptidoglycan. In such abnormal cells division is 'chaotic' but it does take place, showing that the formation of a peptidoglycan septum is not the sole process which can lead to cell division. There are also naturally occurring prokaryotic cells which lack a rigid layer. The 'Mollicutes' (mycoplasmas and spiroplasmas) are the smallest known free-living cells, with between 30 and 50 per cent of the DNA of *E. coli* and, most important, regular shapes and sizes (at least in actively growing cultures; Maniloff & Morowitz, 1972; Garnier, Cleri & Bove, 1984), showing that well-regulated morphogenesis need not require peptidoglycan or other forms of rigid cell wall. Study of the mechanism of division in these organisms may therefore be more useful for understanding the morphogenesis of organelles than is the study of the bacteria. Unfortunately the Mollicutes have been little studied from this point of view.

References

Bachmann, B.N. (1983). Linkage map of *E. coli* K-12, edition 7. *Microbiological Reviews*, **47**, 180–230.

Baker, T.A., Sekimizu, K., Funnell, B.E. & Kornberg, A. (1986). Extensive unwinding of the plasmid template during staged enzymatic initiation of DNA replication from the origin of the *Escherichia coli* chromosome. *Cell*, **45**, 53–64.

Begg, K.J. & Donachie, W.D. (1985). Cell shape and division in *Escherichia coli*: experiments with shape and division mutants. *Journal of Bacteriology*, **163**, 615–22.

Begg, K.J., Spratt, B.G. & Donachie, W.D. (1986). Interaction between membrane proteins PBP3 and RodA is required for normal cell shape and division in *Escherichia coli*. *Journal of Bacteriology*, **167**, 1004–8.

Cohen, S.S. (1970). Are/were mitochondria and chloroplasts microorganisms? *American Scientist*, **58**, 281–9.

Corton, C.J., Ward, J.E. & Lutkenhaus, J.F. (1987). Analysis of cell division gene *ftsZ* (*sulB*) from gram-negative and gram-positive bacteria. *Journal of Bacteriology*, **169**, 1–7.

D'Ari, R. (1985). The SOS system. *Biochimie*, **67**, 343–7.

D'Ari, R. & Huisman, O. (1983). Novel mechanism of cell division inhibtion associated with the SOS response in *Escherichia coli*. *Journal of Bacteriology*, **156**, 243–50.

Di Rienzo, J.M., Nakamura, K. & Inouye, M. (1978). The outer membrane proteins of gram-negative bacterial biosynthesis, assembly and functions. *Annual Review of Biochemistry*, **47**, 481–532.

Donachie, W.D. (1968). Relationship between cell size and initiation of DNA replication. *Nature*, **219**, 1077–9.

Donachie, W.D., Begg, K.J. & Sullivan, N.F. (1984). The morphogenes of *Escherichia coli*. In *Microbial Development*, ed. R. Losick & L. Shapiro, pp. 27–62. Cold Spring Harbor Publications.

Donachie, W.D., Jones, N.C. & Teather, R.M.(1973). The bacterial cell cycle. *Symposium of the Society for General Microbiology*, **23**, 9–44.

Donachie, W.D. & Robinson, A.C. (1987). Cell division of *Escherichia coli* : parameter values and the process. In *The Molecular Biology of Escherichia coli and Salmonella typhimurium*, ed. J. Ingraham, K.B. Low, B. Magasanik, F.C. Neidhardt, M. Schaechter & H.E. Umbarger, pp. 1578–93, American Society for Microbiology .

Garnier, M., Cleri, M. & Bove, J.M. (1984). Growth and division of *Spiroplasma citri*: elongation of elementary helices. *Journal of Bacteriology*, **158**, 23–8.

Hijmans, W., van Boven, C.P.A. & Clasener, H.A.L.(1969). Fundamental biology of the L-phase in bacteria. In *The Mycoplasmatales and the L-phase of Bacteria*, ed. L. Hayflick, pp. 67–143. Amsterdam: North Holland.

Hirota, Y., Ryter, A. & Jacob, F. (1968). Thermosensitive mutants of *E. coli* affected in the process of DNA synthesis and cellular division. *Cold Spring Harbor Symposia on Quantitative Biology*, **33**, 677–93.

Hussain, K., Begg, K.J., Salmond, G.P.C. & Donachie, W.D. (1987). ParD: a new gene coding for a protein required for chromosome partitioning and septum localisation in *Escherichia coli*. *Molecular Microbiology*, **1**, 73–81.

Little, J.W. & Mount, D.W. (1982). The SOS regulatory system of *Escherichia coli*. *Cell*, **29**, 343–7.

Maniloff, J. & Morowitz, H.J. (1972). Cell biology of the mycoplasmas. *Bacteriological Reviews*, **36**, 263–90.

Mitchell, P. & Moyle, J. (1956). Osmotic structure and function in bacteria. *Symposium of the Society for General Microbiology*, **6**, 150–80.

Normark, S., Bowman, H.G. & Bloom, H. (1971). Cell division in a chain forming *envA* mutant of *Escherichia coli* K-12. Fine structure of division sites and effects of EDTA, lysozyme and penicillin. *Acta Pathologica et Microbiologica Scandinavica*, Section B, **79**, 651.

Ochman, H. & Wilson, A.C. (1987). A universal substitution rate: evidence from bacteria. *Journal of Molecular Evolution* (In press).

Spratt, B.G. (1975). Distinct penicillin-binding proteins involved in the division, elongation and shape of *Escherichia coli* K-12. *Proceedings of the National Academy of Sciences, USA*, **72**, 2999–3003.

Steck, T.R. & Drlica, K. (1984). Bacterial chromosome segregation: evidence for DNA gyrase involvement in decatenation. *Cell*, **36**, 1081–8.

Teather, R.M., Collins, J.F. & Donachie, W.D. (1974). Quantal behaviour of a diffusible factor which initiates septum formation at potential division sites in *Escherichia coli*. *Journal of Bacteriology*, **118**, 407–13.

Ward, J.E. Jr. & Lutkenhaus, J.F. (1985). Overproduction of FtsZ induces minicell formation in *E. coli*. *Cell*, **42**, 941–9.

Woese, C.R., Stackebrandt, E., Make, T. & Fox, G.E. (1985). A phylogenetic definition of the major bacterial taxa. *Systematic and Applied Microbiology*, **6**, 143–51.

Wolf–Watz, H. & Normark, S. (1976). Evidence for a role of N–acetylmuramyl–L–alanine amidase in septum separation in *Escherichia coli*. *Journal of Bacteriology*, **128**, 580–6.

Yang, D., Oyaizu, Y., Oyaizu, H., Olsen, G.J. & Woese, C.R. (1985). Mitochondrial origins. *Proceedings of the National Academy of Sciences, USA*, **82**, 4443–7.

C.J. DUNCAN

Mitochondrial division in animal cells

Mitochondrial division in lower organisms

Many cytological and biochemical studies support the hypothesis that new mitochondria arise from older ones by mitochondrial division in protozoans, algae and fungi (see review in Kuroiwa, Hizume & Kawano, 1978). The presence of DNA in mitochondria is firmly established, and two distinct events are involved in mitochondrial division in these lower organisms. One results in the duplication of the mitochondrial nucleoid containing replicated DNA, whilst the other results in the division of the matrix and cristae. Mitochondria synthesise DNA primarily during the S phase of the mitotic cycle.

It seems to be firmly established that the mitochondria in the plasmodium of the slime mould *Physarum* undergo a clear cycle of divisions (Kuroiwa, Kawano & Hizume, 1977). These mitochondria contain a large, rodlike nucleoid situated in the centre of the inner matrix which is composed of a large amount of DNA, RNA and protein and which is of higher electron density than the matrix. The mitochondrial nucleoid elongates longitudinally while the mitochondrion increases in size during the mitochondrial S phase, and it divides by the constriction of the mitochondrion. Such a sequence is similar to that reported for the kinetoplast nucleoid of the Trypanonidae (Simpson, 1972) and for the division of bacteria and *Rickettsiella melolonthae* (De Vauchelle, Meynadier & Vago, 1972). The DNA fibres of the nucleoid are bound closely to fragments of cristae and one end of the elongated nucleoid is associated with the cristae before division, although the association disappears after division, suggesting that the membrane of the cristae plays an important role in the division of the *Physarum* mitochondrial nucleoid (Kuroiwa *et al.*, 1977). Mitochondrial division in *Physarum* is inhibited by cytochalasin B; the dumb-bell shaped dividing organelles become spheroidal and it is suggested that contractile proteins are essential for mitochondrial division (Kuroiwa & Kuroiwa, 1980).

Mitochondrial division in animal cells

Does this cycle of ultrastructural changes of mitochondrial division in the slime mould represent a pattern for the regular division of these organelles of animal cells *in vivo*? There are a number of studies of mitochondrial swelling and septation in vertebrate liver and muscle cells, but the majority of these reports follow various experimental treatments. However, the renewal of the normal mammalian intestinal epithelium involves frequent mitoses in the lower half of the crypt and the migration of the cells to the villus whilst differentiating into non-proliferative absorptive cells. The average number of mitochondria per cell was 21 at the bottom of the crypts, gradually increasing to 42 at the base of the villi. Mitochondria with transverse membranes (similar to those of Fig. 21) and bilateral invaginations (similar to those of Fig. 18) of the outer mitochondrial membrane, characteristic of mitochondria believed to be subdividing, were found, but only in the mid and upper third of the crypts. This migration of the cells and accompanying mitochondrial division continued even when mitosis was inhibited; on the other hand, no mitochondrial division occurs in sections of the intestine that have been surgically isolated (Jeynes & Altmann, 1975). The epithelium of the rat intestine thus appears to be an example of a genuine single mitochondrial division (doubling the number per cell) during the life and migration of the crypt cells.

Mitochondria with a transverse septum have been found in mammalian liver *in situ*, but these are rare; the intracristal spaces were narrow with greatly expanded matrix spaces (Wakabayashi, Asano & Kurono, 1974). Similarly, the occasional development of septa has been described in insect fat body (Larsen, 1970) and insect corpora cardiaca (Normann & Samaranayaka–Ramasamy, 1977).

However, the incidence of mitochondrial division in mammalian liver can be greatly enhanced by such experimental treatments as (i) injection of cuprizone, a copper-chelating agent (Tandler & Hoppel, 1973) or feeding a 0.5 per cent (w/w) cuprizone diet (Flatmark, Kryvi & Tangeras, 1980); (ii) a riboflavin-deficient diet followed by riboflavin injection (Tandler *et al.*, 1969); (iii) inclusion of the non-carcinogenic dye 2–Me–DAB in the drinking water (Lafontaine & Allard, 1964); or (iv) feeding a high-alcohol diet (Koch *et al.*, 1978). These treatments were characterised in the liver by grossly swollen mitochondria (so-called megamitochondria), increases in the number of mitochondria per cell, apparent division of liver mitochondria and, most interestingly, a rapid reversal of the mitochondrial changes when treatment ceased. Such findings strongly suggest that these characteristic divisions of hepatic mitochondria are pathological exacerbations of a process that occurs rarely *in vivo* under normal conditions.

Mammalian muscle

There are a number of accounts of mitochondria with a transverse septum *in situ* in normal mammalian cardiac and skeletal muscles and these have been tabulated in a useful summary by Rudge (1983), but swollen and dividing mitochondria increase conspicuously in pathological conditions. For example, they can be seen following ischaemia in heart (Jennings *et al.*, 1969) and skeletal muscle (Hanzliková & Schiaffino, 1977; Heffner & Baron, 1978). Furthermore many myopathies of human skeletal muscle are characterised by intracristal paracrystalline inclusions or by mitochondrial 'bars' (Heine & Schaeg, 1979; Mastaglia & Walton, 1982). The paracrystalline inclusions are very striking and are normally located in the subsarcolemmal area. The commonest (type I) paracrystalline inclusion appears in thin sections as a rectangle, some 32 nm wide and of varying length, located within the intracristal space or present in the outer compartment space between the inner and outer membranes. The crystalloids appear as a construction of two rows of flattened hexagons with the hexagon-centres of one row alternating with those of the other row (Busch, Jennekens & Scholte, 1981).

Experimental interventions *in vivo* and *in vitro* with either cardiac or skeletal muscles rapidly produce (within 30 min) much more dramatic changes; electron microscopy shows that the mitochondria are swollen, many have developed transverse septa and there is an apparent multiplication of mitochondrial numbers. Mitochondrial bars are also common in skeletal muscle preparations. Treatments that cause such marked changes in the mitochondria include the divalent cation ionophore A23187, caffeine, 2,4–dinitrophenol and anoxia (Publicover, Duncan & Smith, 1977, 1978; Duncan *et al.*, 1980*a*; Duncan, Greenaway & Smith, 1980*b*; Duncan & Smith, 1980; Duncan & Greenaway, 1981; Rudge & Duncan, 1984); similar changes also occur in cardiac mitochondria after mice were fed a cuprizone diet, although over a much longer term (Tandler & Hoppel, 1972).

Rudge (1983) has made detailed quantitative measurements on electron micrographs of frog and mouse heart preparations exposed to all the treatments listed above. She has shown that the area of the muscle occupied by mitochondria increases by 30–40 per cent in frog and by 40–56 per cent in mouse cardiac muscle. The percentage of mitochondria showing a complete transverse septum also increased significantly ($p < 0.001$) above control levels (frog from 0.2 to about 2.0 per cent; mouse from 0.35 to 3.0–5.0 per cent); concomitantly the number of mitochondria per unit area of the electron micrographs increased by 40–50 per cent in frog and by 50–80 per cent in mouse.

Dividing mitochondria produced by experimental means

These configurational changes of mitochondria can be seen in electron micrographs of mouse diaphragm that has been treated with A23187 (6.25 μg ml^{-1}; 37 °C) or with caffeine (10 mM). Control mitochondria have an electron-dense matrix with a moderately swollen intracristal space showing an energised condition (Fig. 1), but some 40 min later the muscle cell is packed with swollen mitochondria, many apparently undergoing subdivision (Figs. 2 and 5). Identical changes can be demonstrated in mouse soleus muscle with dinitrophenol (DNP) (Fig. 3) or A23187 (Fig. 4). A possible sequence of intermediate stages is shown in Figs. 6–11, following exposure to A23187. After 10 min, the mitochondria show changes consistent with their becoming less energised, the intracristal spaces are smaller and parallel, the matrix is less electron dense (Figs. 7 and 8). After 20 min (Fig. 9) the

Fig. 1. Control skeletal muscle from mouse. Contracted mitochondria, relaxed thin (actin) and thick (myosin) filaments. Bar, 0.5 μm.

Fig. 2. Mouse diaphragm exposed to 10 mM caffeine for 40 min. Greatly swollen and subdivided mitochondria, with cristal membranes changing towards the formation of electron-dense bars. Arrows: A, septa apparently forming from electron-dense material; B, septa associated with inpushing of both mitochondrial membranes; C, triradiate septa. Note myofilament damage; hypercontraction and blurred Z-lines. Bar, 1 μm

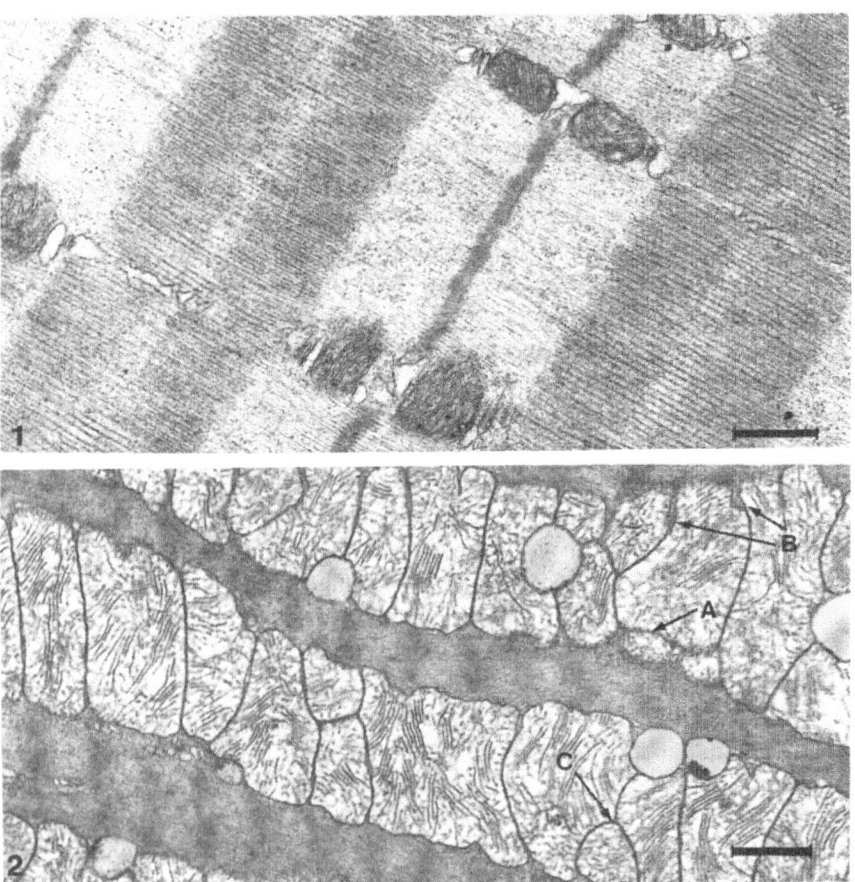

mitochondria swell, may become rectangular in outline, develop transverse septa and develop electron-dense bars, whilst the cristal structure is altered (30 min). Many of the mitochondria elongate before apparent subdivision and multiplication (Figs. 10 and 11). Finally (Figs. 2 and 5) whole areas of the cell become filled with tightly-packed organelles (Publicover, Duncan & Smith, 1977). The long chains of apparently subdivided mitochondria are particularly characteristic of this phase (Figs. 2, 5 and 11). The development of electron-dense bars within these swollen mitochondria is particularly evident in soleus muscle and can be seen in Figs. 2, 3 and 4 and at greater magnification in Figs. 12 and 13. Such bars produced *in vitro* by A23187 or caffeine correspond closely with those described in myopathies *in vivo*, and they apparently form from the cristal membranes, as shown by their laminar structure (Fig. 12).

Similar changes in the mitochondria can be seen in vertebrate cardiac muscle when the intracellular Ca^{2+} concentration is experimentally raised as in the so-called Ca-paradox (Rudge & Duncan, 1984). Fig. 19 shows swollen and packed mitochondria, arranged in long chains and some with complete or developing septa only 10 min after the return of Ca^{2+} in the perfusate. Developing transverse septa, one extending half-way across the mitochondrion and another almost complete, are shown in Fig. 20. In Fig. 22, it can be seen that the septum has been formed from the ingrowth of only the inner mitochondrial membrane.

Closely packed mitochondria from cardiac muscle are shown in Fig. 21 where the newly formed septa interrupt the continuous line of the cristae in the chain of adjacent mitochondria. Such an observation suggests that these mitochondrial chains are not formed from the fusion of separate organelles but are the result of the subdivision of an original swollen megamitochondrion. The mitochondria are frequently subdivided by a septum so that the cristae in each portion are oriented at right angles to each other (Fig. 23), so that one set are in profile and the other en face (Rudge, 1983). Subdivision by septa can also yield mitochondria in which the two (or even three) sections are in different energetic configurations (Wakabayashi *et al.*, 1974; Duncan *et al.*, 1980a), as shown in Figs. 24 and 25 in cardiac muscle and also in two mitochondria from skeletal muscle in Fig. 15.

Development of septa

The available evidence suggests that these mitochondrial septa are genuine and that they subdivide the organelles. Wakabayashi *et al.* (1974) have shown that the septa are not fixation artefacts dependent on such factors as differential rates of penetration of the fixative resulting in one section of the mitochondrion being in the condensed and the other in the orthodox configuration. Tandler & Hoppel (1972) demonstrated that the straight or gently curving septa in isolated cardiac mitochondria (see Figs. 22–24) were true partitions and not merely cristae spanning the width of

Fig. 3. Mouse soleus muscle exposed to 10^{-3} M dinitrophenol, for 60 min. A chain of swollen mitochondria, some with transverse septa and some with mitochondrial bars. Arrow shows septa forming quadripartite organelle. Note myofilament damage with disintegration of actin and Z-line. Bar, 1 μm.

Fig. 4. Mouse soleus muscle exposed to 20 μM A23187 for 60 min. Swollen, adpressed mitochondria with bars. Arrows show triradiate septa, an almost complete tranverse septum and actin and Z-line loss. Note many empty vesicles, probably severely damaged mitochondria. Bar, 1 μm.

Fig. 5. Mouse diaphragm exposed to 6.25 μg ml^{-1} A23187 for 30 min. Arrows A–D indicate increasing density of material laid down to form septa and associated with cristal membranes. Arrow shows lateral intucking in chain of mitochondria. Muscle is hypercontracted with heavily blurred Z-lines. Bar, 1 μm.

the organelle. This is especially evident when the septum separates compartments in different energetic configurations. The quantitative analysis that showed that the frequency of septation increased in parallel with the increase in mitochondrial numbers also supports this view (Rudge, 1983).

Septa do not develop only as a central, gently curving division; triradiate septa (Fig. 14) and even quadripartite organelles (Fig. 3) occur regularly, whilst septa separating a corner of the mitochondrion are common (Figs. 2, 11 and 18).

There appear to be at least three potential ways in which transverse septa might develop:

(1) By a lateral ingrowth of the inner mitochondrial membrane only, which fuses with the opposite side, as illustrated in Figs. 13 and 22.

(2) By a progressive intucking of both inner and outer mitochondrial membranes, the finger-like projection crossing the organelle, as shown in Figs. 2 and 16 to 18. This appears from the electron micrographs to be the most unequivocal method of mitochondrial division.

(3) By the internal deposition of electron-dense material which completely crosses the mitochondrion.

Fig. 5 shows four areas of increasing development of such septa. Initially the material barely traverses the mitochondrion; subsequently there is an increase in density, the membranes become more clearly defined and there are lateral inpushings from the mitochondrial edge. This electron-dense material is probably distinct from the mitochondrial bars (Fig. 13) which are more clearly defined and develop along the length of the cristal membranes (Figs. 12 and 13) whereas the electron-dense material appears at right angles to the membranes, resembling the condition of the fully formed septa of Fig. 21. These apparently internally formed septa may be artefacts, being material associated with the borders of the lateral inpushings (see (2) above), the plane of the section just missing the true septum. Fig. 18 shows how septa

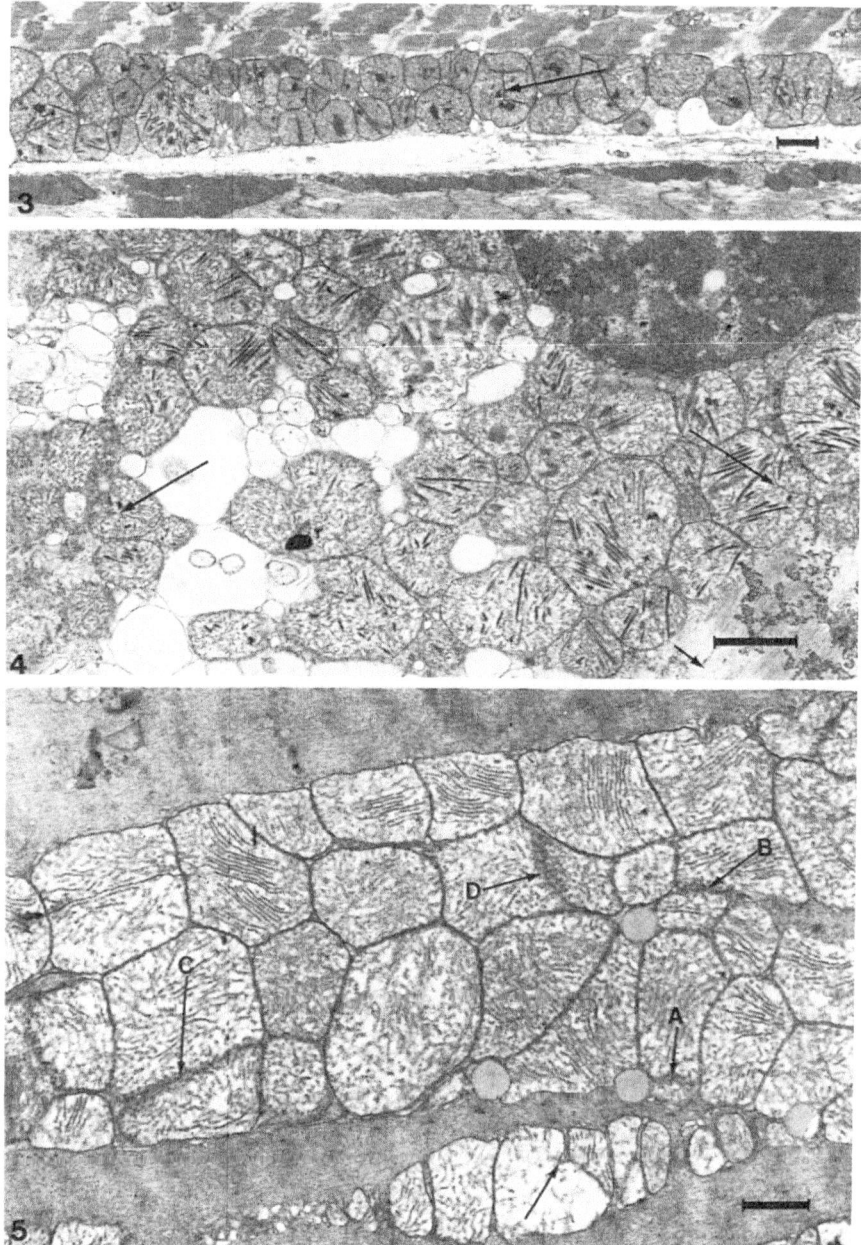

apparently developing from electron-dense material are associated with lateral inpushings of both inner and outer mitochondrial membranes.

These studies with the experimental production *in vitro* and *in vivo* of apparent mitochondrial subdivision in mammalian cardiac and skeletal muscle show the following.

(1) Mitochondrial septa can be produced *in situ* by a variety of experimental techniques.

Figs. 6–11. Possible sequence of ultrastructural changes in mitochondria from mouse diaphragm treated with A23187 (6.25 μg ml^{-1}). Fig. 6, control. Figs. 7 and 8, 10-min exposure. Fig. 9, 20-min exposure. Figs. 10 and 11, 30-min exposure, greatly swollen with a transverse septum (arrow). Bars: Figs. 6–9, 0.1 μm; Figs. 10 and 11, 0.5 μm.

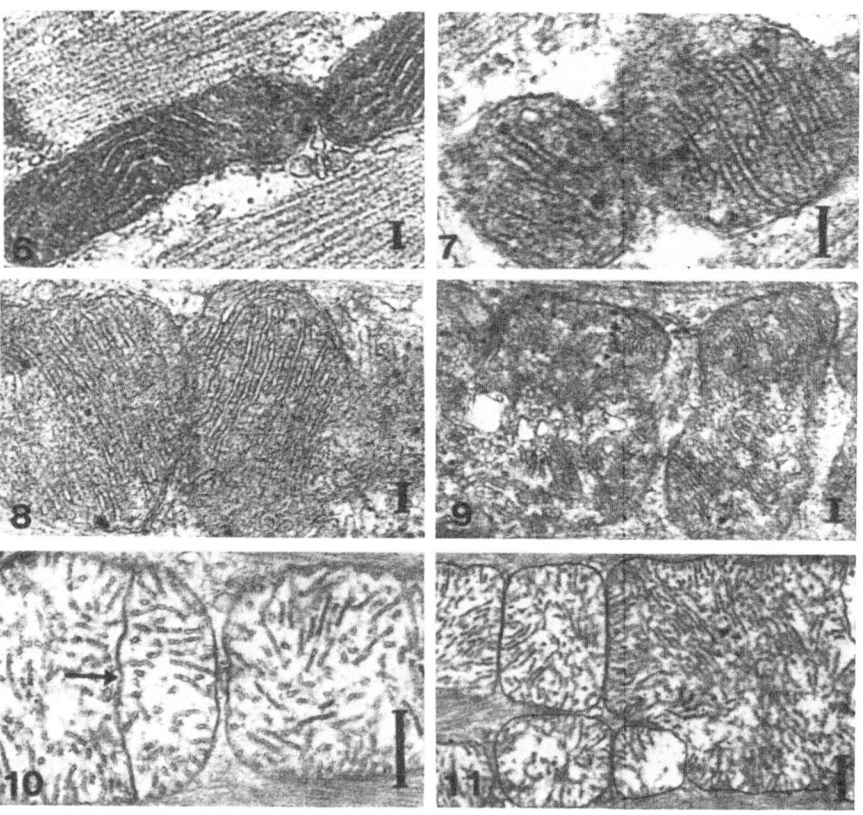

(2) The septa develop quickly (30–45 min, sometimes within 10 min), in contrast with earlier studies where the animals had been maintained on dietary regimes for several weeks.

(3) Earlier studies had shown subdivision into two different energetic configurations only in pellets of isolated mitochondria whereas they occur frequently in muscle *in situ.*

None of the foregoing studies demonstrates unequivocally that true mitochondrial division occurs *in situ* and that the megamitochondria (Fig. 11) and the swollen and adpressed mitochondria (Figs. 2, 5 and 19) are not merely the result of fusion consequent upon the swelling of the organelles. However, the findings are consistent with the hypothesis that experimental intervention can induce (perhaps artificially) rapid division of these organelles.

Are mammalian muscles unique in producing such dramatic mitochondrial changes? Although similar (almost identical) changes have been described in other tissues, such as liver, brown fat and adrenal cortex, the responses in muscle seem to be on a much greater scale and to be much more rapid than reported elsewhere.

Are dividing mitochondria a normal occurrence or are they found only in abnormal or damaged cells? Most of the reports of dividing mitochondria in any numbers follow experimental treatments such as cuprizone or 2–Me–DAB. They are found in normal cells *in vivo*, particularly cardiac muscle (Tandler & Hoppel, 1972; Rudge, 1983), although of rare occurrence and so these experimental interventions may accelerate a normal low-level activity.

Is mitochondrial division associated with cellular damage?

Characteristic mitochondrial division in muscle is usually (perhaps invariably) accompanied by cellular damage, and these observations have led to suggestions that it is an impairment of mitochondrial functioning, which causes a failure in the supply of ATP to the cell, that leads ultimately to muscle cellular damage and cell death. However there are two end-products of cellular damage which, in skeletal muscle, have been shown to be the result of two separate pathways (Duncan & Jackson, 1987):

(1) Sarcolemma damage and the release of cytosolic proteins, e.g. creatine kinase and lactic dehydrogenase.

(2) Severe damage to the myofilament apparatus, either (i) contraction and hypercontraction of the myofilaments with Z-line blurring, shown in Figs. 2, 5, 18 and 19 (compare myofilament structure with Fig. 1), or (ii) breakdown of the Z-line (Fig. 4) and/or actin filaments (Fig. 3) and/or myosin filaments (Fig. 20), culminating in the complete dissolution of the myofilament apparatus (Fig. 13).

It is now evident that in many of these experimental or pathological conditions, such as ischaemia, anoxia, malignant hyperthermia, A23187 or caffeine, there is a marked rise in intracellular Ca^{2+} concentration ($[Ca^{2+}]_i$), caused by the release of Ca^{2+} from the sarcoplasmic reticulum (SR) (caffeine, malignant hyperthermia) or by an increase in the net Ca^{2+} influx at the sarcolemma.

Firstly, this rise in $[Ca^{2+}]_i$ activates a phospholipase A_2 (PLA_2), associated with the sarcolemma, producing arachidonic acid and lysolecithin, both membrane-active agents. Arachidonic acid serves as the substrate for the cyclo-oxygenase enzyme pathway, producing prostaglandins, and for the lipoxygenase enzyme pathway, producing leukotrienes. It is this latter pathway (which can be inhibited by nordihydroguaiaretic acid) that is responsible for the release of CK at the sarcolemma, probably by the generation of oxygen radicals and lipid peroxidation (Duncan & Jackson, 1987).

Secondly, this rise in $[Ca^{2+}]_i$ switches on an unknown process which modifies the −SH groups of the actomyosin of which the myofilament apparatus is composed, causing increased ATPase activity (hypercontraction and rigor-like complexes) and disassembly of the actin and myosin filaments (Duncan, 1987). There is evidence that, associated with the movement of the calcium ions in the cytosol, an NADH oxidase situated on the SR is switched on which generates a transmembrane movement of electrons and the production of superoxides and oxygen radicals. Certainly, the experimental production of superoxides in chemically skinned skeletal muscle cells and in isolated, washed myofibrils simulates the patterns of myofilament damage seen *in vitro* and *in vivo* (C.J. Duncan, unpublished data).

Thus, during *rapid* cellular damage triggered by Ca^{2+} (as distinct from the much slower Ca^{2+}-activated protein turnover, with a time-course of hours, Furono &

Fig. 12. Mitochondrial bars. Note internal laminar structure and alignment with cristal membranes (diaphragm, A23187). Bar, 0.1 μm.

Fig. 13. Mitochondrial bars with gently curved transverse septum formed from internal mitochondrial membrane only (arrow). Soleus, A23187. Bar, 0.5 μm.

Fig. 14. Swollen mitochondria with triradiate septa (arrows). Mouse diaphragm incubated 60 min at 37 $^{\circ}$C and gassed with 95 per cent O_2 plus 5 per cent CO_2. Bar, 1 μm.

Fig. 15. Swollen mitochondria, of which two have a complete, curved septum separating two different energetic configurations. Diaphragm incubated as in Fig. 14. Bar, 1 μm.

Figs. 16–18. Mitochondria undergoing septation and showing progressive intucking in both mitochondrial membranes (arrows) forming septa. Diaphragm, A23187. Bars, 0.25 μm.

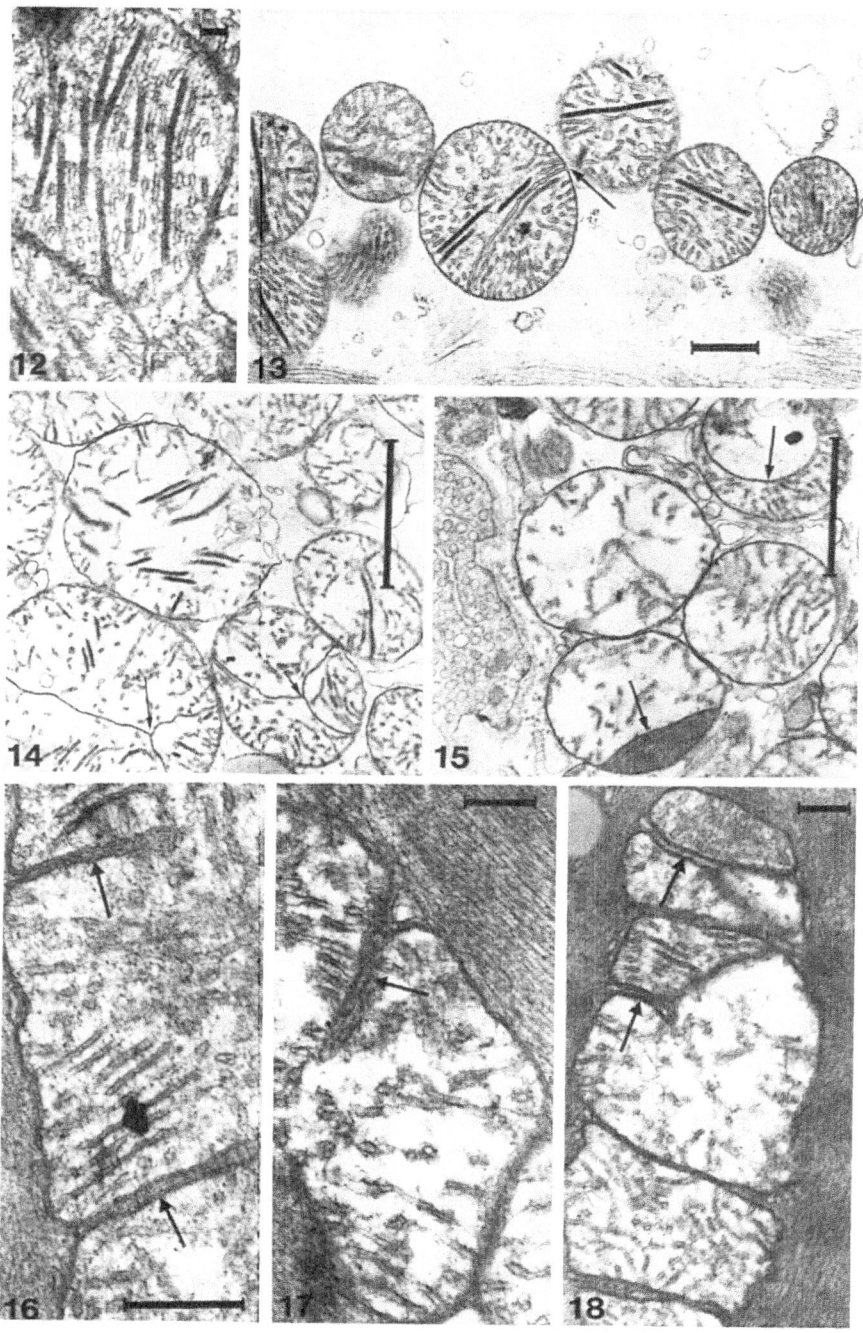

Goldberg, 1986), a number of interacting events occur in muscle which might serve as the stimulus for mitochondrial division and septation:

(1) Ca^{2+} uptake by the mitochondria and SR which, by Ca^{2+}/proton exchange together with massive ATP consumption, causes a fall in pH_i.

(2) Consumption of high-energy phosphate reserves.

(3) Activation of the muscle Ca^{2+}-activated neutral proteases, although mitochondrial division continues when these are inhibited (Duncan, Smith & Greenaway, 1979).

(4) Ca^{2+}-activation of PLA_2, both at the sarcolemma and the mitochondria, generating membrane-active substances and oxygen radicals.

(5) Release of lysosomal enzymes, probably from the sarcotubular system, although it is unlikely that this occurs during the 30 min of rapid muscle damage (Duncan et al., 1980b).

(6) A rise in temperature, particularly evident in malignant hyperthermia.

(7) Modification of key –SH groups on the actomyosin complex by the process that effects myofilament dissolution.

Modification of –SH groups in liver mitochondria

Since the damage to the myofilament apparatus appears to be an attack on –SH groups, perhaps via superoxides, it is possible that modification of –SH groups on mitochondrial membranes might be the stimulus for the formation of septa. Chemical modification of –SH groups on the inner mitochondrial membrane is known to stimulate swelling of the organelles (Lê–Quôc & Lê–Quôc, 1985). Diamide (at 10^{-4} M), a thiol oxidizing agent, rapidly promotes limited septation in isolated frog liver mitochondria, and the effect is partially inhibited by dithioerythritol, a –SH protective agent. Diamide is known to uncouple oxidative phosphorylation in isolated mitochondria (Siliprandi et al., 1975), but DNP does not cause division in isolated mitochondria, and it is concluded that diamide does not promote septation via an uncoupling action. However, after diamide treatment the incidence of divided mitochondria rose to only 1 per cent and the septate mitochondria have a different appearance from the typical dividing mitochondria described above. Thus diamide may be acting to favour the fusion of the internal membrane (Publicover et al., 1979).

We have repeated these experiments with isolated rat liver mitochondria, exposing them at 37 °C to such sulphydryl reagents as N–ethylmaleimide and diamide and also to a superoxide generating system (xanthine/xanthine oxidase). However, the swelling that is characteristic of mitochondria in muscle undergoing cell damage was not seen with isolated liver mitochondria with any of these treatments, although, as with diamide-treated frog liver mitochondria, there was a general shift towards the aggregated state. In parallel, liver mitochondria were exposed to 10^{-4}–10^{-3} M Ca^{2+}, but again active Ca^{2+} uptake (as shown by Ca^{2+} electrode studies) failed to simulate the characteristic septation and division.

Isolated heart mitochondria

In our current experiments, cardiac mitochondria from rat hearts have been isolated by the method of Tyler & Gonze (1967) and then exposed to various treatments for 10 min (a time chosen to correspond with rapid damage and mitochondrial division in mammalian cardiac muscle) at 37 °C before fixation for subsequent electron microscopy. Calcium (10^{-4}M or 5×10^{-3} M) produced a clear modification of the cristal organisation, including configurations resembling myelin figures; and occasionally, straight cristal membranes, reminiscent of mitochondrial bars, were seen. Exposure to a superoxide-generating, xanthine/xanthine oxidase system (with zero [Ca^{2+}] in the incubation medium) produced fewer changes in cristal organisation than did Ca^{2+}, although clear myelin figures were evident in 5–10 per cent of the organelles. However, incubation of the mitochondria with 10^{-4} M Ca^{2+} plus the xanthine/xanthine oxidase produced more clear-cut changes; in particular the cristal organisation resembled the swollen and damaged mitochondria of Figs. 9, 11 and 14, some mitochondria showed intra-organelle developments akin to bars and a few mitochondria showed apparent subdivision. Many mitochondria were

Fig. 19. Mouse ventricle, 10 min after re-perfusion with Ca^{2+} after the Ca-paradox. Note chains and areas of packed, swollen mitochondria. Several mitochondria have gently curving transverse septa (arrows). Myofilament apparatus severely hypercontracted with blurred Z-lines. Compare with skeletal muscle, Figs. 2 and 5. Bar, 1 μm.

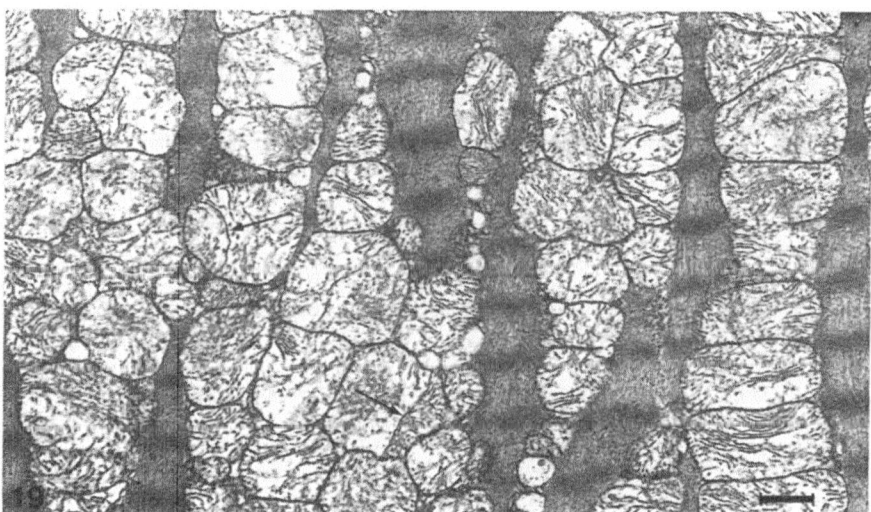

clearly damaged. However, it must be emphasised that these similarities may be superficial and that, so far, it has not been possible to simulate true septation in isolated mitochondria, although a more realistic picture might be achieved by varying the different factors, particularly the Ca^{2+} concentration and time of exposure.

Since the effects of Ca^{2+} and superoxides are synergistic, the most probable explanation of these results is that superoxide attack on –SH groups of the Ca^{2+} channels of the mitochondrion causes Ca^{2+} loading which acts as the trigger for the structural changes in the organelles seen during Ca^{2+}-activation. There are interesting parallels with recent biochemical studies with isolated mammalian renal mitochondria exposed to the synergistic action of oxygen free-radicals plus Ca^{2+}, which caused the inhibition of electron transport, uncoupling of oxidative phosphorylation and 55 per cent inhibition of ATPase activity. The defect was localised to NADH CoQ reductase, since it occurred with site I substrates, but not with site II substrates. The damaging effects were partially mitigated by dibucaine, an inhibitor of Ca^{2+}-activated PLA_2 (Malis & Bonventre, 1986). Thus, the properties and role of mitochondrial PLA_2 (as distinct from PLA_2 associated with the sarcolemma) are of particular interest as a potential trigger for mitochondrial septation. It splits phospholipids, producing a lysolecithin and an unsaturated fatty acid. Mitochondrial PLA_2 is activated by Ca^{2+} or Sr^{2+} (as is muscle damage), and in addition to the PLA_2 of the outer mitochondrial membrane, an active PLA_2 has now been described in the inner mitochondrial membrane in liver cells (Zurini, Hugentobler & Gazzotti, 1981) where it is believed

Figs. 20–25. Mitochondria from cardiac muscle in which $[Ca^{2+}]_i$ has been raised experimentally.

Fig. 20, Mouse. Inpushings of inner mitochondrial membrane (arrows); remnants of myofilament apparatus, showing extreme disintegration; A23187. Bar, 0.5 μm.

Fig. 21. Frog. Cristal membranes apparently continuous across mitochondrial membrane (arrows); A23187. Bar, 0.5 μm.

Fig. 22. Frog. Septum formed from inner mitochondrial membrane only (arrow); 5 mM caffeine. Bar, 1 μm.

Fig. 23. Mouse. The cristae in one section of the subdivided mitochondrion appear in profile and the other en face; 15 min anoxia. Bar, 1 μm.

Fig. 24. Mouse. Two sections of divided mitochondrion in different configurational states. A23187. Bar, 0.25 μm.

Fig. 25. Mouse. Mitochondria subdivided into three sections, the middle one being in a different configurational state from the two outer sections; 5 mM caffeine. Bar, 0.25 μm.

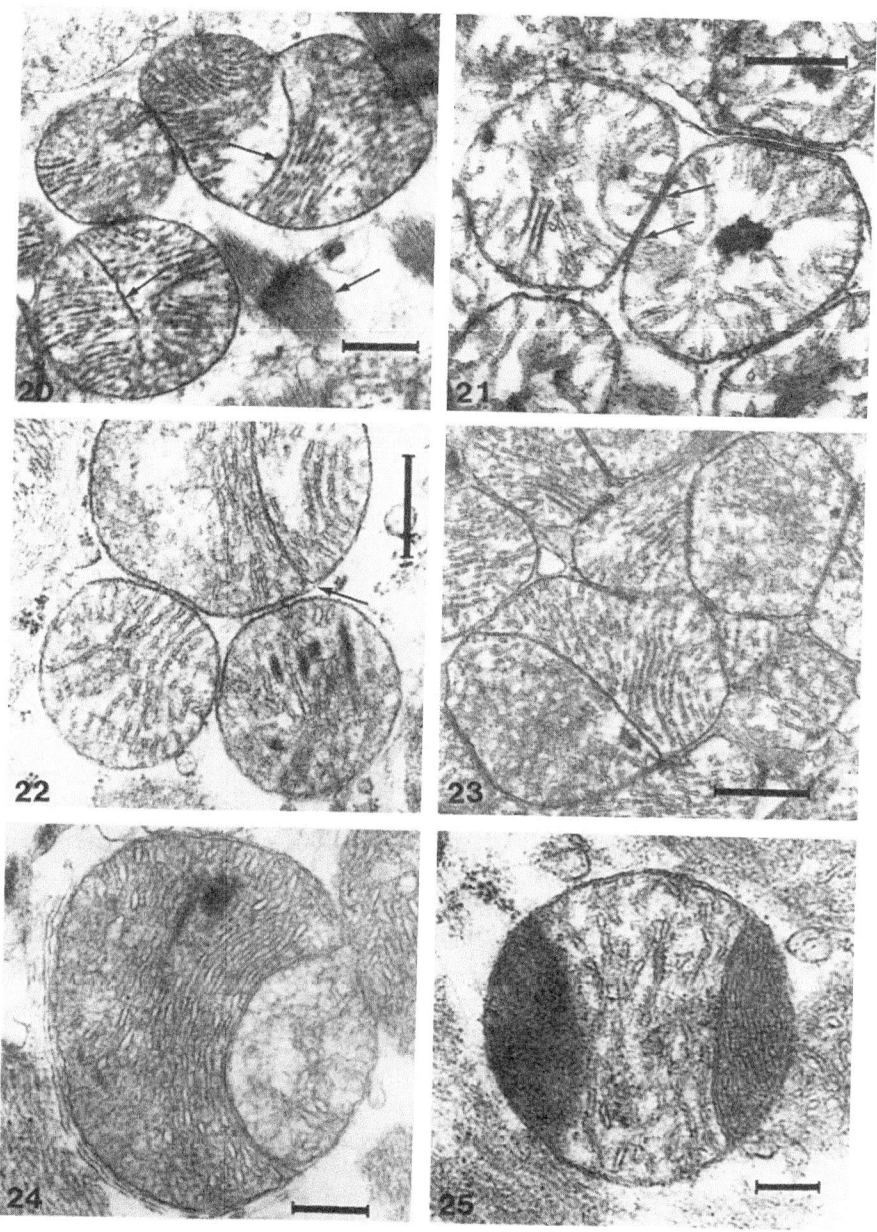

to be involved in determining the functional integrity of the inner membrane and in producing membrane changes in aged mitochondria. Neither the mitochondrial PLA_2 nor muscle damage are modified by calmodulin inhibitors.

It is concluded (Malis & Bonventre, 1986) that the Ca^{2+} potentiation of oxygen free-radical injury in renal mitochondria is because, in part, of activation of PLA_2 located in the mitochondrial membranes; thus an induction of septation in muscle mitochondria by activation of PLA_2 during Ca^{2+}-triggered cell damage would be consistent with the results described above. Thus, rapid cell damage in muscle appears to be associated with elevated $[Ca^{2+}]_i$ and with the production of active radicals that cause disassembly of actomyosin complexes. Muscle mitochondria therefore are vulnerable to damage to membrane $-SH$ groups and to Ca^{2+} overloading, which causes the swelling, whilst activation of a mitochondrial PLA_2 by Ca^{2+} would be responsible for the changes in the outer, inner and cristal membranes, the production of myelin figures and finally the disappearance of all internal membranes leaving empty, swollen organelles (see Fig. 4).

Conclusions

A substantial body of evidence (see survey in Gross, 1971) testifies to the steady turnover of mitochondrial constituents in mammalian tissues in the normal physiological state; also, changes in turnover were observed in thyroidectomized and normal rats following administration of thyroid hormone. The different turnover rates suggest that in liver the pre-existing population of mitochondria is being replaced by another population synthesised under new physiological conditions. However, the problem concerning the origin of this new population is still unresolved; do they originate *de novo* or are they produced by subdivision of pre-existing mitochondria, as shown by the small numbers of mitochondria with a complete transverse septum seen in a number of different normal tissues *in vivo*?

The accounts of dramatic swelling and multiplication of mitochondria seen in mammalian cardiac and skeletal muscle, and also the occurrence of mitochondrial bars, seem to be unique to vertebrate muscle and seem always to be associated with experimental interventions, with pathological conditions, or with severe cell damage. As suggested above, Ca^{2+} activation of mitochondrial PLA_2 may be the trigger for the septation and finally for the destruction of cristal membrane organisation. Whether this mitochondrial septation is unique to cellular damage or an exacerbation of a regular event in a cycle of mitochondrial division remains to be determined.

Acknowledgements

I thank Dr Melinda Rudge for our collaborative studies on cardiac muscle. The assistance of Mr J.L. Smith in electron microscopy and of Miss S. Scott in the preparation of the manuscript is gratefully acknowledged.

References

Busch, H.F.M., Jennekens, F.G.I. & Scholte, H.R. (1981). *Mitochondria and Muscular Diseases*. Beetsterzwaag, Netherlands: Mefar B.V.

De Vauchelle, G., Meynadier, G. & Vago, C. (1972). Etude ultrastructurale du cycle de multiplication de *Rickettsiella melolonthae* (Krieg), Philip, dans les hemocytes de son hate. *Journal of Ultrastructural Research*, **38**, 134–48.

Duncan, C.J. (1987). Role of calcium in triggering rapid ultrastructural damage in muscle: a study with chemically-skinned fibres. *Journal of Cell Science*, **89**, 581–94.

Duncan, C.J. & Greenaway, H.C. (1981). The induction of septation and subdivision in muscle mitochondria. *Comparative Biochemistry and Physiology*, **69A**, 329–31.

Duncan, C.J., Greenaway, H.C., Publicover, S.J., Rudge, M.F. & Smith, J.L. (1980*a*). Experimental production of "septa" and apparent subdivision of muscle mitochondria. *Journal of Bioenergetics and Biomembranes*, **12**, 13–33.

Duncan, C.J., Greenaway, H.C. & Smith, J.L. (1980*b*). 2,4–Dinitrophenol, lysosomal breakdown and rapid myofilament degradation in vertebrate skeletal muscle. *Naunyn-Schmiedeberg's Archives of Pharmacology*, **315**, 77–82.

Duncan, C.J. & Jackson, M.J. (1987). Different mechanisms mediate structural changes and intracellular enzyme efflux following damage to skeletal muscle. *Journal of Cell Science*, **87**, 183–8.

Duncan, C.J. & Smith, J.L. (1980).Action of caffeine in initiating myofilament degradation and subdivision of mitochondria in mammalian skeletal muscle. *Comparative Biochemistry and Physiology*, **65C**, 143–5.

Duncan, C.J., Smith, J.L. & Greenaway, H.C. (1979). Failure to protect frog skeletal muscle from ionophore-induced damage by the use of the protease inhibitor leupeptin. *Comparative Biochemistry and Physiology*, **63C**, 205–7.

Flatmark, T., Kryvi, H. & Tangeras, A. (1980). Induction of megamitochondria by cuprizone biscyclo hexanone oxaldihydrazone. Evidence for an inhibition of the mitochondrial division process. *European Journal of Cell Biology*, **23**, 141–8.

Furuno, K. & Goldberg, A.L. (1986). The activation of protein degradation in muscle by Ca^{2+} or muscle injury does not involve a lysosomal mechanism. *Biomedical Journal*, **237**, 859–64.

Gross, N.J. (1971). Control of mitochondrial turnover under the influence of thyroid hormone. *Journal of Cell Biology*, **48**, 29–40.

Hanzlikova, U. & Schiaffino, S. (1977). Mitochondrial changes in ischemic skeletal muscle. *Journal of Ultrastructural Research*, **60**, 121–33.

Heffner, R.R. & Baron, S.A. (1978). The early effects of ischaemia upon skeletal muscle mitochondria. *Journal of Neurological Science*, **38**, 295–315.

Heine, H. & Schaeg, G. (1979). Origin and function of 'rod-like structures' in mitochondria. *Acta Anatomica*, **103**, 1–10.

Jennings, R.B., Sommers, H.M., Herdson, P.B. & Kaltenbach, J.P. (1969). Ischaemic injury of the myocardium. *Annals of the New York Academy of Sciences*, **156**, 61–78.

Jeynes, B.J. & Altman, G.G. (1975). A region of mitochondrial division in the epithelium of the small intestine of the rat. *Anatomical Record*, **182**, 289–96.

Koch, O.R., Roatta de Conti, L.L., Balanos, L.P. & Stoppani, O.M. (1978). Ultrastructural and biochemical aspects of liver mitochondria during recovery from ethanol-induced alterations. *Americal Journal of Pathology*, **90**, 325–37.

Kuroiwa, T., Hizume, M. & Kawano, S. (1978). Studies on mitochondrial structure and function in *Physarum polycephalum* IV. *Cytologia*, **43**, 119–36.

Kuroiwa, T., Kawano, S. & Hizume, M. (1977). Studies on mitochondrial structure and function in *Physarum polycephalum*. V. Behaviour of mitochondrial nucleoids throughout mitochondrial division cycle. *Journal of Cell Biology*, **72**, 687–94.

Kuroiwa, T. & Kuroiwa, H. (1980). Inhibition of *Physarum polycephalum* mitochondrial division by cytochalasin B. *Experientia*, **36**, 193–4.

Lafontaine, J.G. & Allard, C. (1964). A light and electron microscope study of the morphological changes induced in rat liver cells by the azo dye 2–Me–DAB. *Journal of Cell Biology*, **22**, 143–72.

Larsen, W.J. (1970). Genesis of mitochondria in insect fat body. *Journal of Cell Biology*, **47**, 373–83.

Lê–Quôc, K. & Lê–Quôc, D. (1985). Crucial role of sulfhydryl groups in the mitochondrial inner membrane structure. *Journal of Biological Chemistry*, **260**, 7422–8.

Malis, C.D. & Bonventre, J.V. (1986). Mechanism of calcium potentiation of oxygen free radical injury to renal mitochondria. *Journal of Biological Chemistry*, **261**, 14201–8.

Mastaglia, F.L. & Walton, J. (1982). *Skeletal Muscle Pathology*. London: Churchill Livingstone.

Normann, T.C. & Samaranayaka–Ramasamy, M. (1977). Secretory hyperactivity and mitochondrial changes in neurosecretory cells of an insect. *Cell and Tissue Research*, **183**, 61–9.

Publicover, S.J., Duncan, C.J. & Smith, J.L. (1977). Ultrastructural changes in muscle mitochondria in situ, including the apparent development of internal septa, associated with the uptake and release of calcium. *Cell and Tissue Research*, **185**, 373–85.

Publicover, S.J., Duncan, C.J. & Smith, J.L. (1978). The use of A23187 to demonstrate the role of intracellular calcium in causing ultrastructural damage in mammalian muscle. *Journal of Neuropathology and Experimental Neurology*, **37**, 544–57.

Publicover, S.J., Duncan, C.J., Smith, J.L.& Greenaway, H.C. (1979). Stimulation of septation in mitochondria by diamide, a thiol oxidising agent. *Cell and Tissue Research*, **203**, 291–300.

Rudge, M.F. (1983). 'Comparative Studies on the Experimental Induction of Ultrastructural Damage in Vertebrate Cardiac Muscle.' Ph.D. thesis, University of Liverpool.

Rudge, M.F. & Duncan, C.J. (1984). Comparative studies on the role of calcium in triggering subcellular damage in cardiac muscle. *Comparative Biochemistry and Physiology*, **77A**, 459–68.

Siliprandi, D., Toninello, A., Zoccarato, F., Rugulo, M. & Siliprandi, N. (1975). Synergic action of calcium ions and diamide on mitochondrial swelling. *Biochemical and Biophysical Research Communications*, **66**, 956–61.

Simpson, L. (1972). The kinetoplast of the hemoflagellates. *International Review of Cytology*, **32**, 139–209.

Tandler, B., Erlandson, R.A., Smith, A.L. & Wynder, E.L. (1969). Riboflavin and mouse hepatic cell structure and function. II. Division of mitochondria during recovery from simple deficiency. *Journal of Cell Biology*, **41**, 477–93.

Tandler, B. & Hoppel, C.L. (1972). Possible division of cardiac mitochondria. *The Anatomical Record*, **173**, 309–24.

Tandler, B. & Hoppel, C.L. (1973). Division of giant mitochondria during recovery from cuprizone intoxication. *Journal of Cell Biology*, **56**, 266–72.

Tyler, D.D. & Gonze, J. (1967). The preparation of heart mitochondria from laboratory animals. In *Methods of Enzymology*, vol. 10, ed. R.W. Estabrook & M.E. Pullman. New York: Academic.

Wakabayashi, T., Asano, M. & Kurono, C. (1974). Some aspects of mitochondria having a "septum". *Journal of Electron Microscopy (Tokyo)*, **23**, 247–54.

Zurini, M., Hugentobler, G. & Gazzotti, P. (1981). Activity of phospholipase A$_2$ in the inner membrane of rat-liver mitochondria. *European Journal of Biochemistry*, **119**, 517–21.

R. A. E. TILNEY-BASSETT

Inheritance of plastids in *Pelargonium*

The evolution of the genus *Pelargonium* has probably included many mutations in the unusually large 217–kb plastid genome (Palmer, 1985). Several mutations were recognised by the disharmony observed after hybridisation between cultivated zonal pelargoniums (*Pelargonium × hortorum*) and the species *P. zonale* (Metzlaff *et al.*, 1982), and after hybridisation between the species *P. zonale* and *P. inquinans* (Pohlheim, 1986). An interaction between the hybrid nuclei and the plastids from the two parents gives rise to the phenomenon of hybrid variegation (Kirk & Tilney–Bassett, 1978). In the variegated seedlings the plastid inheritance is biparental. The plastids from one parent are compatible with the hybrid nucleus and they develop normally into green chloroplasts within the cotyledon and leaf cells; the plastids from the other parent are incompatible, fail to develop fully, and remain colourless. Other seedlings are fully green or completely colourless showing that their plastids have a purely maternal or purely paternal origin. This evidence of dissimilar plastids, which develop normally in their original nuclear background, but which are incompatible in a hybrid nuclear background, is confirmed by differences in the electrophoretic banding patterns of their respective DNAs following digestion with restriction enzymes (Metzlaff, Börner & Hagemann, 1981; Metzlaff *et al.*, 1982).

A different approach to the analysis of plastid inheritance is to make crosses between normal green plants and the ornamental, white-margined, variegated-leaf chimeras (Tilney–Bassett, 1986), which are an important source of spontaneous, mutant white plastids. These studies have a long history, which will be summarised briefly in order to develop a firm foundation for the evidence of an important genetic switch controlling the manner of plastid inheritance in the zonal pelargoniums.

The maternal inheritance model

By crossing green with variegated cultivars, a female gamete containing wild-type green plastids (G) is effectively being fertilised with a male gamete containing mutant white plastids (W), and so, for the sake of brevity, these can be referred to as G × W crosses, and the reciprocal as W × G crosses. These crosses produce progeny with highly variable mixtures of maternal zygotes (MZ), biparental zygotes (BPZ), and paternal zygotes (PZ), as defined by the presence or absence of green or white plastids in the young embryos, and later seedlings, into which the zygotes develop. Although only a portion of the embryos or seedlings from a cross receive plastids from both parents, the segregation pattern as a whole is referred to as biparental in contrast to the uniparental patterns in which usually only the maternal parent, but occasionally only the paternal parent, is the source of the plastids for all the progeny (Sears, 1980).

The classical experiments of Baur (1909) led him to confront the generally held view that the pollen could not transmit plastids. He suggested that in zonal pelargoniums the plastids were inherited from both parents. As a result of this biparental inheritance, the zygotes contained two types of plastid, which sorted-out during embryogenesis into pure cells with only one type of plastid – green or white. In principle, the results of later workers confirmed those of Baur. In detail, they revealed considerable heterogeneity in the relative proportions of maternal, biparental and paternal zygotes among the progeny of different crosses, and especially between reciprocal crosses. Thus, sometimes, plastid inheritance was maternal plus biparental (Noack, 1924, 1925; Imai, 1936; Tilney–Bassett, 1963; Hagemann & Scholze, 1974), paternal plus biparental (Baur, 1909; Noack, 1924, 1925; Imai, 1936; Tilney–Bassett, 1963), maternal (Roth, 1927; Imai, 1936; Tilney–Bassett, 1973), paternal (Imai, 1936; Tilney–Bassett, 1964), or maternal plus paternal with no biparental (Tilney–Bassett, 1973, 1974b). There was no satisfactory explanation for this extraordinary variability (Hagemann, 1964) and the problem was looked at again, first in Oxford and later in Swansea.

The first step was to look at the behaviour of a range of cultivars, and then to choose for further analysis some similar and some dissimilar variegated cultivars with good crossibility and fertility (Tilney–Bassett, 1963). From the two white-margined, variegated-leaf chimeras 'Dolly Varden' and 'Flower of Spring', pure green shoots were isolated and, for each cultivar, two isogenic clones were therefore established one with white and the other with green plastids in the germ cells (Tilney–Bassett, 1965). By making G × W and W × G crosses within the isogenic clones, it was possible to input the normal or mutant plastids from either parent without varying the nuclear input. These comparisons showed that reciprocal crosses did not give reciprocal results. Whether the input was through the female or the male parent, the mutant plastids were less successfully transmitted than the normal plastids. One striking result of the inequality between the plastids was that, averaged across the progeny, a G × W cross often showed a predominantly maternal inheritance, whereas the reciprocal W × G cross showed a predominantly paternal inheritance (Tilney–Bassett, 1975).

An important practical step was to score young embryos by dissecting them out of the seed coats and so avoid the necessity of always maturing and germinating seed (Tilney–Bassett, 1970a). This made it possible to demonstrate that the differing frequencies of maternal, biparental and paternal zygotes from dissimilar cultivars, and from reciprocal crosses, could not be attributed to selection against white embryos containing wholly mutant plastids (Tilney–Bassett 1970a, c). The segregation

Table 1. *Examples of varying segregation frequencies following G ×W and W × G plastid crosses between the type I 'Miss Burdette–Coutts' (MBC) and the type II 'Flower of Spring' (FS)*

Cross			% Embryos				% Zygotes			
			G	V	W	Total	MZ	BPZ	PZ	% MZ + BPZ
MBC	G × W	FS	96.8	3.2	0	155	96.8	3.2	0	100.0
MBC	W × G	FS	4.9	69.2	25.9	266	25.9	69.2	4.9	95.1
FS	G × W	MBC	45.4	17.3	37.3	359	45.4	17.3	37.3	62.7
FS	W × G	MBC	73.5	25.2	1.3	155	1.3	25.2	73.5	26.5

Data from Tilney–Bassett (1976).

patterns also proved to be well buffered against the fluctuating environment, and so were quite repeatable (Tilney–Bassett, 1970*b*). Hence the significant differences between crosses had to be attributed to an underlying genetic control.

Once embryo selection and environmental effects had been ruled out as major causes of variation, the way was open for a deeper understanding of the genetic factors controlling plastid inheritance. Using the green clones of 'Dolly Varden' and 'Flower of Spring', alternately as female or as male parents, in crosses with six variegated cultivars, it became clear that the segregation pattern was highly repeatable with a constant female cultivar and varying males, and highly variable with differing females and a constant male (Tilney–Bassett, 1970*c*). This conclusion was confirmed in a larger experiment involving the isogenic green and variegated clones of six cultivars in an analysis of 36 G × W and 36 W × G crosses (Tilney–Bassett, 1976). After G × W plastid crosses these six cultivars had progeny with very distinctive segregation patterns. With three cultivars, 'Miss Burdette Coutts', 'Lass O'Gowrie' and 'Dolly Varden', most zygotes were maternal, rather less biparental and a few paternal (MZ > BPZ > PZ). With the other three cultivars, 'J.C. Mapping', 'Flower of Spring' and 'Foster's Seedling', maternal and paternal zygotes were about the same frequency and biparental few (MZ > BPZ < PZ) (Table 1). The constancy of their distinctive segregation patterns implied that the cultivars were genetically different. Just how different was shown by selfing and crossing these cultivars (Tilney–Bassett, 1983).

'Dolly Varden' was true-breeding. All its selfed progeny, when grown up and tested for their behaviour in G × W plastid crosses, had segregation frequencies in a pattern similar to itself. This was referred to as the type I pattern. By contrast, 'Flower of Spring' was not true-breeding. Approximately half its selfed progeny had segregation frequencies in a pattern similar to itself. This was referred to as the type II pattern. The other half of its selfed progeny had segregation frequencies in a type I pattern. It thus appeared that 'Dolly Varden' was homozygous for a gene controlling the pattern of plastid segregation and 'Flower of Spring' heterozygous. Moreover, as befitting crosses between homozygotes and heterozygotes, crosses between the two cultivars produced progeny which also segregated into those producing a type I and those producing a type II segregation pattern. The segregation into type I or type II was independent of the origin of the wild-type plastids, which ruled out any possibility of the plastids themselves controlling the alternative patterns of plastid inheritance. Hence the two patterns were considered to be under the control of a major nuclear gene. The gene was symbolised as Pr, with alternative alleles $Pr1$ and $Pr2$ on the assumption that the gene controlled plastid segregation through an effect, direct or indirect, on plastid replication.

These discoveries led to the realisation that a good model for the control of plastid inheritance in zonal pelargoniums was that of the left- and right-hand coiling of the shell in the snail *Limnaea peregra* (Gurdon, 1978). This classic example of maternal inheritance is determined by alternative alleles of a gene in which the direction of coiling is controlled, not by the genotype of the mollusc housed inside the shell, but by the nuclear genotype – not the phenotype – of its mother. This is a good model for pelargonium because the fate of the plastids is determined in the zygote or very early embryo just as is the coiling of the snail shell and, indeed, some other phenomena exhibiting maternal effects (Newth & Balls, 1979). Under the maternal inheritance model, the segregation pattern, as revealed by the scoring of a large sample of embryos, is the phenotype of the maternal parent from which its genotype is deduced.

The inheritance of the *Pr* gene

The method of analysing the Pr genotype is rather unusual because, after all selfs and crosses, the progeny has to be tested by scoring the segregation patterns that each of them produce a generation later. To use a human analogy: after mating the parents, a large sample of the infant grandchildren have to be observed in order to tell the phenotypes, and hence deduce the genotypes, of the children.

When homozygous type I plants, $Pr1Pr1$, were crossed with heterozygous type II plants, $Pr1Pr2$, the progeny generally segregated in good agreement with the expected 1 : 1 backcross ratio, although there were some unaccountable exceptions in which the expected type II plants were rare or absent (Tilney-Bassett, 1973, 1974*b*). When heterozygous type II plants, $Pr1Pr2$, were selfed or intercrossed, instead of the progeny consisting of three genotypes in the expected monohybrid ratio of 1 $Pr1Pr1$: 2 $Pr1Pr2$: 1 $Pr2Pr2$, the progeny again appeared in the ratio of 1 $Pr1Pr1$: 1 $Pr1Pr2$, and this behaviour has since been repeated many times. The absence of the $Pr2Pr2$ homozygote could not be attributed to zygotic selection as this would give rise to a

Table 2. *Independence of red : non-red petal colour and type I : type II plastid segregation patterns: after (A) Red I × White II and (B) Red II × White II crosses; between red flowers of 'Dolly Varden', 'Miss Burdette–Coutts', 'Flower of Spring' and 'J.C. Mapping', and the white flowers of 'J.C. Mapping'*

Red ×White Crosses	Red I	Red II	Non-red I	Non-red II	Total	df	χ^2	P
A								
DV × JCM	14	11	13	10	48	3	0.833	0.9–0.5
MBC × JCM	10	18	21	19	68	3	4.118	0.5–0.1
B								
FS × JCM	16	18	18	10	62	3	2.774	0.5–0.1
JCM × JCM	9	9	10	4	32	3	2.750	0.5–0.1
Ratio	1	1	1	1				
Total						12	10.475	
Pooled:	49	56	62	43	210	3	3.905	0.5–0.1
Heterogeneity:						9	6.570	0.5–0.1

In the "Progeny" header spanning columns: Red I, Red II, Non-red I, Non-red II, Total.

1 : 2 ratio and not the 1 : 1 ratio observed. Gametic selection against the *Pr2* allele appeared too simple an explanation as crosses between type I and type II plants revealed that the *Pr2* allele could be transmitted by both male and female parents. Hence, as a tentative explanation, it was suggested that *Pr1* behaves as a self-compatible, recessive allele, *Pr2* as a self-incompatible, dominant allele, and *Pr1* and *Pr2* as cross compatible alleles (Tilney–Bassett & Abdel–Wahab, 1982). Self-fertilisation was probably determined by sporophytic control on the male side and gametophytic control on the female side. Under sporophytic control all pollen grains derived from a type II parent behave as if they are *Pr2*, even though half are actually *Pr1*, and are therefore accepted by *Pr1* eggs and opposed by *Pr2* eggs. The resulting progeny ratio is 1 *Pr1Pr1* : 1 *Pr1Pr2* in which the *Pr2* allele is derived from the male parent. It was also possible to envisage sporophytic control on the female side and gametophytic control in the male, in which case the *Pr2* allele would be derived from the female parent.

The consistency of these ratios has been verified in an extensive new crossing programme in which the behaviour of other genes has been monitored at the same time. For example, in crosses involving plants with red versus non-red flower colours and type I or type II genotypes, progeny have been obtained in good agreement with a 1 : 1 : 1 : 1 ratio indicative of the Mendelian segregation of both genes with independent assortment (Table 2). As yet no linkage has been found between *Pr* and any other gene. The occasional exceptions to the expected 1 : 1 ratios have not yet been explained, but suggest that in some instances a second gene might be suppressing the expression of the type II genotype.

The variation in gene expression

A search for type I or type II plants among a dozen green cultivars which were unfamiliar to the author, showed them all to be type I (Tilney–Bassett, 1974*a*). Like the many cultivars examined in the past, they showed a wide range of variation within this pattern. In order to determine whether this variation had a genetic basis, or was purely environmental, an analysis of variance was performed on the estimated maternal or paternal allelic frequencies among the progeny after 36 G × W and 36 reciprocal W × G crosses within and between three type I and three type II cultivars. One of the main effects showed that the variance caused by differences between female cultivars was highly significant. Orthogonal contrasts within the main effects confirmed the highly significant differences between individual cultivars, as a result of which some cultivars were regarded as more strongly maternal than others. The variation in the expression of the two segregation patterns suggested the influence of polygenic effects, or there might be multiple alleles of the *Pr* gene. These results were from six cultivars taken at random from the much larger population that exists and so did not reveal the nature of the variability as a whole.

In order to obtain a better insight into the nature of the population variability, the variation between 214 families was looked at and signs were found of a skewed or normal distribution among type I or type II plants respectively (Tilney–Bassett, 1984). This line of investigation has now been expanded by taking 11 cultivars and selfing and intercrossing them in 111 different ways, both within, and between, type I and type II plants, to generate 2567 families. These were all classified as type I or II by crossing them, as green females, with the variegated 'Flower of Spring' as the pollen parent. The resulting segregation patterns were assessed from an average of 76 embryos per plant and a minimum of 20 embryos.

For the majority of plants the assessment of their segregation pattern as type I or II was straightforward. In some progeny, however, there was a high frequency of maternal zygotes, no biparental zygotes, and just a few paternal ones. In some other progeny the frequencies of biparental and paternal zygotes were identical or were very close to each other. In both cases, the overall maternal transmission placed the plants well within the type I range and on the edge of the type II range, and so they were more likely to be type I than type II. It was therefore decided to add the rider that for a type II classification after G × W crosses, the progeny had to express the pattern MZ > BPZ < PZ and either have at least three paternal zygotes (white embryos), or have at least three more paternal than biparental zygotes.

The variable frequencies of maternal, biparental and paternal zygotes were all converted into percentages for ease of comparison. The progeny were then scored for the percentage of maternal plus biparental zygotes combined. This tells us the percentage of zygotes among the progeny of a family that received at least some maternal plastids. The progeny were then grouped into either 100 per cent zygotes with maternal plastids, or in 4 per cent intervals down to a minimum of an average of 20 per cent zygotes with at least some maternal plastids, and the frequencies for each group were plotted as a histogram (Fig. 1). The figure shows that the type I plants fall into a very skewed distribution with such a strong maternal bias that over 70 per cent of the families have at least some maternal plastids in every zygote. Thereafter

Fig. 1. Histogram of the frequencies of families with the percentages of zygotes having at least some maternal plastids, in 4 per cent intervals. Analysis of G × W plastid crosses with the variegated 'Flower of Spring' as constant pollen parent. Frequencies of the families of type I plants shaded, and of type II plants unshaded.

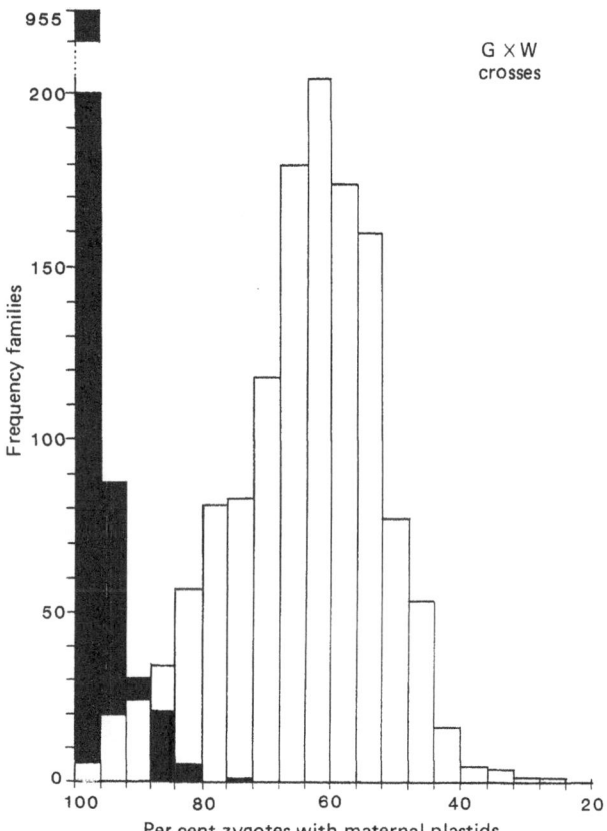

the frequencies of families with less than 100 per cent zygotes containing maternal plastids falls off steeply. By contrast, the type II plants exhibit a normal-looking distribution ranging from over 96 per cent zygotes containing maternal plastids down to 20 per cent with a mode at 60–64 per cent.

These results clearly demonstrate the wide variation in the expression of the *Pr* gene, and just how much its expression is modified by the genetic background, as well as the error variance around each and every genotype. Within each segregation type there is so much variance that the effects of individual modifying genes are submerged and the continuous variation so produced becomes just like many other metrical characters.

Fig. 2. Histogram of the frequencies of families with the percentages of zygotes having at least some maternal plastids, in 10 per cent intervals. Analysis of W × G plastid crosses with variable pollen parents. There is no separation of type I and type II plants, which are scattered across the range.

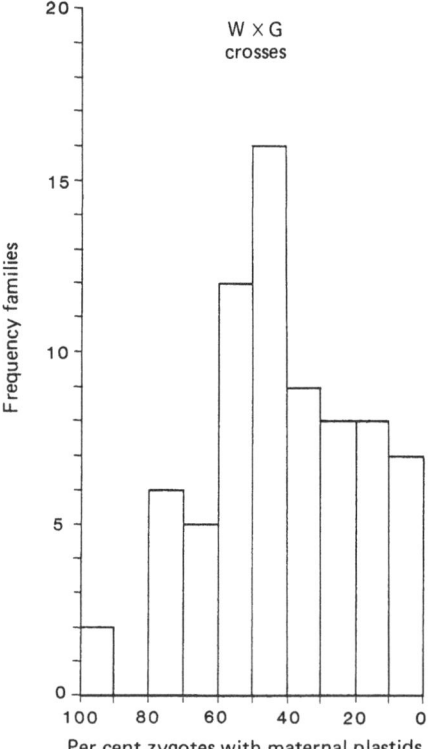

The examination of W × G crosses on a large scale is difficult. Most work has involved G × W crosses, which tend not to generate high frequencies of variegated progeny of the kind that develop into stable white-margined chimeras. Furthermore, the number of such plants in cultivation is rather limited and not all of these are fertile. Nevertheless, by including the published results of earlier workers, I have been able to assemble data on over 70 W × G crosses. These reveal an extraordinary range of variation (Fig. 2). At one extreme one family had 100 per cent of the zygotes with some maternal plastids (white plus variegated embryos), and at the other extreme one family had only 0.5 per cent of the zygotes with some maternal plastids. There was such a continuous range of variation in between that the mode at 40–50 per cent was not very strong. Where data were available on both, the W × G crosses showed no correlation with their reciprocal G × W crosses; the type I and type II individuals were scattered across the range throwing doubt on the significance of an apparent correlation when only a few families had been compared (Tilney–Bassett, 1976, 1984). Caution is required, however, as there may be different mutant plastids among the variegated cultivars, and these might vary in their interactions with the maternal nuclei creating additional heterogeneity among the families (Abdel–Wahab & Tilney–Bassett, 1981) and masking important relationships.

The plastid segregation patterns

Tilney–Bassett & Birky (1981) examined the problem of how to quantify the segregation patterns by considering each embryo from a cross as a population of plastids with a particular gene frequency. They assumed that each cell, and each plastid, was homoplasmic for one allele or the other, and hence the percent of tissue, in an embryo with the maternal or paternal phenotype (G or W) was an estimate of the frequency of the corresponding allele in the entire embryo. The different embryos from a cross, classified according to their allelic frequencies, then constituted an ensemble of populations. The varying shapes of these ensembles could then be assessed by plotting the frequency distributions of their constituent embryos. The maternal allelic frequencies of the type I progeny fall within a characteristic L-shaped distribution, and the type II progeny within a characteristic U-shaped distribution (Fig. 3).

Among the type I progeny, where maternal alleles are more frequent than paternal alleles, there is also a bias in favour of the maternal alleles among the biparental zygotes. As maternal transmission decreases the initial L-shaped curves become increasingly U-shaped. The only modes are at 100 per cent and at 0 per cent maternal alleles. There is no mode corresponding to the population mean and no sign of a Gaussian distribution.

The U-shaped distribution of type II progeny arises because most zygotes are either maternal or paternal and few are biparental. Yet, if on average the ratio of maternal to paternal alleles is approximately equal, one would expect the majority of zygotes to be biparental. The rapid appearance of pure embryos does not support Baur's classical hypothesis of the random sorting-out of plastids, and so it was suggested (Tilney–Bassett, 1970c) that of the two types of plastid, frequently only one is replicated. This hypothesis of competition between the parental organelles, or their DNAs, for replication in the zygote was later described by Birky (1983) as a

Fig. 3. Two histograms from G × W plastid crosses.

Above: Segregation pattern in the progeny of a type I mother showing the typical L-shaped distribution with a strongly maternal, intermediate and low biparental, and rare paternal plastid gene transmission (MZ > BPZ > PZ).

Below: Segregation pattern in the progeny of a type II mother showing the typical U-shaped distribution with approximately equal maternal and paternal and rare biparental plastid gene transmission (MZ > BPZ < PZ). Uniparental-maternal and paternal progeny open, biparental progeny shaded.

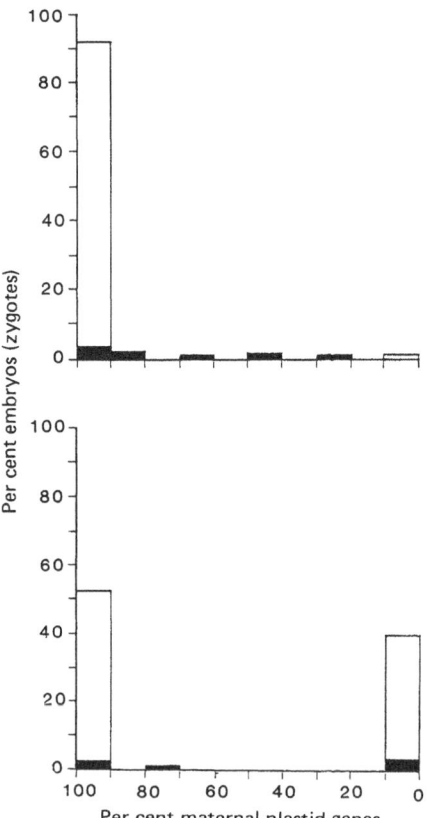

forerunner of his more general stochastic hypothesis (Birky *et al.*, 1982) which he has developed over many years.

The frequency distributions strongly resemble those seen in random drift of finite populations in Mendelian population genetics. They resemble those seen for chloroplast genes in *Chlamydomonas* (Birky *et al.*, 1981) and for mitochondrial genes in yeast (Thrailkill *et al.*, 1980). The distributions suggest that there is some repeated stochastic event or events occurring in zygotes and early embryos which change gene frequencies. Such events would contribute to uniparental inheritance by fixing one allele or the other in some embryos, and would also eliminate the Gaussian distribution expected around the input frequencies.

Tilney–Bassett & Birky (1981) considered several possible mechanisms to account for the rapid fixation of one or other allele. After arguing against the less likely mechanisms, they concluded that a random element in the replication or degradation of plastids or plastid DNA molecules is a serious possibility. If there was a turnover of plastids, and the plastids selected for replication or degradation were chosen randomly, this would cause the frequencies of green and white plastids, respectively alleles, to drift. The rate of drift would depend on the number of plastids, or plastid DNA molecules, in the zygote – the fewer the organelles or molecules the more rapid the drift. The rate would also be speeded up if groups of genetically identical molecules were selected for replication or degradation lowering the effective population size. This could happen if, for example, a polymerase enzyme replicated a plastid DNA molecule and then preferentially replicated the daughter DNA molecules because the polymerase was still in the vicinity.

In the case of type II plants, the probability of replicating a maternal or paternal plastid after G × W crosses appears, on average, only marginally in favour of the maternal plastid; there is a clear stochastic choice. But among the progeny of type I plants the probability appears to be weighted much more strongly in favour of the maternal plastid, introducing an element of selection. The cause of such selection is not known, but one possibility would be if in the homozygous recessive type I plants the paternal, mutant plastid DNA was preferentially degraded, perhaps by a restriction enzyme, while the maternal plastid DNA within the same zygote was protected. The apparent absence of selection in the heterozygous type II plants would then be indicative of the suppression of the activity of the restriction enzyme.

The drift and selection may have elements of several different mechanisms. The important principle is that, in contrast to the behaviour of chromosomes, the segregation and/or replication of plastids and plastid DNA appears to have a strong random element which results in random drift of gene frequencies within the zygote and subsequent mixed cells. This, coupled with selection, leads to a fixation of alleles giving many purely maternal or paternal embryos, and even whole progenies having maternal or paternal inheritance patterns, or mixed progenies with extremely high variance of gene frequencies.

The *Pr* switch gene

The choice of cultivars used in our earliest experiments with G × W crosses led to the impression that limited male plastid transmission was a characteristic of type I plants (Kirk & Tilney–Bassett, 1978). We now see that such plants were towards the edge of a range of variation. In over 70 per cent of the families tested either there was no male transmission whatsoever, or a very few male plastids were recovered as biparental zygotes. As the skewed histogram (Fig. 1) shows, the decreasing maternal plastid contribution shows a steep decline in frequency. So it appears that the homozygous maternal parents produce *Pr1* eggs, containing normal plastids, that do not permit the retention of most of the mutant plastids brought in by the male gamete at fertilisation. Assuming there is plastid replication and turnover, this could happen if the nuclear encoded plastid DNA polymerase had a greater affinity for the normal than for the mutant plastid DNA. Such a selective force would ensure that in most zygotes there was a rapid drift towards fixation of the maternal plastid. Nevertheless, a few mutant plastids might be preserved by chance and, even if the probability was quite low, occasionally replicated. The effect of different genetic backgrounds might be sometimes to enhance and sometimes to reduce the affinity of the polymerase for the normal plastid DNA. This might be a direct effect, or indirect owing, for example, to variations in the ratio of the input of plastids from the two parents or in the initial numbers of copies of the plastid DNAs.

Whatever the true explanation for the huge success of the maternal plastids in homozygous plants, there is a dramatic switch in behaviour in the heterozygous maternal plants. The heterozygous mothers produce eggs that behave as if they all contain the dominant *Pr2* allele, even though half contain the recessive *Pr1* allele. Under the sporophytic control of the *Pr2* allele, the selection against the incoming mutant male plastids appears to be switched off. Continuing with the polymerase model, it appears that in the *Pr2* eggs the enzyme has equal affinity for either normal or mutant plastid DNA. The overall average position slightly favours the maternal parent, but this might reflect nothing more than a greater initial plastid input. Again, as for type I plants, differences in the genetic background may increase or decrease the probability for replicating maternal or paternal plastid DNA, and so shift the direction of drift either side of the mean resulting in the normal distribution observed.

When one looks at the W × G crosses, the effect of the switch gene is still far from clear. At one end of the range there are families in which the normal plastids are again extremely successful. These are less frequent than with type I G × W crosses, probably because the input of green plastids from the male parent is likely to be less than from the female parent. In other words, the stronger affinity of the polymerase for the normal plastids might rarely be sufficient to overcome completely the disadvantage of an initially greater number of mutant plastids. In general, one can expect much greater variance if the initial input from the maternal plastids and the selective advantage of the paternal plastids are counter forces pulling in opposite directions than if they are joint forces pulling in the same direction. This is probably the situation in most type I W × G crosses, which have high variance in comparison with the reciprocal G × W crosses, which have relatively low variance. With the type II plants, even though the selective advantage of the normal plastids may have gone, variation in the genetic background might still create enough variability for sometimes the maternal, sometimes the paternal parent to have the upper hand. It is significant that the proportion of biparental zygotes is higher among W × G than G × W crosses, reflecting the greater equality of the alternative plastid types.

Another possible hypothesis is that the switch gene only interacts with the normal plastids in the egg cell and not with mutant plastids. In which case, although there would be a big difference in the response of *Pr1* and *Pr2* eggs with G × W crosses, there would be no differential response with W × G crosses. On the other hand, some groups of W × G crosses show evidence of a maternal effect (Tilney–Bassett, 1976). The real difficulty is that still more W × G crosses need analysing before one can expect to thoroughly understand the basic causes of their excessive variability – a daunting task for the future.

Acknowledgements
I should like to thank Mr A.B. Almouslem and Mr H.H. Kabwazi for allowing me to include some of their results in Figs. 2 and 3 respectively.

References
Abdel–Wahab, O.A.L. & Tilney–Bassett, R.A.E. (1981). The role of plastid competition in the control of plastid inheritance in the zonal *Pelargonium. Plasmid*, 6, 7–16.

Baur, E. (1909). Das Wesen und die Erblichkeitsverhältnisse der 'Varietates albomarginate hort' von *Pelargonium zonale. Zeitschrift für Vererbungslehre*, 1, 330–51.

Birky, C.W. Jr. (1983). Relaxed cellular controls and organelle heredity. *Science*, 222, 468–75.

Birky, C.W. Jr., Acton, A.R., Dietrich, R. & Carver, M. (1982). Mitochondrial transmission genetics: Replication, recombination and segregation of mitochondrial DNA and its inheritance in crosses. In *Mitochondrial Genes*, ed. G. Attardi, P. Borst & P.P. Slonimski, pp. 333–48. Cold Spring Harbor, NY: Cold Spring Harbor Laboratory.

Birky, C.W. Jr., VanWinkle–Swift, K.P., Sears, B.B., Boyton, J.E. & Gillham, N.W. (1981). Frequency distributions for chloroplast genes in *Chlamydomonas* zygote clones: Evidence for random drift. *Plasmid*, 6, 173–92.

Gurdon, J.B. (1978). *Gene Expression during Cell Differentiation*, 2nd edn. In *Carolina Biology Readers* 25, ed. J.J. Head, pp. 1–32. Carolina Biological Supply Company, North Carolina.

Hagemann, R. (1964). *Plasmatische Vererbung.* Jena: Veb. Gustav Fischer Verlag.

Hagemann, R. & Scholze, M. (1974). Struktur und Funktion der genetischen Information in den Plastiden. VII. Vererbung und Entmischung genetisch unterschiedlicher Plastidensorten bei *Pelargonium zonale* Ait. *Biologisches Zentralblatt*, 93, 625–48.

Imai, Y. (1936). Geno- and plasmotypes of variegated *Pelargoniums*. *Journal of Genetics*, **33**, 169–95.

Kirk, J.T.O. & Tilney–Bassett, R.A.E. (1978). *The Plastids: Their Chemistry, Structure, Growth and Inheritance*, 2nd edn. Amsterdam: Elsevier/North Holland.

Metzlaff, M., Börner, T. & Hagemann, R. (1981). Variations of chloroplast DNAs in the genus *Pelargonium* and their biparental inheritance. *Theoretical and Applied Genetics*, **60**, 37–41.

Metzlaff, M., Pohlheim, F., Börner, T. & Hagemann, R. (1982). Hybrid variegation in the genus *Pelargonium*. *Current Genetics*, **5**, 245–9.

Newth, D.R. & Balls, M. (eds.) (1979). *Maternal Effects in Development*. British Society for Developmental Biology Symposium 4. Cambridge University Press.

Noack, K.L. (1924). Vererbungsversuche mit buntblättrigen *Pelargonien*. *Verhandlungen physikalisch-medizinischen Gesellschaft Würzb.*, **49**, 45–93.

Noack, K.L. (1925). Weitere Untersuchungen über das Wesen der Buntblättrigkeit bei *Pelargonium*. *Verhandlungen physikalisch-medizinischen Gesellschaft Würzb.*, **50**, 47–97.

Palmer, J.D. (1985). Comparative organization of chloroplast genomes. *Annual Review of Genetics*, **19**, 325–54.

Pohlheim, F. (1986). Hybrid variegation in crosses between *Pelargonium zonale* (L) l'Herit. ex Ait. and *Pelargonium inquinans* (L) l'Herit. ex Ait. *Plant Breeding*, **97**, 93–6.

Roth, L. (1927). Untersuchungen über die periklinal bunten Rassen von *Pelargonium zonale*. *Zeitschrift für Vererbungslehre*, **45**, 125–9.

Sears, B.B. (1980). Elimination of plastids during spermatogenesis and fertilization in the Plant Kingdom. *Plasmid*, **4**, 233–55.

Thrailkill, K.M., Birky, C.W. Jr., Lückemann, G. & Wolf, K. (1980). Intracellular population genetics: Evidence for random drift of mitochondrial allele frequencies in *Saccharomyces cerevisiae* and *Schizosaccharomyces pombe*. *Genetics*, **96**, 237–62.

Tilney–Bassett, R.A.E. (1963). Genetics and plastid physiology in *Pelargonium*. *Heredity*, **18**, 485–504.

Tilney–Bassett, R.A.E. (1964). Failure to transmit mutant plastids in a *Pelargonium* cross. *Heredity*, **19**, 516–18.

Tilney–Bassett, R.A.E (1965). Genetics and plastid physiology in *Pelargonium* II. *Heredity*, **20**, 451–66.

Tilney–Bassett, R.A.E. (1970a). Genetics and plastid physiology in *Pelargonium* III. Effect of cultivar and plastids on fertilization and embryo survival. *Heredity*, **25**, 89–103.

Tilney–Bassett, R.A.E. (1970b). Effect of environment on plastid segregation in young embryos of *Pelargonium × hortorum* Bailey. *Annals of Botany*, **34**, 811–16.

Tilney–Bassett, R.A.E. (1970c). The control of plastid inheritance in *Pelargonium*.. *Genetical Research, Cambridge*, **16**, 49–61.

Tilney–Bassett, R.A.E. (1973). The control of plastid inheritance in *Pelargonium* II. *Heredity*, **30**, 1–13.

Tilney–Bassett, R.A.E. (1974a). A search for the rare type II (G > V < W) plastid segregation pattern among cultivars of *Pelargonium × hortorum* Bailey. *Annals of Botany*, **38**, 1089–92.

Tilney–Bassett, R.A.E. (1974b). The control of plastid inheritance in *Pelargonium* III. *Heredity*, **33**, 353–60.

Tilney–Bassett, R.A.E. (1975). Genetics of variegated plants. In *Genetics and Biogenesis of Mitochondria and Chloroplasts* (The First Colloquium of the

College of Biological Sciences of the Ohio State University), eds. C.W. Birky Jr., P.S. Perlman & T.J. Byers. pp. 268–308. Ohio State University Press.

Tilney–Bassett, R.A.E. (1976). The control of plastid inheritance in *Pelargonium* IV. *Heredity*, **37**, 95–107.

Tilney–Bassett, R.A.E. (1984). The genetic evidence for nuclear control of chloroplast biogenesis in higher plants. In *Chloroplast Biogenesis*, ed. R.J. Ellis, pp. 13–50. Cambridge University Press.

Tilney–Bassett, R.A.E. (1986). *Plant Chimeras*. London: Edward Arnold.

Tilney–Bassett, R.A.E. & Abdel–Wahab, O.A.L. (1982). Irregular segregation at the *Pr* locus controlling plastid inheritance in *Pelargonium*: Gametophytic lethal or incompatibility system? *Theoretical and Applied Genetics*, **62**, 185–91.

Tilney–Bassett, R.A.E. & Birky, C.W. Jr. (1981). The mechanism of the mixed inheritance of chloroplast genes in *Pelargonium*: Evidence from gene frequency distributions among the progeny of crosses. *Theoretical and Applied Genetics*, **60**, 43–53.

H. G. DICKINSON AND F.L. LI

Organelle behaviour during higher plant gametogenesis

The three genomes of the higher plant cell remain surprisingly conserved for a particular species. This is all the more extraordinary in view of the fact that both mitochondrial (mt) DNA and chloroplast (cp) DNA recombine readily (Palmer & Shields, 1984) and, in recent years, DNA sequences have been shown not only to move between chloroplast and mitochondria, but also between these organelles and the nucleus (Scott & Timmis, 1984; Kemble et al., 1983). A clue to the stability of organellar (org-) DNA was provided by Wilkie (1973; Fonty et al., 1978) who demonstrated that in Saccharomycetes cerevisiae, the nucleus exerts a very strong control over organellar genotypes. This was most clearly exemplified by work with newly formed diploids where the first buds produced from these cells contained mitochondria with recombinant DNA, whilst subsequent buds reverted to a 'parental' mitochondrial genotype. In more recent years, this relationship between the nuclear and organellar genomes has been further clarified by studies of protoplast 'cybrids', where a mixture of two cytoplasms can be offered to a single nucleus (Pelletier et al., 1985). The nucleus is, however, unlikely to be capable of causing the generation of novel genotypes of org-DNA de novo, but merely of favouring selection of particular base sequences.

When considering the selective pressures operating upon plants in the field, a conflict of interest must clearly exist between the requirements of the population, which demand that as many individuals as possible be generated, and the needs of the 'species' which require the selection of individuals best fitted for the environment. When considered at the levels of org-DNA, the 'population; would demand the survival of as many organelles as possible, irrespective of their genomic construction, whilst the 'species' would require that there should be an active selection for good genomic arrangements. Since the major part of the variation acted upon by selection is generated during sexual reproduction, it might thus be reasonable to expect a 'sorting out' of organelles at this point in the life cycle. It would matter little were deleterious rearrangements of org-DNA to arise during development of the plant body, but it would be particularly unfortunate were such lesions or chimeras to be propagated through sexual reproduction. A massive imbalance obviously exists between the numbers of male and female gametes produced in plants and, since there are no clear figures available to indicate the role that pollen : ovule ratios play in the evolutionary process, it is impossible to comment as to whether pollen is over-produced. However, were this to be the case, pollen development would provide a unique opportunity for 'weeding out' unacceptable organellar genomes, without much cost to the reproductive strategy of the plant (Dickinson, 1986).

At first sight, a proposal that organelle selection might take place during male gametogenesis in angiosperms would appear foolish. That the majority of organelles contained in the zygote are derived from the egg cell cytoplasm is widely recognised, and for only a few species (e.g. *Pelargonium*, Hagemann, 1976) has transmission of plastids firmly been established. However, these facts concerning plastid transmission are known solely for the reason that the fate of these organelles may easily be followed. Very little is known of mitochondrial transmission, although it has tacitly been assumed that these organelles are degraded in the synergid, prior to karyogamy. Recent elegant studies by Russell (1986), and others (McConchie *et al.*, 1985), on the structure and behaviour of the male germ unit as it approaches the egg cell, show this asumption to be totally unjustified, and that male organelles are intimately involved in the fertilisation process.

Having established a somewhat insubstantial *prima facie* case for some form of selection operating during gametogenesis, do any past studies of these events indicate the presence of such a process? The literature is particularly encouraging in this respect for, from the time of Guillermond (1924), fundamental changes have been reported as occurring during angiosperm gametogenesis. In those early days, individual organelles could not be resolved, let alone their genomes, but the evidence was such that Guillermond (1924) proposed that continuity of organelle line was maintained throughout gametogenesis, despite the spectacular changes affecting the 'chondriome' during this period. Working with more or less the same data, Wagner (1927), on the other hand, suggested the converse proposing the chondriome was regenerated *de novo* in both pollen and egg cells. Continuity of organelle line continued to be debated until the 1960s where, in one of the first electron microscopic studies, Bal & De (1961) showed the mitochondrial line to be maintained through pollen development in *Tradescantia*. Plastids were however, reported to be eliminated from the cells, only to be regenerated from large pleomorphic bodies. Marumaya (1968), on the other hand, was able to show the presence of plastids throughout pollen development in *Lilium*. In subsequent investigations (e.g. Dickinson & Heslop–Harrison, 1970a) the maintenance of both plastid and mitochondrial lines has been confirmed, but evidence has also emerged of a spectacular cycle of dedifferentiation and redifferentiation in both these organelle populations; events which are closely linked with other, even more spectacular, changes that are taking place in the meiocyte cytoplasm at this time.

The cellular environment during gametogenesis

Painter demonstrated as long ago as 1943 that the character of the meiocyte cytoplasm differs radically from that of normal somatic tissue for, as both male and female cells enter prophase, their affinity for basic dyes is lost. This feature remained only a cytochemical observation until these cells were examined using the electron microscope, when it became clear that this loss of basophilia results from the near-eradication of the cytoplasmic ribosome population (Mackenzie, Heslop–Harrison & Dickinson, 1967). Subsequent investigations (Dickinson & Heslop–Harrison, 1970b; Williams, Heslop–Harrison & Dickinson, 1973; Dickinson & Heslop–Harrison, 1977) have established that a cycle of ribosomal RNA metabolism accompanies meiosis in male cells of higher plants. While the precise timing and intensity of the cycle varies from species to species, there is normally a pronounced drop in extractable RNA, and indeed the numbers of ribosomes visible, during early prophase. Restoration of ribosome numbers occurs almost immediately, and normal 'somatic' levels are regained by the tetrad stage. The mechanics of this restoration of ribosomes has yet to be fully elucidated, but there is some evidence that the nucleolar-like bodies (nucleoloids) seen in the cytoplasm at this time (McClintock, 1934; Gavaudan & Chih–Chen, 1936; Williams, Heslop–Harrison & Dickinson, 1973) may be aggregations of ribosomal RNA, which then disperse into the cytoplasm to form the new ribosome population. The origin of these cytoplasmic nucleoloids is far from clear; since the nucleolar organisers of these cells are particularly active during meiotic prophase, it is tempting to propose that it is the product of this activity which composes the nucleoloids. However, these structures are not visible until the anaphase/telophase stage of meiosis I and, for this reason, there remains an interval during late prophase and the metaphase stages during which this material is undetectable. It has been assumed that the nascent RNA becomes associated with the metaphase chromosomes, as is held to be the case for nucleolar RNA during mitosis. Although the cytoplasmic nucleoloids resemble nucleoli very closely and react positively with some cytochemical probes for RNA, their composition has yet to be fully determined. Certainly, they do not contain any messenger (m) RNA, for they will not bind radioactively labelled polyuridylic acid (Dickinson & Willson, 1985).

The cycle of RNA metabolism is not restricted to male cells, for a decline in levels has been demonstrated to take place in megaspore mother cells, and the presence of cytoplasmic nucleoloids has also been recorded (Dickinson & Potter, 1975). More surprisingly, similar events have also been observed in lower plants, and it is the study of these groups which provides us with the evidence that these cycles of RNA metabolism are probably associated with meiosis, rather than alternation of generation or, indeed, any other developmental stage. For, in species as diverse as *Saccharomyces cerevisiae* (Hopper *et al.*, 1974), *Coprinus cinerea* (Moore, 1984) and *Pteridium aquilinum* (Sheffield & Bell, 1979) substantial decreases in RNA levels can be shown to take place during meiotic prophase. The only exception to this rule is in the alga *Chlamydomonas reinhardii*, where ribosomal metabolism takes place during gametogenesis, rather than at meiosis (Siersma & Chiang, 1971).

The biochemical mechanism responsible for the elimination of cytoplasm RNA from these cells is far from clear. In some plants (e.g. *Cosmos bipinnatus*, Knox *et al.*, 1973) elevated levels of acid phosphatase have been demonstrated to precede the decline in RNA. However, detailed analysis of the enzymes present has revealed increased general levels of acid hydrolase, but no synthesis of new isozymes. Electron microscopic autoradiography has shown the levels of RNA synthesis, both at the nuclear organiser and elsewhere in the chromatin (Porter, Bird & Dickinson, 1982) to be very low throughout meiosis, and it remains possible that the decline in RNA results more from this lack of synthesis and by normal turnover, than from the activity of specific enzymes.

Experiments with the probe [^{3}H]–polyuridylic acid indicate conclusively that messenger RNA is also affected during this period (Porter, Parry & Dickinson, 1983) declining to minimal levels throughout the zygotene, pachytene and diplotene stages. As has been mentioned earlier, it does not appear that new message is transferred to the cytoplasm via the cytoplasmic nucleoloids, but more likely by the normal route through the nuclear pores. In a number of plant groups, particularly the gymnosperms, conspicuous modifications overcome the nuclear envelope in the immediate post-meiotic period, with the formation of finger-like invaginations into the nucleoplasm (Dickinson & Bell, 1970; Dickinson & Potter, 1975; Li and Dickinson, 1987). It has been suggested that these modifications permit the increased flow of messenger RNA from the nucleus to the cytoplasm while the new haploid nucleus is establishing its influence in the cytoplasm. The distribution of these post-meiotic nuclear envelope modifications within the angiosperms has yet to be investigated, but their presence has already been demonstrated in the Compositae (Dickinson, H.G., unpublished observation).

In both male and female cells, the control of immediate post-meiotic development is clearly very complex. For example, it is known that the patterning of the pollen grain wall that forms around the young microspore is under sporophytic control, whilst many of the aspects of its manufacture are gametophytically determined (Heslop–Harrison, 1971; Sheldon & Dickinson, 1983). Such hybrid regulation obviously must occur in both male and female cells and, until more is known about the expression of both sporophytic and gametophytic sets of genes, the control of post-meiotic development will remain far from clear.

The dynamics of organelle populations during gametogenesis

All the early reports on changes in the chondriome during gametogenesis, including those of Guillermond (1924), Py (1932) and Wagner (1927), comment upon numerical changes in the chondriome 'granules' – structures which we must identify now with organelles. Although it has subsequently been shown that many of the changes observed result from aggregation and disaggregation of organelle populations, perhaps occasioned by changes in pH (Heslop–Harrison & Dickinson, 1967) there is little doubt that very comprehensive changes occur in organelle number during this period of development. As both male and female cells enter meiotic prophase, mitochondria and plastids decrease in size. Fig. 1 shows the size of mitochondria in pollen mother cells of *C. bipinnatus*, and it can be seen that the diameter of these organelles decreases by at least 50 per cent between the leptotene and pachytene stages. These small plastids and mitochondria now enter a period of very

Fig. 1. Average diameters of mitochondria and spherical inclusions present in the male meiocytes of *Cosmos bipinnatus*. Developmental stages are as follows: PM: premeiosis, L/Z: leptotene/zygotene, P: pachytene, D/D: diplotene/diakinesis, M: metaphase, T: tetrad. Vertical bars indicate 95 per cent confidence limits.

Average diameter (μm)

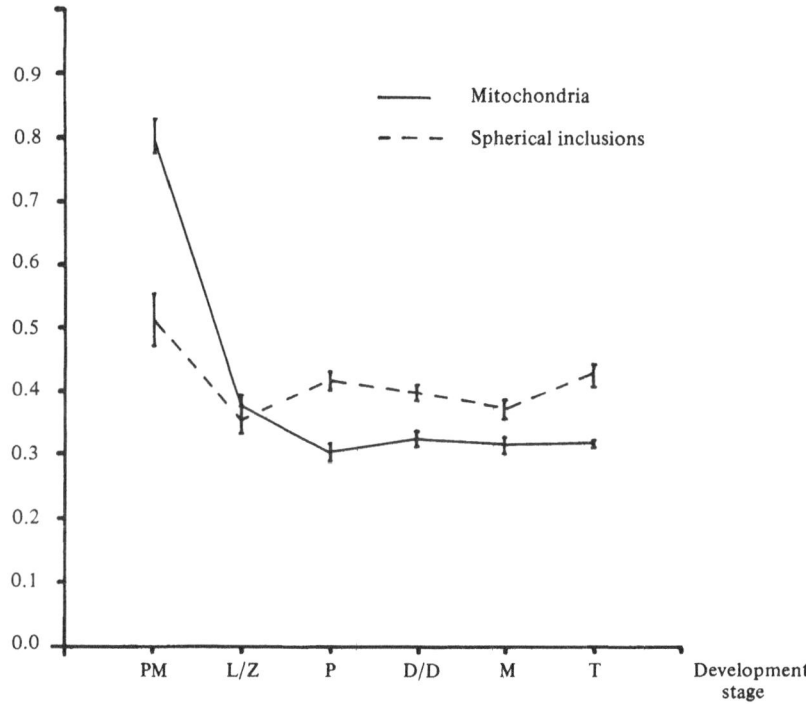

rapid replication, the like of which is not observed elsewhere in the plant life cycle. Division is particularly marked in the mitochondria, and Fig. 2 shows the change in mitochondrial number during meiotic prophase in *C. bipinnatus*. This dramatic increase is also reflected in the plastid population, but to a far lesser extent. Equally, although some organelle replication does occur in female cells with, in particular, large numbers of mitochondria being observed, the rate of replication is far less than in the pollen mother cell.

Interpretation of electron micrographs from these stages is confused by the presence of a new population of organelles, termed for convenience spherical inclusions (SI). These are present in both male and female cells, and represent a range of structures, extending from inclusions resembling mitochondria to simple single

Fig. 2. Relative numbers of mitochondria and spherical inclusions, expressed in arbitrary units, present in developing meiocytes and young microspores of *Cosmos bipinnatus*. Developmental stages as in Fig. 1, and vertical bars indicate 95 per cent confidence.

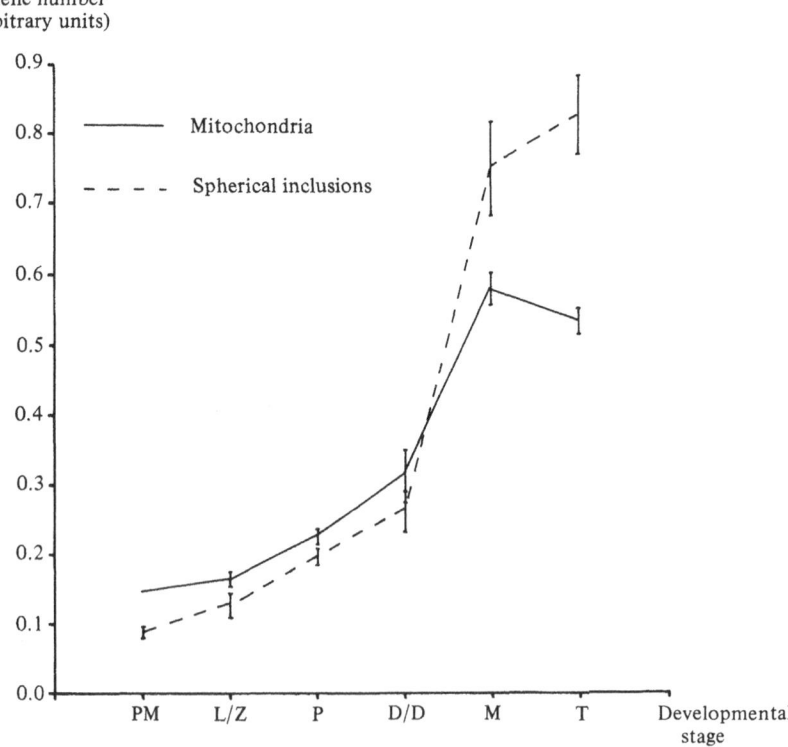

Fig. 3. Large spherical mitochondria (m) in the cytoplasm of pre-meiotic archesporial tissue of *Cosmos bipinnatus*. ($\times$ 12 $\times$ 10^3).

Fig. 4. Transverse of section of entire meiocyte of *Cosmos bipinnatus*. The nucleus (n), plastids (p), mitochondria (m) and spherical inclusions (arrows) are visible. Tapetal cells: t. ($\times$ 2.3 $\times$ 10^3).

Fig. 5. Prophase cytoplasm of a male meiocyte of *Cosmos bipinnatus*, showing profiles (s) identified with degenerating mitochondria. Many of these inclusions possess the internal vesicle (arrows) characteristic of mitochondria at these stages. ($\times$ 24 $\times$ 10^3).

Fig. 6. The cytoplasm of a young microspore of *Cosmos bipinnatus* at the early tetrad stage. Intermediates between spherical inclusions (s) and mitochondria (m) are no longer visible. The majority of the spherical inclusions contain electron opaque material (arrows). ($\times$ 30 $\times$ 10^3).

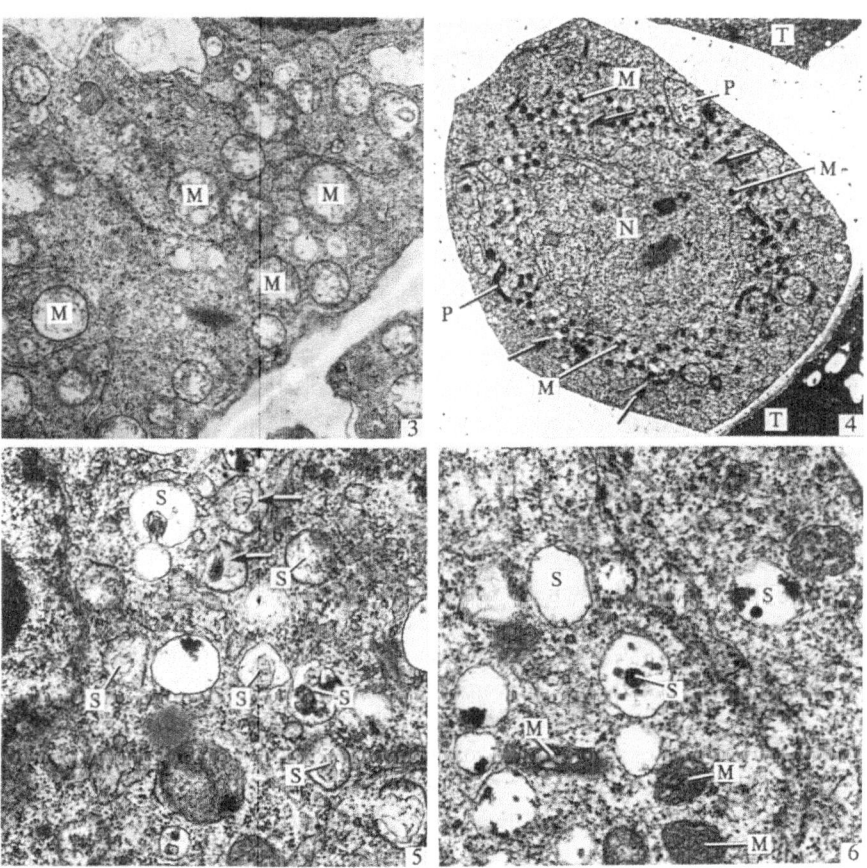

membrane-bound vesicles, containing electron-opaque material (see Figs. 3—6). This similarity to mitochondria has led previous workers to propose that SIs are degenerating organelles, most probably mitochondria (Bell & Muhlethaler, 1964; Dickinson & Potter, 1978; Bird, Porter & Dickinson, 1983).

The numbers of SIs present in the cytoplasm is also plotted on Fig. 2, where they can be seen to appear early in meiosis during the period of decrease in mitochondrial size. Increase in SI number parallels the enlargement of the mitochondrial population through early prophase, their number increasing even more rapidly than the mitochondria such that, by the time the mitochondrial population has peaked, the number of SIs in the cytoplasm is greater than the sum of all other organelles present. The balance of inclusions within these cells remains unaltered during the division stages of meiosis I and II. As the new haploid nuclei reform and cross walls are formed between members of the tetrad, the numbers of SIs begin to fall. The mitochondrial and plastid populations remain constant over this period and, indeed, do so until the release of the microspore from the tetrad. Although plastid and mitochondrial numbers do increase during maturation of the pollen, this is never accompanied by the production of SIs.

Stereological investigations have yet to be carried out on female cells, but it is clear from simple inspection that while mitochondrial number does increase in the megaspore mother cell, SIs are not produced in large numbers (Dickinson & Potter, 1978). Instead, the mitochondrial population behaves very much as would be anticipated in actively metabolising somatic tissue.

The numbers and the nature of the organelles present in mature pollen is dependent upon species. However, in the development of all pollen, the young microspore divides asymmetrically by mitosis to give rise to the vegetative and generative cells. In plants which produce 'trinucleate' pollen grains, the generative cell itself divides again to form the two sperm cells. In plants with 'binucleate' pollen, this second mitotic division occurs in the pollen tube after pollination. It is the first pollen mitosis which determines which organelles are included in the sperm cells (Hagemann, 1976), and thus be available for transmission at fertilisation. Although complex hypotheses have been proposed to explain the lack of plastids in the generative cell (Vaughan et al., 1980), it is now commonly accepted that plants which do not transmit plastids simply exclude them from the generative cell during its formation (Hagemann, 1976), presumably by the action of the asymmetric spindle. No further exclusion takes place at the second mitosis, but evidence is now emerging that this division also may be asymmetric, with regular, unequal partitioning of organelles between the two sperm cells (Russell, 1986). Reconstructions from serial sections indicate that both sperm cells and the vegetative cell constitute an interlinked structure which moves to the tip of the pollen tube where it remains until fertilisation. This assembly, now known as the male germ unit (McConchie et al., 1985) carries the entire complement of both nuclear and organellar genomes which the male parent transmits to the next generation.

Organellar differentiation and behaviour during gametogenesis

The male and female archesporial tissue of angiosperms contains an organelle complement resembling that of normal somatic meristems (see Fig. 3). Mitochondria are generally elongate, measuring some 1.5 μm in maximum dimension, while the plastids, which may contain small starch grains, measure about 2.5 μm. As the cells enter meiosis a period of division ensues, as a result of which the general size of each type of organelle diminishes considerably (see Figs. 1, 4 and 5). In many plants, coincident with this phase of division, starch is lost from the plastids. Whether this reserve is simply utilised in metabolism, or transformed into the lipid droplets which begin to populate the cytoplasm is not known. Thus, as both male and female meiocytes enter meiotic prophase, the mitochondrial population is represented by small spherical organelles, termed 'promitochondria' by Bal & De (1961). In many plants, the plastid population is also similarly dedifferentiated, comprising simply large pleomorphic bodies containing neither starch nor lamellar systems (Bal & De, 1961; Dickinson & Heslop–Harrison, 1970a). It is noteworthy, however, that the plastids of some plants do continue to hold starch reserves throughout meiosis (Marumaya, 1968).

During mid-prophase, the mitochondria of male cells enter a phase of very rapid division. As might be anticipated, this active division is accompanied by very rapid synthesis of organellar DNA (see Fig. 7) throughout this period (Bird, Porter & Dickinson, 1983). Some synthesis of DNA also occurs in the plastid population in these cells, but stereological investigations suggest that these organelles do not divide with the rapidity of the mitochondria. During this period of rapid division, the first SIs become visible (see Figs. 5 and 6), increasing in number in the regions where mitochondria are concentrated. The range of structures present within this group of inclusions very strongly suggests that SIs are mitochondria undergoing progressive degeneration. For example, they contain internal lamellar systems, their population kinetics follows that of the mitochondria very closely, and the more highly differentiated contain cytochrome-*c* oxidase (Willson & Dickinson, unpublished observations). SIs are observed within the female cytoplasm, but at a much lower frequency than in male cells. Male and female material has been examined very carefully for degeneration products of the plastid population, but these have not been found in any number.

In a number of plants, particularly those belonging to the Liliaceae, redifferentiation of the plastids starts towards the end of meiotic prophase (Willson & Dickinson, 1984). Here, enzymes responsible for the synthesis of starch accumulate on small sections of membrane within the stroma, sometimes assuming quite complex configurations and reminiscent on occasions of the prolamellar body of the etioplast.

Small (*c.* 100 nm) starch grains then soon develop around them, presenting a very characteristic image under the electron microscope (Dickinson, 1986; Dickinson & Willson, 1983). The entire assemblies, which have been termed membrane particle associations (MPAs) (Dickinson & Heslop–Harrison, 1970*a*), generally occur singly within an organelle, although exceptions do occur. Enzymic digestion of MPAs reveals a close association with the org-DNA (Dickinson, 1981).

In male cells, the mitochondrial population stabilises late in prophase, and the meiocytes enter the division stages of meiosis containing small mitochondria, large numbers of SIs, and plastids commencing starch synthesis. The picture is very much the same for female cells, except in that fewer SIs are present.

Conclusions are difficult to draw concerning immediate post-meiotic development, since radically different forms of organelle differentiation seem to take place dependent on the sex, and the species involved. The majority of reports describe both

Fig. 7. Incorporation of tritiated thymidine into plastids and mitochondria during meiosis in *Lilium*. The method used for the calculation of these levels is quite complex, and set out in Porter, Bird & Dickinson (1982). The confidence limits of these measurements are discussed more fully in Bird, Porter & Dickinson (1983), from which this figure has largely been redrawn. Developmental stages are as follows: PL: preleptotene, L: leptotene, LL: late leptotene, EZ: early zygotene, Z: zygotene, P: pachytene, D/D: diplotene/diakinesis, T: tetrad.

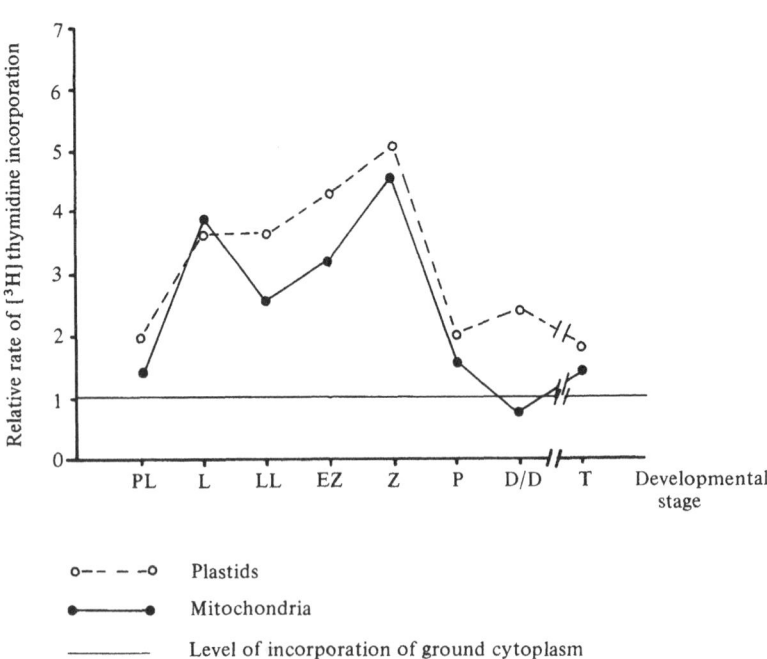

plastids and mitochondria remaining relatively unchanged during the period the young microspore is retained within the tetrad in male cells (Bal & De, 1961; Dickinson & Heslop–Harrison, 1970a; Dickinson & Willson, 1983) and, indeed, a similar situation pertains within the newly formed meiotic products of the ovule. However, in the young spores of some species, a very striking association takes place between the promitochondria and the surface of the haploid nucleus. Such an event was first described in female cells of *Myosurus minimus* by Woodcock & Bell (1968), and has since been extended to the microspores of a number of species, particularly those within the Compositae (Dickinson & Potter, 1979; Dickinson, 1986). Indeed, this association is not restricted to plants, for very similar events have been reported by Baker & Franchi (1969) as occurring during female gamete formation in primates. The association is very intimate, in that the organelles cannot be detached by osmotic shock (Dickinson & Potter, 1979) and, in favourable micrographs, they can be seen adhering to the nuclear surface by means of small granules or rodlets. Frequently, the mitochondria attach in register with small areas of chromatin appressed to the inner face of the nuclear envelope (Dickinson, 1986).

Reasonably, it has been suggested that this association permits the direct injection of ATP into the young haploid nucleus, which clearly must be involved in very active syntheses. In an attempt to derive some support for this hypothesis, young microspores were either exposed to a nitrogen atmosphere for an extended period, or to physiologically active levels of dinitrophenol (DNP). Both these treatments should halt ATP production immediately, the nitrogen atmosphere by depriving cells of oxygen, and the DNP by uncoupling electron transport from phosphorylation. In neither of these circumstances did the relationship between the mitochondria and the nuclear envelope alter in *Cosmos bipinnatus* (see Figs. 8–11). There is no doubt that the treatments did affect the cells, in that signs of necrosis set in after some 30 h of either treatment; the mitochondria, however, remained firmly attached to the nuclear envelope. In another experiment, the possibility was explored that the relationship between the nucleus and mitochondria involved transcription of DNA, with the transfer of mRNA into the organelles. After 48 h in the presence of actinomycin D, the mitochondria were seen to remain adhering to the nuclear surface.

This particular nuclear–cytoplasmic interaction is certainly spectacular, but clearly no commentary can be made as to its significance until its distribution within the plant and animal kingdom has been established. We do not even know whether it occurs generally in female cells of higher plants, although the earliest report of this phenomenon (Woodcock & Bell, 1968) was for young megaspores. We have examined female development in *Cosmos bipinnatus* very thoroughly and, so far, found no sign of this type of nuclear–mitochondrial interaction. The material is, however, very difficult to stage and it is very possible that we have so far failed to encounter the exact stage at which this interaction happens.

Fig. 8. Nucleus (n) of a young microspore of *Cosmos bipinnatus* following 30 hours exposure to a nitrogen atmosphere. Whilst there is some evidence of tissue degeneration (arrows) mitochondria (m) remain identifiable and associated with the nuclear envelope, ($\times 12.5 \times 10^3$).

Fig. 9. The nucleus (n) of a young microspore taken from a bud which had been fed for 30 h with an aqueous solution of 10^{-4} M dinitrophenol. Mitochondria (arrows) are clearly visible associated with the nuclear envelope (e) ($\times 16 \times 10^3$).

Fig. 10. A high power detail of material similar to that shown in Fig. 9. The nuclear envelope (e) is evident, as are the mitochondria (m) affixed to it. There is some evidence (arrows) of physical connection between the organelles and the envelope ($\times 37.2 \times 10^3$).

Fig. 11. Tangentially sectioned nucleus of a young microspore of *Cosmos bipinnatus* taken from a bud fed with an aqueous solution of actinomycin D at a concentration of 1 µg ml. Although the envelope is not easily seen, the chromatin (c) stains darkly, and an association of the mitochondria (m) is clearly visible ($\times 19 \times 10^3$).

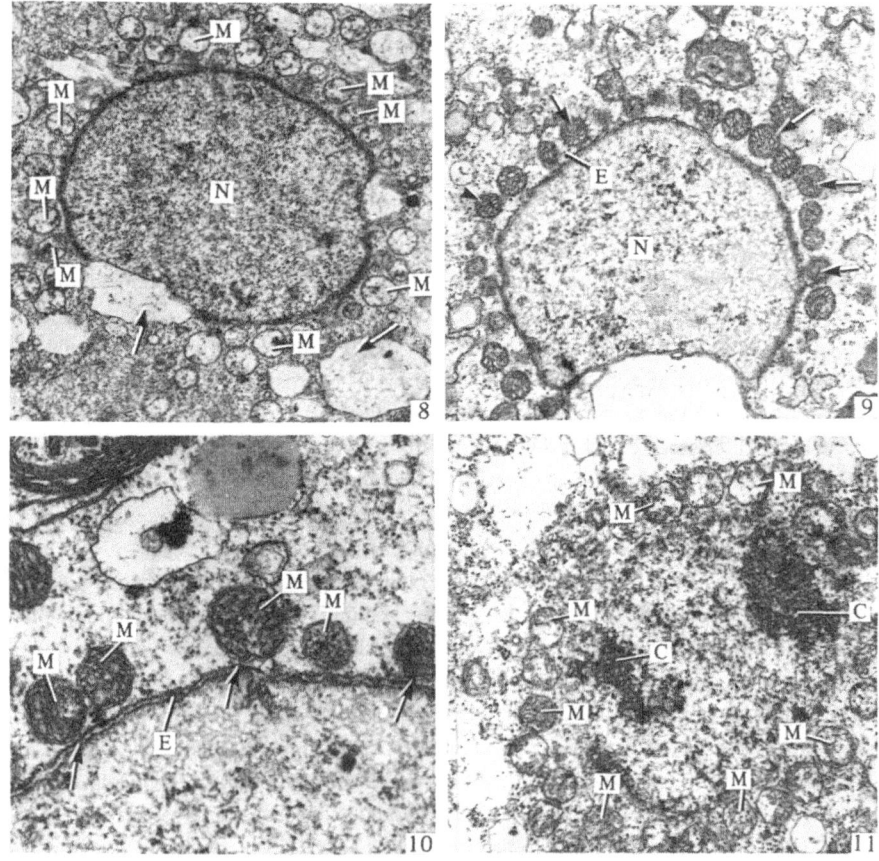

Whether or not nuclear–organellar interactions occur during the tetrad stage in all species, major redifferentiation of the organelle populations always occurs following release of the microspores from the tetrad. Mitochondria undergo a short period of division and enlarge considerably to assume a conventional conformation. On some occasions electron opaque material is observed within these newly differentiated organelles (Dickinson & Potter, 1978, 1979), but nothing is known of its chemistry save that it is osmiophilic. In the plastids, starch accumulation occurs in earnest and, depending upon the species, amyloplasts may continue to increase in size and number until they fill the major part of the pollen grain (Dickinson & Willson, 1983). Both mitochondria and plastids exist in very large numbers within mature pollen, largely being contained within the vegetative cell. Although both plastids and mitochondria play an important role in the establishment of the embryo sac and egg apparatus within the ovule, no unusual differentiation occurs. The mature egg cell is strikingly undifferentiated, containing populations of both organelles (Cass & Karas, 1974), but with no particular structural modification except for a little starch in the plastids.

The consequences of organellar interactions taking place during plant gametogenesis

When considering the phase of organellar replication, and the association between the mitochondria and the nuclear envelope, it is helpful to examine any short-term advantage they may confer on the cells involved, and any long-term advantage they may hold for the species or population. It is not surprising that large numbers of mitochondria are formed during pollen development; both meiocytes and tapetum have been shown to exhibit parallel phasic syntheses of proteins, lipids, and the complex polymer sporopollenin, all of which require active mitochondrial participation (Dickinson & Heslop–Harrison, 1970a, b). In this connection, it is interesting that the numbers of mitochondria in the tapetum also rise in a spectacular fashion, almost paralleling that seen in the young meiocyte. Since it is likely that mitochondrial replication will be under nuclear control, we must assume that this event is regulated by the expression of genes specific to sporogenesis in both the meiocytes and tapetum. It has been suggested (Bird et al., 1983) that sudden changes in organelle structure and number may result from the lack of nuclear regulation caused by the low levels of messenger RNA present in the cytoplasm at this time. Certainly this may account for the rapid and synchronous organellar dedifferentiation, but it is not easy to see how lack of controlling elements from the nucleus could result in such a rapid increase in mitochondrial number. Certainly, organelles do exhibit some most unusual morphologies over this period, and there is evidence of junctions being formed between individual plastids (Dickinson, 1981). It is tempting to propose that this period parallels that found following diploid formation in yeast, when organelles are apparently free to be involved in recombination of DNA (Wilkie,

1973), but far firmer evidence is required before inferences of this type may be drawn.

We have good evidence that a large proportion of the mitochondria replicated degenerate, and the conclusion is inescapable that those surviving must, in some way, be better fitted to the prevailing conditions. A *de facto* selection system must therefore exist and it is possible that it weeds out recombinant and mutant mitochondrial genomes. At first sight, this might provide the basis of an hypothesis to explain some types of cytoplasmic male sterility (CMS), which normally occur in plants possessing mitochondria with defective genomes (Forde & Leaver, 1980). Unfortunately, close examination of microsporogenesis in CMS plants reveals that the first perturbations of development occur often in the tapetal cells (Bino, de Hoop & van der Neut, 1985). It thus seems more likely that CMS is caused by a requirement for an unusual synthesis by the mitochondria in both the tapetum and meiocytes.

Female cells do not have the same intense requirement for the synthesis of lipids and sporopollenin, and it is not surprising that rapid mitochondrial replication does not occur. It is interesting, however, that the changes in dedifferentiation and morphology seen in the organelles, attributed to low levels of RNA, do occur in the megaspore mother cells (Dickinson & Potter, 1978). If we consider these events simply in terms of changes to the organelle populations, we then have a selection system operating in the male but not the female. Whether or not this would be of advantage to the plant depends very much on any over-production of males under normal breeding conditions and further evidence must clearly await detailed work on pollen flow in these species.

The association of mitochondria with the nuclear envelope has yet to be shown in all angiosperms, but it is striking that it can be present in male and female cells, and also during gametogenesis in animals (Baker & Franchi, 1969). It appears a very positive interaction, and is clearly not a tenous attachment via electrostatic forces, or a consequence of cytoplasmic movement. Despite the fact that plastids may often be seen attached to the nuclear envelope in male cells of members of the Compositae (Dickinson, 1986), the most attractive explanation for these events remains the provision of ATP to the haploid nucleus. However, it is surprising neither anoxia, nor the uncoupling of respiration from phosphorylation was capable of detaching the organelles. Even were organelles to provide ATP for the nucleus, it is difficult to conceive the mechanism which directed them so precisely to sites on the nuclear surface, ensuring their register with the chromatin. More heretical explanations for this interaction might include the transfer of nuclear-encoded message to the organelles, but the evidence from the actinomycin D experiments would point to this not being the case. Even the exchange of DNA between the organellar and nuclear genomes has been considered; clearly this does happen (Scott & Timmis, 1984) and its mechanism has yet to be demonstrated. Nevertheless, it has always been assumed that this exchange of DNA sequences occurs slowly during evolution, and probably involves viral or other intermediates.

By the time the tetrads of microspores or megaspores commence their next stages of development, all organellar interactions specific to gametogenesis are complete. From this point onwards both populations behave as they would in active meristematic cells elsewhere in the plant.

Conclusion

Events taking place during male and female gametogenesis in higher plants have been examined with a view to their effect on the number and nature of the cell organelles, and on transmission of these organelles to the next generation. While events differ both between species and between the sexes within a species, the organelles of the micro- and megasporocytes of most angiosperms exhibit spectacular changes during development. In both male and female cells, mitochondrial and plastid populations undergo comprehensive dedifferentiation, such that the possibility of discontinuity of organelle line has been considered. In male cells, mitochondrial number increases dramatically during early meiotic prophase, with the accompanying production of profiles which indicate that many of these newly produced organelles degenerate. At a later stage in the male cells of some plants, up to 30 per cent of the mitochondrial population become firmly affixed to the nuclear envelope. Using electron microscopic autoradiography, DNA synthesis is shown to increase rapidly within the dedifferentiated organelles. Since both male and female meiocytes have been demonstrated to contain very low levels of ribosomal and messenger RNA, it is suggested that some of the changes observed may result from the lack of nuclear-encoded regulatory molecules reaching the organelles. This may account for some of the morphological differences, but it is suggested that the phases of rapid division, and indeed the nuclear–organellar interaction, are more positively regulated. The degeneration of a proportion of the mitochondrial population during the phase of rapid division in male cells would suggest that those surviving form the 'fittest' group within the population. Likewise, the phase of dedifferentiation seen in all the organelles may also act as a selection mechanism. The association of mitochondria and the nuclear envelope in post-meiotic cells of some species is most likely evidence of direct ATP injection into the nucleus. Unfortunately, experiments using nitrogen atmospheres, weak acid uncouplers and other inhibitors suggest that this may not necessarily be the case.

Acknowledgements

The authors wish to thank the Agricultural and Food Research Council of the United Kingdom for financial support during this work. The assistance of Sue Mitchell in the drawing of diagrams, and of Simon Brookes in the preparation of micrographs is also most gratefully acknowledged.

References

Baker, T.G. & Franchi, L.L. (1969). The origin of cytoplasmic inclusions from the nuclear envelope of mammalian oocytes. *Zeitschrift für Zellforschung und Mikroscopische Anatomie*, **93**, 45–55.

Bal, A.K. & De, D.N. (1961). Developmental changes in the sub-microscopic morphology of cytoplasmic components during microsporogenesis in *Tradescantia*. *Developmental Biology*, **3**, 341–54.

Bell, P.R. & Muhlethaler, K. (1964). The degeneration and reappearance of mitochondria in the egg cells of a plant. *Journal of Cell Biology*, **20**, 235–248.

Bino, R., De Hoop, S.J. & Van der Neut, A. (1985). Cytochemical localisation of cytochrome C oxidase in anthers of cytoplasmic male sterile *Petunia hybrida*. In *Sexual Reproduction in seed plants, ferns and mosses*, ed. M.T.M. Willemse & J.L. van Went, pp. 44–46, Pudoc (Wageningen).

Bird, J., Porter, E.K. & Dickinson, H.G. (1983). Events in the cytoplasm during male meiosis in *Lilium*. *Journal of Cell Science*, **59**, 27–42.

Cass, D. & Karas, I. (1974). Ultrastructural organisation of the egg of *Plumbago zeylandica*. *Protoplasma*, **81**, 49–62.

Dickinson, H.G. (1981). The structure and chemistry of plastid dedifferentiation during male meiosis in *Lilium henryi*. *Journal of Cell Science*, **52**, 223–41.

Dickinson, H.G. (1986). Organelle selection during flowering plant gametogenesis. In *The Chondriome: chloroplast and mitochondrial genomes*. ed. S.H. Mantell, G.P. Chapman & P.F.S. Street, pp. 37–60. Longmans Scientific and Technical, Harlow, UK.

Dickinson, H.G. & Bell, P.R. (1970). Nucleo-cytoplasmic interaction at the nuclear envelope in post-meiotic microspores of *Pinus banksiana*. *Journal of Ultrastructural Research*, **33**, 356–60.

Dickinson, H.G. & Heslop–Harrison, J. (1970a). The behaviour of plastids during meiosis in the microsporocyte of *Lilium longiflorum* Thunb. *Cytobios*, **6**, 103–18.

Dickinson, H.G. & Heslop–Harrison, J. (1970b). The ribosome cycle, nucleoli and cytoplasmic nucleoloids in the meiocytes of *Lilium*. *Protoplasma*, **69**, 187–200.

Dickinson, H.G. & Heslop–Harrison, J. (1977). Ribosomes, membranes and organelles during meiosis in angiosperms. *Philosophical Transactions of the Royal Society of London, Series B*, **277**, 327–42.

Dickinson, H.G. & Potter, U. (1975). Post-meiotic nucleo-cytoplasmic interactions in *Pinus banksiana*: the secretion of RNA by the nucleus. *Planta*, **122**, 99–104.

Dickinson, H.G. & Potter, U. (1978). Cytoplasmic changes accompanying female meiosis in *Lilium longiflorum* Thunb. *Journal of Cell Science*, **29**, 147–69.

Dickinson, H.G. & Potter, U. (1979). Post-meiotic nucleo-cytoplasmic interaction in *Cosmos bipinnatus*. *Planta*, **145**, 449–57.

Dickinson, H.G. & Willson, C.E. (1983). Two stages in the redifferentiation of amyloplasts in the microspores of *Lilium*. *Annals of Botany*, **52**, 803–10.

Dickinson, H.G. & Willson, C.E. (1985). Behaviour of nucleoli and cytoplasmic nucleoloids during meiotic divisions in *Lilium henryi*. *Cytobios*, **43**, 349–65.

Fonty, G., Goursot, R., Wilkie, D. & Bernardi, G. (1978). The mitochondrial genome of wild-type yeast cells. VII. Recombination in crosses. *Journal of Molecular Biology*, **119**, 213–35.

Forde, B.G. & Leaver, C.J. (1980). Nuclear and cytoplasmic genes controlling synthesis of variant mitochondrial polypeptides in male-sterile maize. *Proceedings of the National Academy of Sciences of the USA*, **77**, 418–22.

Gavaudan, P. & Chih–Chen, Y. (1936) Centrosomes et extrusions chromatiques chez les angiosperms. *Actualités Scientifiques Industrielles* 319, Paris, Herman et Cie.

Guillermond, A. (1924). Recherches sur l'evolution du chondriome pendant le developpement du sac embryonnaire et des cellules mères des grains de pollen dans les Liliaceaces et sur la signification des formations ergastoplasmiques. *Annales de Sciences Naturelles, Botanique et Biologie Vegetale*, 6, 1–52.

Hagemann, R. (1976). Plastid distribution and plastid competition in higher plants and the induction of plastome mutations by Nitroso-urea compounds. In *Genetics and Biogenesis of Chloroplasts and Mitochondria*. ed. Bucher, F., *et al.*, Amsterdam, North Holland, pp. 331–8.

Heslop–Harrison, J. (1971). Wall patterning in angiosperm microsporogenesis. In *Control mechanisms of growth and development, Symposium of the Society for Experimental Biology*, 25, 277–300.

Heslop–Harrison, J. & Dickinson, H.G. (1967). A cycle of sphaerosome aggregation and disaggregation contrasted with the meiotic divisions in *Lilium*. *Phytomorphology*, 17, 195–9.

Hopper, A.K., Magee, P.T., Welch, S.K., Friedman, M. & Hall, B.D. (1974). Macromolecule synthesis and breakdown in relation to sporogenesis and meiosis in yeast. *Journal of Bacteriology*, 119, 619–28.

Kemble, R.J., Mans, R.J., Gabay–Laughnan, S. & Laughnan, J.R. (1983). Sequences homologous to episomal mitochondrial DNAs in the maize nuclear genome. *Nature*, 304, 744–7.

Knox, R.B., Dickinson, H.G. & Heslop–Harrison, J. (1973). Cytoplasmic RNA and enzyme activity during the meiotic prophase in *Cosmos bipinnatus*. In *Pollen development and physiology*. ed. J. Heslop–Harrison, pp. 32–5, Butterworths (London).

Li, F.L.& Dickinson, H.G. (1987). The structure and function of nuclear invaginations characteristic of microsporogenesis in *Pinus banksiana. Annals of Botany*, (in press).

Mackenzie, A., Heslop–Harrison, J. & Dickinson, H.G. (1967). Elimination of ribosomes during meiotic prophase. *Nature*, 215, 997–9.

McLintock, B. (1934). The relation of a particular chromosome element to development of the nucleoli in *Zea mays. Zeitschrift für Zellforschung und Mikroskopische Anatomie*, 21, 294–328.

McConchie, C.A., Jobson, S. & Knox, R.B. (1985). Computer assisted reconstruction of the male germ unit in pollen of *Brassica campestris. Protoplasma*, 127, 57–63.

Marumaya, L. (1968). Electron microscopic observation of plastids and mitochondria during pollen development in *Tradescantia palidosa. Cytologia*, 33, 482–97.

Moore, D. (1984). Developmental biology of the *Coprinus cinerea* carpophore; metabolic regulation in relation to cap morphogenesis. *Experimental Mycology*, 8, 283–97.

Painter, T.S. (1943). Cell growth and nucleic acids in the pollen of *Rhoeo discolor. Botanical Gazette*, B105, 58–68.

Palmer, J.D. & Shields, C.R. (1984). Tripartite structure of the *Brassica campestris* chloroplast genome. *Nature*, 307, 437–40.

Pelletier, G., Vedel, F. & Belliard, G. (1985). Cybrids in genetics and breeding. *Hereditas Supplement*, 3, 49–56.

Porter, E.K., Bird, J. & Dickinson, H.G. (1982). Nucleic acid synthesis in microsporocytes of *Lilium* cv cinnabar: events in the nucleus. *Journal of Cell Science*, 57, 229–46.

Porter, E.K., Parry, D. & Dickinson, H.G. (1983). Changes in poly(A)$^+$ RNA during male meiosis in *Lilium*. *Journal of Cell Science*, **62**, 177–86.

Py, G. (1932). Recherches cytologiques sur l'assis nourriciere des microspores et les microspores des plantes vasculaires. *Revue General de Botanique*, **44**, 316–413, 450–62.

Russell, S.D. (1986). Dimorphic sperm cell, cytoplasmic transmission and preferential fertilisation in *Plumbago zeylandica*. In *The Chondriome: chloroplast and mitochondrial genomes*. ed. S.H. Mantell, G.P. Chapman & P.F.S. Street, pp. 69–116, Longman Scientific, Harlow, UK.

Scott, N.S. & Timmis, J.N. (1984). Homologies between nuclear and plastid DNA in spinach. *Theoretical and Applied Genetics*, **67**, 279–88.

Sheldon, J. & Dickinson, H.G. (1983). Determination of patterning in the pollen wall of *Lilium henryi*. *Journal of Cell Science*, **63**, 191–208.

Sheffield E. & Bell, P.R. (1979). Ultrastructural aspects of sporogenesis in a fern, *Pteridium aquilinum* (L) Kuhn. *Annals of Botany*, **44**, 393–405.

Siersma, P.W. & Chiang, K.S. (1971). Conservation and degradation of cytoplasmic and chloroplast ribosomes in *Chlamydomonas reinhardi. Journal of Molecular Biology*, **58**, 167–85.

Vaughan, K., De Bonte, L.R., Wilson, K.G. & Schaffer, G.W. (1980). Organelle alteration as a mechanism for maternal inheritance. *Science*, **208**, 196–8.

Wagner, N. (1927). Evolution du chondriome pendant la formation de pollen des angiosperms. *Biologia Generalis*, **3**, 15–66.

Williams, E., Heslop–Harrison, J. & Dickinson, H.G. (1973). The activity of the nucleolus organising region and the origin of cytoplasmic nucleoloids in meiocytes of *Lilium. Protoplasma*, **77**, 79–93.

Wilkie, D. (1973). Cytoplasmic genetic systems of eukaryotic cells. *British Medical Bulletin*, **29**, 263–8.

Woodcock, C.L.F. & Bell, P.R. (1968). Features of the ultrastructure of the female gametophyte of *Myosurus minimus. Journal of Ultrastructural Research*, **22**, 546–63.

E. B. GINGOLD

The replication and segregation of yeast mitochondrial DNA

The central role played by the yeast *Saccharomyces cerevisiae* in studies of organelle heredity is no accident but arises directly from its properties as a facultative anaerobe. Put simply, this means that this yeast can survive without active mitochondria, the cells then obtaining their energy purely by fermentation of a sugar such as glucose. As the mitochondria are thus dispensable, it is possible to conduct studies using mutations that abolish all or part of their activity. Such mutations would be lethal in obligatory aerobes, a category that includes most other eukaryotic species!

The mitochondrial genome of yeast (a term which will be taken to mean *Saccharomyces cerevisiae* for the purposes of this review) is one of the most studied pieces of eukaryotic DNA and is almost entirely sequenced. But it is not the purpose of this discussion to review the rapid progress that has been made in understanding the organisation of this genome and the many unusual features found in the expression of its coding sequences; the reader is referred elsewhere for reviews of these topics (Evans, 1983; Dujon, 1981). Rather, it is the intention here to look at the information that has become available on the control of the transmission of the mitochondrial genome to progeny cells. We will observe a situation that is very different from the highly organised processes found for chromosomal genes, with their mechanisms of mitosis and meiosis to ensure orderly transmission of each piece of genetic information to daughter cells, and which finds its pattern of inheritance described by the classical Mendelian laws. The replication and segregation of mitochondrial genomes will be seen to rely far more on chance than on a deterministic mechanism. Yet despite the vast amount of effort that has gone into the study of this process, there is a lot that is still unclear. It is the intention of this review to describe both the strengths and shortcomings of our current understanding.

The cytoplasmic petite mutation

The study of yeast mitochondrial genetics owes its origin to the discovery in 1949 by Ephrussi and his co-workers of the induction of a novel class of mutant by the agent acriflavine (Ephrussi, Hottinguer & Chimenes, 1949; Ephrussi, Hottinguer & Tavlitzki, 1949). Such mutants, clearly recognisable by their small colony size on glucose media (hence their name 'petite'), were found to arise spontaneously at a frequency of around 1 per cent in most strains, but to be inducible at frequencies approaching 100 per cent after a period of growth in acriflavine. On non-fermentable substrates these petite mutants were totally incapable of growth and were thus concluded to be respiratory deficient. This was confirmed with the demonstration that the mutant cells had inactive mitochondria, with an absence of cytochrome oxidase, cytochrome b and modification of other features characteristic of wild-type mitochondria (Slonimski & Ephrussi, 1949; Slonimski, 1953).

It was the genetic behaviour of this new class of mutant that was to attract the most interest. The petite state itself was found to be stably inherited by progeny cells during vegetative growth without any reversion to wild-type. On crossing such mutants to a wild-type strain, however, the petite characteristic disappeared and did not reappear in either the vegetative diploid progeny of the zygotes or, most significantly, in any of the haploid products of subsequent sporulation. With a yeast chromosomal mutation it is not unusual to obtain diploids from a cross with wild-type cells that show only the wild-type appearance. This is simply the classical genetic phenomenon of recessiveness. But, in contrast with the petite mutation, sporulation would for a chromosomal mutation lead to the reappearance of the mutant phenotype in two of the four spores in the ascus. Ephrussi & Slonimski (1955) concluded from this work that the petite mutation was due to the loss or alteration of a self-replicating cytoplasmic factor, possibly residing in the mitochondria. This factor has since become known as the *rho* factor, with wild-type strains designated rho^+ and the petite mutants rho^- (Sherman & Ephrussi, 1962).

Later work from the same laboratory revealed that the initially studied acriflavine induced petites were just one of a number of classes of respiratory deficient petite mutations. Mutants which are unable to transmit their petite characteristic to crosses of the progeny are now known to be a special case generally referred to as *neutral petites*. Amongst spontaneously arising petites, however, *suppressive petites*, first described by Ephrussi, de Margerie–Hottinguer & Roman (1955), are the most commonly found class.

The phenomenon of suppressiveness is best illustrated by an extreme example and such an example is shown contrasted with a neutral petite in Fig. 1. Petites of this type will, on crossing to a wild-type strain, yield zygotes which in almost every case give rise to clones consisting entirely of petite cells. It is thus the normal rather than the mutant phenotype that never appears in the progeny from these zygotes. It is clear that such suppressive petites must originate from an alteration in the postulated *rho* factor, making it able to dominate the progeny, rather than from a simple loss of this factor.

Most newly isolated petites are neither neutral nor entirely suppressive but fall somewhere between these two extremes. Such intermediate suppressive petites give both wholly petite zygotic clones and other zygotic clones consisting of both petite and wild-type progeny. The percentage of wholly petite zygotic clones, referred to as the per cent suppressiveness, is a characteristic of a given mutant, and can take values

Fig. 1. The inheritance of neutral and suppressive petite mutations. Cells containing the petite genome are shown as 0 and the wild-type genome as ●. Note that the zygote is initially shown as mixed in both cases.

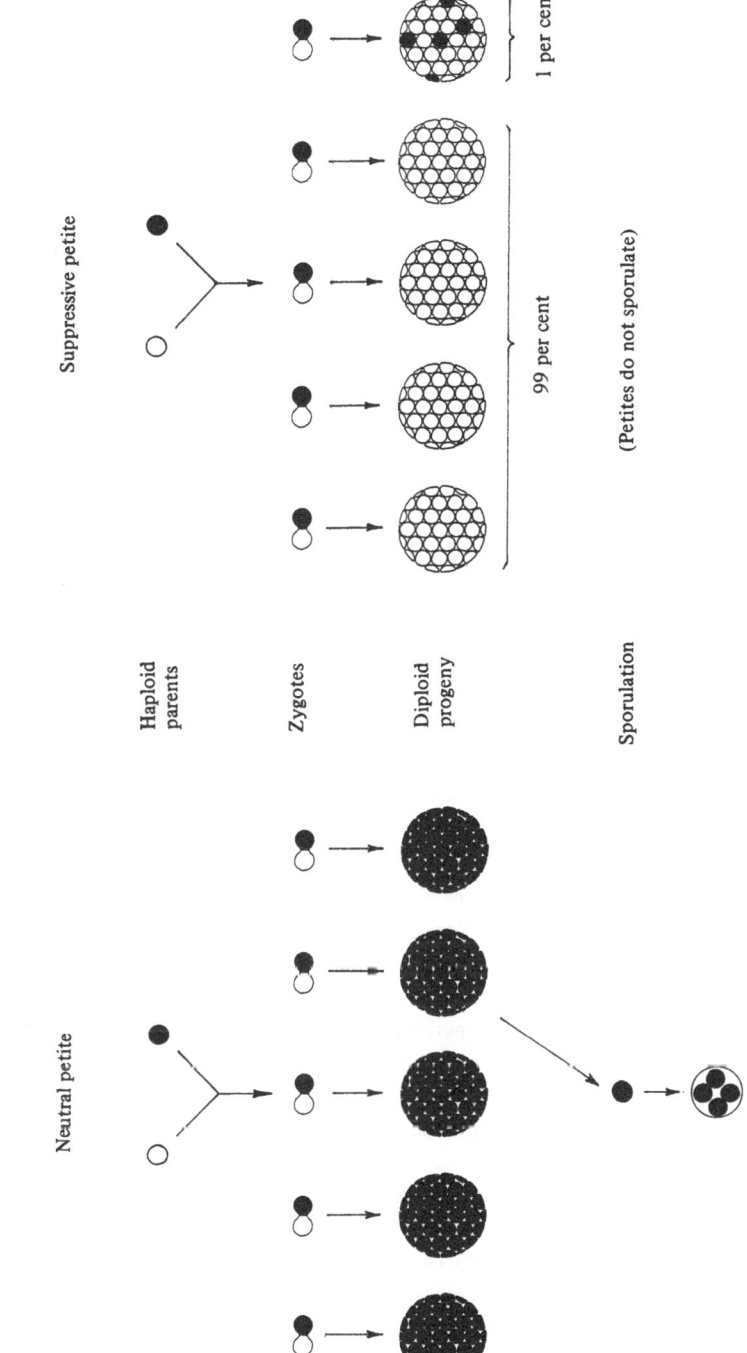

from 1 per cent through to over 99 per cent (Ephrussi & Grandchamp, 1965). It should be emphasised that segregation of two classes of progeny during vegetative growth from the zygote would not be expected with a chromosomally determined trait as all zygotes and their progeny will have an identical nuclear constitution.

An extensive investigation of the transmission of the suppressive petite was reported by Ephrussi, Jakob & Grandchamp (1966) but a fuller understanding of its mode of action had to await the discovery of the nature of the *rho* factor. It will be seen later that the highly suppressive petite mutants, with their extreme transmission bias, have played a major part in building up our understanding of the control of replication and segregation of the mitochondrial genome.

Although it is not central to the aims of this review, it is appropriate to mention that there are a large number of chromosomally inherited mutations with respiratory deficient phenotypes similar to those of the cytoplasmic petites (Sherman, 1963; Tzagoloff, Akai & Needleman, 1975). The existence of these mutants emphasises that the control of the transmission and expression of the mitochondrial genome is a product of the interaction of both nuclear and cytoplasmically inherited factors.

Yeast mitochondrial DNA

A physical basis for the *rho* factor was provided by the discovery that mitochondria from yeast, in common with those from other species, contain their own DNA (Schatz, Haslbrunner & Tuppy, 1964; Tewari, Jayaraman & Mahler, 1965). Despite difficulty in isolating this yeast mitochondrial DNA (mtDNA) intact, it was shown by observation of the contents of lysed mitochondria under the electron microscope that the mtDNA was a closed circular molecule of about 25 μm circumference (Hollenberg, Borst & Van Bruggen, 1970). This figure is in agreement with results obtained from restriction analysis suggesting a strain-dependent size of between 68 and 75 kbp (Prunell *et al.*, 1977; Sanders *et al.*, 1977). More recently, sequencing data have enabled the size of the mtDNA to be more accurately estimated as between 78 and 85 kbp (de Zamaroczy & Bernardi, 1985).

The link between the yeast mtDNA and the genetic determinant *rho* was clearly established by the work of Mounolou, Jakob & Slonimski (1966) who demonstrated that the mtDNA from certain petite mutants had a grossly altered buoyant density. Furthermore, a group of neutral petites (now designated rho^0) were later shown to be completely devoid of mtDNA (Nagley & Linnane, 1970). Thus the mtDNA, like the postulated *rho* factor, was seen to be altered in suppressive petites and absent in many neutral petites.

An important factor to be considered in any theory of segregation of yeast mtDNA is the large number of molecules present in each cell. It is generally agreed that, under normal growth conditions, mtDNA accounts for around 15 per cent of the total yeast DNA, a figure that leads to an estimate of about 50 copies in a haploid yeast cell and 100 in a diploid (Williamson, Moustacchi & Fennell, 1971; Grimes, Mahler &

Perlman, 1974). This level has been reported to be reduced by a number of factors including glucose repression (Goldthwaite, Cryer & Marmur, 1974) and anaerobic growth in the absence of ergosterol and unsaturated fatty acids (Nagley & Linnane, 1972). It has also been suggested that the level of mtDNA is controlled by the cell volume (Lee & Johnson, 1977) or perhaps even the mitochondrial volume (Conrad & Newlon, 1982). From a genetic study of strains with differing mtDNA contents Hall, Nagley & Linnane (1976) concluded that at least two nuclear genes were involved in determining the mtDNA level. Despite all such variation, however, it is clear that each cell contains a large number of copies of the mitochondrial genome.

The replication of yeast mtDNA

Classical biochemical techniques have not shed a great deal of light on the process of replication of mtDNA in yeast. It has not even been possible to demonstrate that mtDNA replication proceeds via the expected semi-conservative mechanism. Experiments in which cells grown in the presence of ^{15}N are shifted to ^{14}N media have not produced a peak of intermediate density as expected from conventional theory but a continuous shift in density suggestive of dispersive replication (Williamson & Fennell, 1974; Sena *et al.*, 1975). This unexpected result was not taken to indicate that yeast mtDNA replicates by a different mechanism to all other DNA molecules but, rather, that frequent recombination between different mtDNA molecules distributes the label in a continuous fashion. In a recent electron microscopic study of petite mtDNA molecules obtained from vegetatively growing cells the level of recombination intermediates observed was sufficiently high to support this explanation (Sena, Revet & Moustacchi, 1986).

Yeast mitochondria have been shown to contain their own DNA polymerase with properties that distinguish it from both the yeast nuclear polymerases and the *gamma* type polymerases found in higher eukaryotes (Wintersberger & Wintersberger, 1970; Wintersberger & Blutsch, 1976). Petite cells are unable to engage in mitochondrial protein synthesis and yet can replicate their mtDNA; it is thus clear that this DNA polymerase must be coded for by the nuclear system. Genga, Bianchi & Foury (1986) have described a nuclear mutant deficient in mtDNA replication and suggest that the gene involved might code for the polymerase itself.

Despite the chromosomal location of the gene coding for the polymerase, it is clear that mtDNA replication is not closely coupled with chromosomal DNA replication. It is generally accepted that yeast mtDNA replication takes place continuously throughout the cell cycle and is not confined to the S phase (Williamson & Moustacchi, 1971; Sena *et al.*, 1975), although some workers have suggested a more synchronised control (Cottrell & Lee, 1981). It has also been reported that inhibitors of cytoplasmic protein synthesis (and hence of initiation of new rounds of nuclear DNA replication) do not affect mtDNA replication (Grossman, Goldring & Marmur, 1969), a result consistent with the continuation of mtDNA replication observed in cell cycle mutants blocked in the initiation of nuclear DNA synthesis (Newlon & Fangman, 1975). On the other hand, mutants deficient for continued replication during the S phase itself not only inhibit mtDNA replication but also lead to the formation of petite mutants (Newlon, Ludescher & Walter, 1979).

It can be concluded from these studies that nuclear genes play a major role in the control of mtDNA synthesis, but that they act for the most part independently from the system controlling the nuclear chromosomes. Further progress in understanding the control of mtDNA replication and its role in genome transmission was made possible primarily as a result of genetic studies on the mitochondrial genome itself.

Point mutations of the mitochondrial genome

The cytoplasmic petite mutation has played an invaluable role in the study of the yeast mitochondrial genome but could not in itself be used to study the normal segregational patterns of mtDNA. Petite cells are at a growth disadvantage on all media and this leads to a situation in which the final composition of a culture cannot be taken as any measure of the proportion of petite and wild-type segregants. And, as already outlined, the petite mutation itself can dramatically affect the segregational pattern. To investigate the normal segregational patterns of the mitochondrial genome it was thus necessary to obtain mutations of this genome which differed less drastically from the wild-type.

The most useful mutants for studies on such transmission have proved to be the point mutations conferring resistance to drugs that selectively inhibit the functioning of yeast mitochondria. Linnane *et al.* (1968) and Thomas & Wilkie (1968) both reported mutants resistant to erythromycin, an antibiotic that inhibits mitochondrial but not cytoplasmic yeast ribosomes. When analysed genetically, these mutations displayed a mode of inheritance suggesting a non-chromosomal location. Later, similarly inherited mutations conferring resistance to other mitochondrial ribosome inhibitors such as chloramphenicol, to agents such as antimycin, which inhibits electron transport, and to oligomycin, which inhibits oxidative phosphorylation, were also obtained (see Nagley, Sriprakash & Linnane, 1977).

The outcome of a typical cross involving mutants of this type is illustrated in Fig. 2. It can be seen that, unlike the situation in crosses involving suppressive petites, diploid cells of both parental types are found amongst the progeny arising from most zygotes. (The small proportion of exceptional uniparental clones will be returned to later.) It should also be noted that the progeny diploids breed true for their characteristic, thus on sporulation sensitive diploids yield only sensitive spores whilst resistant diploids yield only resistant spores. This is, of course, in sharp contrast to the absence of segregation during vegetative growth and the 2:2 ratio on sporulation found for nuclear-coded genes.

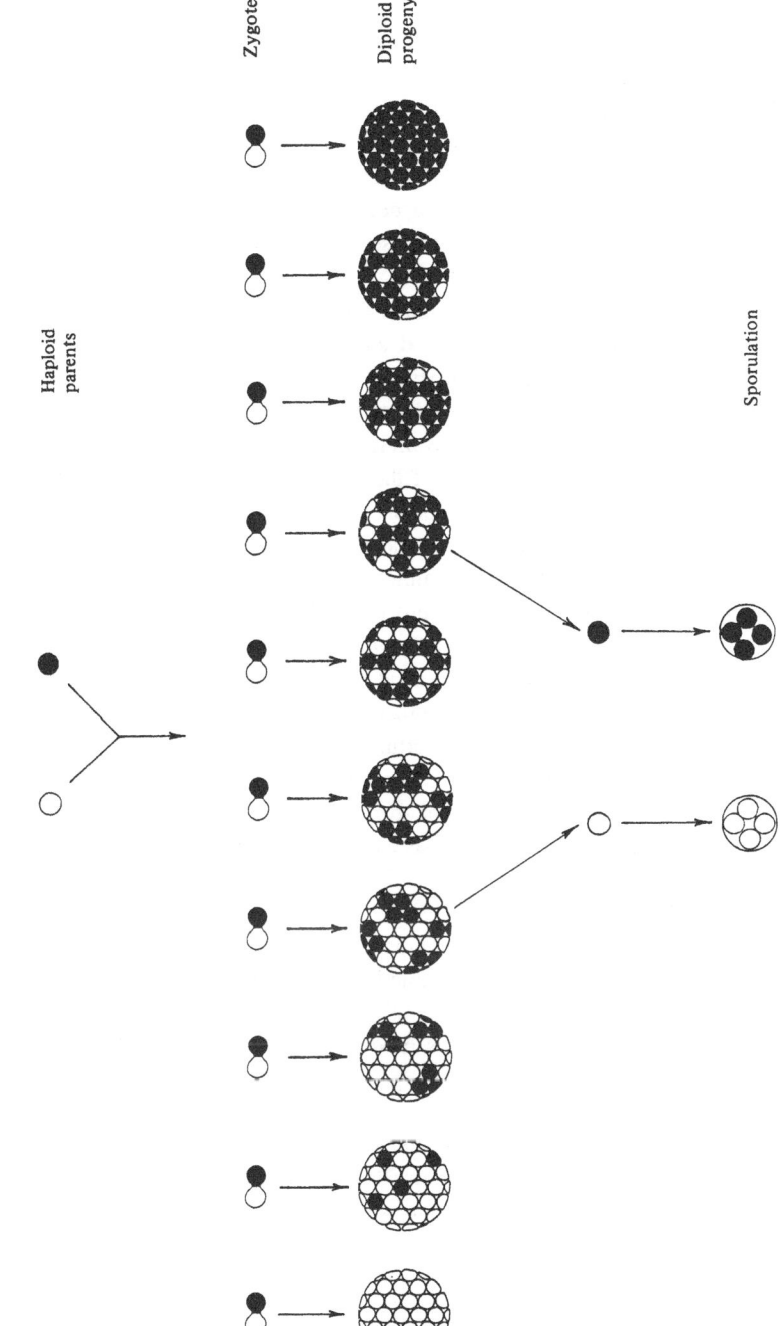

Fig. 2. The inheritance of the erythromycin resistance mutation. Cells containing the resistant genome are shown as ● and the sensitive genome as 0.

Subsequent genetic analysis revealed that the resistance alleles were frequently, but not always, lost when the cells mutated to *rho⁻* (Gingold *et al.*, 1969). From this evidence Saunders *et al.* (1971) deduced that the mutations were located on the *rho* factor itself, that is, the mitochondrial DNA. Whether an allele was lost or retained in a petite cell was seen as a reflection of the region of the wild-type genome retained in the petite cell. This phenomenon of loss or retention of mitochondrial markers in particular petite isolates has since become the basis of an important method of mapping the mitochondrial genome (see Molloy, Linnane & Lukins, 1975; Dujon, 1981).

The antibiotic resistance mutations made it possible to perform crosses differing in two or more mitochondrial encoded loci and thus to study mitochondrial genome recombination and segregation. Early work in this area was dominated by the observation that certain crosses were highly 'polar'. In such crosses the great majority of the progeny carried the alleles from one parent and particular recombinant classes were found far in excess of the reciprocal product (Coen *et al.*, 1970). It is now apparent that this so-called 'polarity' arose because of the location of genes for chloramphenicol and erythromycin resistance in a region of the mtDNA with unusual properties. In some yeast strains, although not all, a sequence of DNA known as the *omega* factor is found inserted into the mtDNA within this particular region. In crosses between strains which carry *omega* and those which lack it, this sequence promotes high frequency unidirectional gene conversion leading to the *omega* factor, along with closely linked genes, spreading to most originally *omega* genomes (see Dujon, 1981). It is somewhat unfortunate that early studies were so dominated by this region, as, although polarity is a topic of great interest, conversion events of this type are by no means typical of the great majority of the mitochondrial genome and certainly did much to obscure the basic segregational pattern.

The Dujon model

The model for mitochondrial genome inheritance presented by Dujon, Slonimski & Weill (1974) undoubtedly represented a major breakthrough in our understanding of mitochondrial genetics and remains the central element in present-day thought on the subject. This model drew its inspiration from studies on the infection of bacteria by virulent phage and the genetic nature of the phage progeny. The authors drew an analogy between the multiplying and freely recombining phage DNA molecules in the infected bacteria and the mtDNA molecules in a yeast cell. In essence, it was suggested that the mtDNA molecules in a zygote were free to engage in multiple rounds of recombination with each other. The choice of partner was seen as totally random, with pairings between like and unlike molecules both occurring. Segregation into the daughter buds was also visualised as an entirely random process.

Probably the greatest strength of the phage analogy or Dujon model was its realisation of polarity as a special case and the prediction that in crosses involving genes distant from the *omega* factor (or indeed even in crosses involving chloramphenicol and erythromycin resistance alleles, but in which both parents either carry or both do not carry *omega*) a very simple relationship should exist between the input and output allele frequencies. If the phage analogy model holds, it should be possible to demonstrate that the frequency of the allele types in the progeny of a cross is directly dependent on the contribution of the two parents.

In a normal cross the two parents will, of course, contribute approximately equal numbers of mtDNA molecules to the zygotes and hence one would expect, and indeed one finds, a 50 : 50 ratio of the two allele types amongst the zygotes. But Dujon *et al.* (1974) took this further by reducing the contribution of one parent by UV irradiation and demonstrating a consequential reduction in the frequency of the alleles contributed by that parent amongst the progeny. Furthermore, the effect was the same for each individual allele carried by the irradiated parent. Other, less drastic, methods of altering the mtDNA contribution of one parent such as glucose repression (Goldthwaite *et al.*, 1974) and the use of a mutant strain with an elevated mtDNA level (Perlman *et al.*, 1976) have also produced results in line with the predicted relationship.

The picture which emerges from the Dujon model is one of mtDNA molecules in a cell being free to mix, randomly recombine and segregate to the buds. On the level of the population of zygotes, the model provides an adequate explanation for the results obtained. It will be seen in the next section, however, that when the results from individual zygotes and their progeny are examined there is still much to explain.

The ploidy paradox and other difficulties

When the purpose of crossing strains with different mitochondrial markers is to obtain sufficient data to investigate recombination between mitochondrial genes, the most suitable approach is the 'standard cross' (Coen *et al.*, 1970). In this procedure the two parental strains are mated, the zygotes plated on a suitable medium to allow propagation of the diploid progeny (but not the parents), and allowed to grow for about 20 generations. After this time the *total* diploid population is harvested and plated to obtain single colonies. These can then be analysed for the mitochondrial markers by replica plating onto media containing the relevant antibiotics. While such experiments give results in excellent agreement with the phage analogy model (Dujon *et al.*, 1974), by their very nature they only give information about the population as a whole and not the individual zygotic clones. To obtain a picture of the segregational patterns at a clonal level, and hence of the actual events occurring in individual zygotes, it is necessary to use different, and more time-consuming, methods.

In the first approach, *zygotic clone analysis*, each individual zygotic clone is *separately* harvested, plated and analysed. This procedure, of course, requires far more work and will be less likely to give as accurate a picture of the overall segregation than the standard cross. By such a separate analysis of each individual clone, however, it is possible to investigate variation between the events occurring in different zygotes.

The second approach, *pedigree analysis*, involves the use of micromanipulation to separate initial buds from the zygotes, thus allowing the growth of the progeny from each bud into an individual colony. The analysis of the cells in each colony enables the building of a detailed picture of the rate of segregation of the different mitochondrial genotypes emerging from the individual zygotes in the cross. Needless to say, this method is by far the most time consuming and is generally limited by the small size of the total sample!

The most striking result from such analyses is the rate at which cells pure for a single allele or even a complete genotype emerge. Such cells, described as homoplasmic, are frequently found even amongst first buds. For example, in an early analysis, Dujon *et al.* (1974) found that a cross between an oligomycin resistant and an oligomycin sensitive strain resulted in 29 out of the 74 first buds being pure for the oligomycin sensitive allele. Such a result could, of course, simply indicate a lack of cytoplasmic mixing and thus a tendency for early buds to receive their cytoplasm from only one parent. The observation that buds forming from the ends of the zygotes more frequently lead to homoplasmic clones than those from the central region where mixing of the cytoplasms is more likely (Strausberg & Perlman, 1978; Waxman & Birky, 1982) might be seen to add weight to this suggestion. The demonstration, however, that the segregation of such homoplasmic buds actually increases in subsequent generations (Waxman & Birky, 1982) calls for a more fundamental explanation.

Given a process of random distribution of the mitochondrial genomes to buds, it is to be expected that buds pure for one genotype would eventually arise as a result of chance partitioning. The probability of such events depends, however, on the number of segregating units involved. In an analysis using an intentionally simplified model, Dujon *et al.* (1974) found that their results were consistent with a figure of three genetic units entering each bud. Using a more sophisticated model to analyse a range of data, Birky *et al.* (1978) and Waxman & Birky (1982) arrived at estimates of between 2 to 20 segregating units in the mother cell and of between 1 to 4 in the buds. In contrast to this, measurements of the numbers of mtDNA molecules have produced figures of around 100 for the zygotes and 40 for their buds (Sena *et al.*, 1976). Random partitioning of molecules in numbers such as these would require many generations of growth before any significant segregation of homoplasmic cells occurred.

This contradiction between the physical number of mtDNA molecules and genetic data, the so-called 'ploidy paradox', had been observed much earlier when considering the kinetics of induction of petite mutations. Attempts to estimate the number of targets involved generally lead to figures of between 1 and 6

(Wilkie, 1963; Slonimski, Perrodin & Croft, 1968). Thus here too, a far lower number of genetic units than mtDNA molecules was inferred from the results.

A phenomenon even more difficult to explain is the degree of variation between individual zygotic clones. When viewed as a total population individual crosses are seen to have characteristic output frequencies for each of the input alleles. However, the use of zygotic clone analysis has revealed that the frequencies for individual zygotic clones are distributed over a wide range and even include a number of clones pure for the allele contributed by one of the two parents (Linnane *et al.*, 1968; Coen *et al.*, 1970; Birky, 1975). Such 'uniparental' zygotes generally represent only a few percent of the cross output, but in crosses with a mtDNA input imbalance this can rise to almost 50 per cent of all zygotes (Birky *et al.*, 1978). Birky *et al.*. (1978) also demonstrated a strong tendency in crosses involving a number of mitochondrial markers for all the alleles from a single parent to become uniparental together, thus eliminating any explanation for this phenomenon based on local events such as gene conversion.

Even with small numbers of segregating units, it would be difficult to see how a mechanism based purely on random partitioning of the input genomes into buds could lead to zygotic clones in which the genomes from only one of the two parents was represented amongst the progeny. The Dujon model clearly needs modification to deal with these difficulties.

Intracellular genetic drift

One of the many contributions made by William Birky Jr. to mitochondrial genetics has been the concept of random mtDNA replication and genetic drift (see Birky, 1983). In essence, it is proposed that instead of each mtDNA molecule replicating once per cell cycle, the choice of molecules for replication is more random such that an *average* of one round of replication takes place, but with some molecules replicating more than once and others not at all. Such random choice would be expected, purely as a result of chance, to increase the frequency of some genomes (and hence alleles) and decrease the frequency of others. Thus, across a population as a whole, a far greater degree of variation between clones would be found than would be the case if each genome replicated once. If it is assumed that some degradation of mtDNA also occurs, it becomes possible to explain how a zygote could come to pass only one genotype on to its progeny. This situation is closely analogous to the population genetics phenomenon of genetic drift.

If a process of genetic drift is taking place in the cell, it would be expected that the partitioning of genomes by cell division would be a limiting factor. Thrailkill *et al.* (1980) tested this idea using various periods of starvation of newly formed zygotes to delay segregation and found that, as predicted, both the inter-clonal variation and the proportion of uniparental zygotes increased in line with starvation time. Thus delaying segregation clearly gave genetic drift more time to operate and allowed individual clones of more extreme composition to arise.

The problem of mitochondrial genome ploidy still remains. Mathematical analysis and computer simulations have shown that genetic drift could only be effective if the copy number of the genetic unit was far below the number of mtDNA molecules in the cell (Birky & Skavaril, 1976). Once again, it would appear that the number of genetic units is the major stumbling block to a coherent model.

As a way around this dilemma it has been suggested that the actual genetic units are not the individual mtDNA molecules themselves but rather clusters of a number of associated mtDNA molecules, perhaps corresponding to the clusters of mtDNA molecules observed by fluorescent staining and termed 'chondriolites' (see Williamson et al., 1977). It has been proposed that while variation could occur between such units, the mtDNA molecules within each cluster would be mainly homozygous and hence act as a single unit. It is not at all clear, however, how mtDNA molecules within these clusters would all come to have the same genotype at not just one locus, but over their entire genome.

A somewhat different solution has been suggested by Gingold (1981) and Birky (1983). Both authors have considered the possibility that a molecule which has already replicated may have an enhanced probability of engaging in further rounds of replication as a consequence of already being in a position of access to the replication machinery. Such a situation would *amplify* the effects of an initially random selection for replication and thus increase the possibility of extreme variation. It must be pointed out, however, that all physical measurements go against any suggestion of one group of mtDNA molecules replicating more frequently than others (Williamson et al., 1977).

In conclusion, it must be stated that the genetic studies using mitochondrial point mutations have not totally clarified the mode of segregation of the mtDNA genome but have at least emphasised the importance of random recombination, replication and partitioning to buds.

The mode of action of the suppressive petite genome

One of the most useful contributions that genetics can make to the study of cell biology is the supply of mutants that are grossly altered in the control of key systems. The suppressive petite mutants, with their ability to pass on the petite characteristic to the overwhelming majority of cross progeny, might thus seem to be precisely the sort of mutant required for the study of the controls on mitochondrial genome transmission. But before they could be used in this way, it was first necessary to establish that the suppressive petite mutation did indeed act by perturbing the normal transmission of mtDNA and not by some other mechanism.

To understand the possible mechanisms by which suppressiveness could act, it is necessary to examine the nature of the mtDNA in petite cells. This topic is well reviewed by Dujon (1981) and only a brief outline will be given here. Petite genomes consist of massive deletions from the wild-type mtDNA such that only small regions of the original genome remain. The retained sequence is, however, found at an elevated copy number such that the total mtDNA content is generally not greatly altered from that of a wild-type cell. Furthermore, the basic retained unit is found repeated in multimeric molecules as is shown in Fig. 3. The so-called 'repeat unit' can vary from a third of the wild-type genome down to less than 100 base pairs. While it is usual to represent the molecule as being of normal wild-type size (as in Fig. 3), the evidence in fact suggests that a range of circle sizes from one repeat unit upwards is found in the petite cell (Lazowska & Slonimski, 1976).

For a long period it was believed that the suppressive genome acted by destroying the integrity of the wild-type genome by a process of 'destructive recombination' (see, for example, Michaelis, Petrochilo & Slonimski, 1973; Perlman & Birky, 1974). Supported by the success of high frequency gene conversion in explaining polarity, and in conjunction with the genetic and physical evidence of petite genomes taking part in recombination, a plethora of theories arose which had as a common element the integration of the suppressive petite genome into the wild-type mtDNA followed by its subsequent degeneration into the petite state. A consequence of such models was that the mtDNA of the petite progeny would for the most part be expected to result from the breakdown of the wild-type genome rather than the efficient transmission of the original petite mtDNA. In other words, it would have been expected that the progeny cells would contain a variety of petite genomes drawn from all over the wild-type mtDNA.

Fig. 3. An illustration of the process of petite formation. Note that while ● in this case represents an *ori* sequence, it could be taken to represent any particular mitochondrial region.

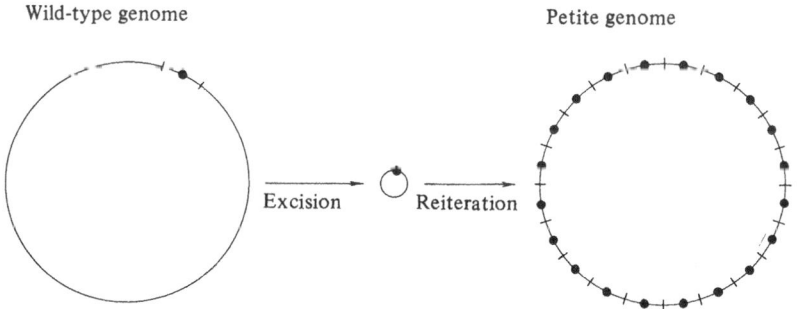

Wild-type genome   Petite genome

Excision   Reiteration

● – Replication origin

Two types of experiments demonstrated that, contrary to this expectation, the progeny from such petite by wild-type crosses had genomes derived directly from the petite parent. Goursot *et al.* (1980) and Blanc & Dujon (1980) both studied crosses of very high suppressive petites (so-called 'super' or 'hyper-suppressives') with wild-type strains and were able to demonstrate by restriction analysis that the mtDNA from petite progeny cells was identical to that of the petite parent. Supporting this conclusion, Gingold (1981) demonstrated by genetic analysis that the petite progeny of a cross of a somewhat more moderately suppressive mutant carried genetic markers only from the region of the genome present in the petite parent.

It was thus clear that, rather than recombining into the wild-type mtDNA, the suppressive genome was indeed acting by ensuring its preferential transmission to progeny cells. This new evidence led to the revival of an earlier hypothesis to explain suppressiveness, that of out-replication (Carnevali, Morpurgo & Teece, 1969). The petite genome was hypothesised to be able to replicate with a greater efficiency than the wild-type genome and thereby to enhance its transmission to progeny.

Mitochondrial replication origins

When the repeat units from a number of super-suppressive petites were compared it was found that they all carried one of a small number of closely related sequences (de Zamaroczy, Baldacci & Bernardi, 1979; de Zamaroczy *et al.*, 1981; Blanc & Dujon, 1980; 1982). It was suggested that these sequences, designated *ori* and *rep* by the two groups of workers, represented mtDNA origins of replication. Some dispute has existed about the number of such sequences on wild-type genomes, although current evidence suggests the presence of four active *ori* sequences and at least three inactive but similar sequences in most strains (Faugeron–Fonty *et al.*, 1984).

Initial evidence for the *ori* sequences being true replication origins was somewhat circumstantial. Clearly, a molecule containing a reiteration of a replication origin (as in Fig. 3) would be expected to be able to out-compete normal wild-type molecules for replication factors and hence be preferentially replicated and passed on to progeny cells. Furthermore, de Zamaroczy *et al.* (1981) were able to show that the suppressiveness of a mutant varied with the density of *ori* sequences along its genome, while deletion or rearrangement of the *ori* sequences led to a drastic drop in suppressiveness. Blanc & Dujon (1982) demonstrated that these sequences when added to plasmids of a type unable to replicate in yeast enabled them to do so (albeit in the nucleus). This result was somewhat diminished, however, by the demonstration that many other regions of the mtDNA, possibly as many as 40, could also act in this way (Hyman, Cramer & Rownd, 1983). While it might be considered unlikely that all these regions function as replication origins in wild-type yeast, they could represent the weaker 'surrogate origins' that have been hypothesised to enable low suppressive petites without *ori* sequences to replicate their mtDNA (Goursot, Mangin & Bernardi, 1982).

An examination of the properties of the *ori* sequences has provided more direct evidence for the role of such sequences as replication origins. Fig. 4 shows the common features of *ori* sequences as revealed by de Zamaroczy *et al.* (1981). A and B are two short GC clusters (regions of DNA composed almost entirely of the bases guanine and cytosine) separated by a 25 base pair AT stretch (a DNA region almost entirely composed of the bases adenine and thymine). A 200 base pair AT stretch (designated *l*) is followed by a third GC cluster designated C. These authors noted that the structure of clusters A and B had the capacity to be folded in a similar way to a sequence found within the mitochondrial replication origins from mammalian cells. Furthermore, it was noted that the sequence of cluster C was highly homologous to that of a similarly located GC cluster in the mammalian origins.

Baldacci & Bernardi (1982) reported that the *ori* sequences could promote initiation of RNA transcription in petite mutants. By hybridisation of the transcripts with denatured DNA strands followed by S1 nuclease digestion of the unprotected DNA, they were able to demonstrate that the RNA synthesis initiates within an AT rich region (designated *r*) next to cluster C. Baldacci, Chérif–Zahar & Bernardi (1984) used similar techniques to demonstrate that DNA synthesis itself is initiated within the *ori* sequences of petite strains of yeast. As illustrated in Fig. 4, newly synthesised DNA strands were shown to begin within the C cluster itself, with RNA primers for each strand initiating in the AT regions *r* and *l*.

It can thus be concluded that strong support exists for the idea that the *ori* sequences found in suppressive petites represent elements with the ability to initiate mtDNA synthesis in petite cells. It seems reasonable, though as yet unproven, to suggest that these sequences act in a similar way in wild-type genomes. What remains to be demonstrated is whether all four *ori* sequences are active in a wild-type yeast and indeed if these are the only points at which mtDNA synthesis can originate. It is, of course, feasible that the surrogate origins mentioned above are also capable of initiating replication of the wild-type genome even if such events occur at a lower frequency than those from the stronger *ori* sequences.

Fig. 4. The structure of an *ori* sequence. A, B and C are GC clusters, while *l* and *r* represent AT stretches. The arrows indicate the positions of initiation of RNA priming and DNA synthesis on each strand.

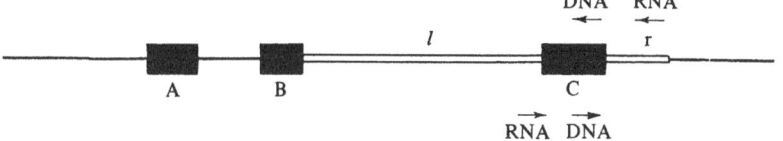

Replication and the control of mtDNA transmission

The concept of suppressive petite genomes acting by means of a reiteration of replication origins fits in very well with the ideas already discussed on the random choice of mtDNA molecules for replication. One must hypothesise that some factor involved in the initiation of the replication is in limiting supply, a reasonable assumption given that mtDNA levels are clearly under some form of control. Whilst it is probably easiest to think of this in terms of a limited number of soluble initiation factors, there is no reason why the limiting factor could not be something like a membrane bound DNA replication complex.

While the choice of which origin obtains access to the limiting replication factor can be viewed as entirely random, it is obvious that a molecule with a high number of replication origins would be more likely to be replicated than a molecule with only a few. Thus in the case of a zygote containing suppressive petite mtDNA and wild-type DNA, it is the petite genome that would have the highest probability of binding the initiation factor to one of its many *ori* sequences and of being replicated. Even when, by chance, it is a wild-type genome which replicates and hence increases its chances of being included in a bud, it will still face the same unbalanced competition from the petite mtDNA in the bud. It is only when a bud arises in which no petite genomes have been included, an unlikely event given the replicative imbalance, that the wild-type genome can establish itself. It is of interest to note that Ephrussi *et al.* (1966) demonstrated that wild-type cells could be induced to appear in zygotic clones otherwise destined to be wholly petite by a change in media over 24 hours into colony formation, an indication of just how long the wild-type mtDNA can survive in a clone and yet not lead to wild-type progeny.

In experiments from our own laboratory we have shown that a period of zygote starvation greatly increases the degree of suppressiveness measured for a number of crosses (Chambers & Gooding, unpublished). Presumably, during the period of starvation, some mtDNA synthesis continues and brings with it the opportunity for the petite genome to gain an even greater advantage in the absence of partitioning. This result emphasises the similarity between the underlying mechanisms responsible for suppressiveness and uniparentalism in non-petite crosses.

In a cross between two *rho+* strains, there would not normally be an imbalance of *ori* sequences, but the random competition for initiation factors would still be expected to lead by chance to some molecules replicating more than others. Thus, this competition would provide a basis for genetic drift, and in a similar though less dramatic fashion to the situation seen with suppressiveness, the possibility of the fixation of one or other genome type in the zygote and its progeny. When an input imbalance exists, the probability of fixation of the majority allele will, of course, increase.

Is there segregational control?

While recent results have clearly supported the view that control of mtDNA replication is a major factor in determining transmission, other evidence from studies on suppressive petites suggests it is not the only one. In a series of experiments reported by Chambers & Gingold (1986), we set out to test directly if suppressive petite genomes did indeed replicate in the zygotes at a faster rate than the wild-type mtDNA. The method involved radio-labelling the replicating DNA in newly formed zygotes, isolation of the mtDNA, and separation of the products of restriction on a gel. The enzymes were chosen such that the petite and wild-type mtDNA digestion products formed clearly separate bands. By means of fluorography of the gel it was thus possible to ascertain the relative levels of synthesis of the two types of mtDNA in the zygote. When a 79 per cent suppressive petite was tested, we found a clear replicative advantage for the petite mtDNA, thus confirming that this genome was out-replicating the wild-type species. But when other petites with suppressiveness ranging from 12 per cent to 43 per cent were tested, no such replication advantage was seen.

It must be emphasised that a 43 per cent suppressiveness indicates that 43 per cent of the zygotic clones contain *no* wild-type progeny, with many of the remaining clones being mainly petite as well. Thus such a petite genome clearly has a competitive advantage in the zygote. This advantage, however, was not demonstrated in terms of replication.

Fangman & Dujon (1984) reported on a petite strain with an 89 base pair repeat unit consisting entirely of A and T. Yet despite the lack of any *ori* sequences, the mutant had a suppressiveness of 30 per cent. Here too, the simple replication theory seems inadequate.

It must be considered possible that factors other than just replication efficiency determine the likelihood of a mitochondrial genome being transmitted to the progeny. Segregation to buds is generally seen as a passive process, with partitioning being determined purely by genome position. If, however, the segregation of the mtDNA molecules has a more active basis, some form of competition for transmission could develop. Clearly, a petite genome with an advantage in this process would, like the replication efficient genomes, be able to predominate in the progeny. It can thus be hoped that the study of petite genomes with suppressiveness but without a replication advantage might offer new insights into the factors controlling mitochondrial genome segregation.

References

Baldacci, G. & Bernardi, B. (1982). Replication origins are associated with transcription initiation sequences in the mitochondrial genome of yeast. *The EMBO Journal*, **1**, 987–94.

Baldacci, G., Chérif–Zahar, B. & Bernardi, B. (1984). The initiation of DNA replication in the mitochondrial genome of yeast. *The EMBO Journal*, **3**, 2115–20.

Birky, C.W. Jr. (1975). Zygote heterogeneity and uniparental inheritance of mitochondrial genes in yeast. *Molecular and General Genetics*, **141**, 41–58.

Birky, C.W. Jr. (1983). Relaxed cellular controls and organelle heredity. *Science*, **222**, 468–75.

Birky, C.W. Jr., Demko, C.A., Perlman, P.S. & Strausberg, R. (1978). Uniparental inheritance of mitochondrial genes in yeast: dependence on input bias of mitochondrial DNA and some preliminary investigations of the mechanisms. *Genetics*, **89**, 615–51.

Birky, C.W. Jr. & Skavaril, R.V. (1976). Maintenance of genetic homogeneity in systems with multiple genomes. *Genetic Research (Cambridge)*, **27**, 249–65.

Blanc, H. & Dujon, B. (1980). Replicator regions of the yeast mitochondrial DNA responsible for suppressiveness. *Proceedings of the National Academy of Sciences, USA*, **77**, 3942–46.

Blanc, H. & Dujon, B. (1982). Replicator regions of the yeast mitochondrial DNA active *in vivo* and in yeast transformants. In *Mitochondrial Genes*, ed. P. Slonimski, P. Borst & G. Attardi, pp. 279–94. Cold Spring Harbor, New York: Cold Spring Harbor Laboratory.

Carnevali, F., Morpurgo, G. & Teece, G. (1969). Cytoplasmic DNA from petite colonies of *Saccharomyces cerevisiae*. A hypothesis on the nature of the mutation. *Science*, **163**, 1331–3.

Chambers, P. & Gingold, E. (1986). A direct study of the relative synthesis of petite and grande mitochondrial DNA in zygotes from crosses involving suppressive petite mutants of *Saccharomyces cerevisiae*. *Current Genetics*, **10**, 565–71.

Coen, D., Deutsch, J., Netter, E., Petrochilo, E. & Slonimski, P.P. (1970). Mitochondrial genetics I. Methodology and phenomenology. In *Control of Organelle Development. Symposium of the Society for Experimental Biology, No. 24*, ed. P.L. Miller. pp. 449–96. Cambridge: Cambridge University Press.

Conrad, M.N. & Newlon, C.S. (1982). The regulation of mitochondrial DNA levels in *Saccharomyces cerevisiae*. *Current Genetics*, **6**, 147–52.

Cottrell, S.F. & Lee, L.H. (1981). Evidence for the synchronous replication of mitochondrial DNA during the yeast cell cycle. *Biochemical and Biophysical Research Communications*, **101**, 1350–6.

De Zamaroczy, M., Baldacci, G. & Bernardi, B. (1979). Putative origins of replication in the mitochondrial genome of yeast. *FEBS Letters*, **108**, 429–32.

De Zamaroczy, M. & Bernardi, B. (1985). Sequence organisation of the mitochondrial genome of yeast – a review. *Gene*, **37**, 1–17.

De Zamaroczy, M., Marotta, R., Fauregon–Fonty, G., Goursot, R., Mangin, M., Baldacci, G. & Bernardi, B. (1981). The origins of replication of the yeast mitochondrial genome and the phenomenon of suppressivity. *Nature*, **292**, 75–8.

Dujon, B. (1981). Mitochondrial genetics and functions. In *Molecular Biology of the Yeast Saccharomyces: Life Cycle and Inheritance*, ed. J.N. Strathern, E.W. Jones & J.R. Broach, pp. 505–635. Cold Spring Harbor, New York: Cold Spring Harbor Laboratory.

Dujon, B., Slonimski, P.P. & Weill, L. (1974). Mitochondrial genetics. IX. A model for the recombination and segregation of mitochondrial genes in *Saccharomyces cerevisiae*, *Genetics*, **78**, 415–37.

Ephrussi, B. & Grandchamp, S. (1965). Études sur l'suppressivité des mutants à deficience repiratoire de la levure I. Existence au niveau cellulaire de divers degrés de suppressivité.*Heredity*, **20**, 1–7.

Ephrussi, B., Hottinguer, H. & Chimenes, Y. (1949). Action de l'acriflavine sur les levures. I. La mutation 'petite colonie'. *Annales de l'Institut Pasteur*, **76**, 351–67.

Ephrussi, B., Hottinguer, H. & Tavlitzki, J. (1949). Action de l'acriflavine sur les levures. II. Étude génétique du mutant 'petite colonie'. *Annales de l'Institut Pasteur*, **76**, 419–50.

Ephrussi, B., Jakob, H. & Grandchamp, S. (1966). Études sur la suppressivité des mutants à déficience repiratoire de la levure II. Étapes de la mutation grande en petite provoquée par le facteur suppressif. *Genetics*, **54**, 1–29.

Ephrussi, B., de Margerie–Hottinguer, H. & Roman, H. (1955). Suppressiveness: a new factor in the genetic determinism of the synthesis of respiratory enzymes in yeast. *Proceedings of the National Academy of Sciences, USA*, **41**, 1065–71.

Ephrussi, B. & Slonimski, P.P. (1955). Yeast mitochondria: Subcellular units involved in the synthesis of respiratory enzymes in yeast. *Nature*, **176**, 1207–8.

Evans, I.H. (1983). Molecular genetic aspects of yeast mitochondria. In *Yeast Genetics. Fundamental and Applied Aspects*, ed. J.F.T. Spencer, D.M. Spencer & A.R.W. Smith, pp. 269–370. New York: Springer–Verlag.

Fangman, W.L. & Dujon, B. (1984). Yeast mitochondrial genomes consisting of only A–T base pairs replicate and exhibit suppressiveness. *Proceedings of the National Academy of Sciences, USA*, **41**, 1065–71.

Faugeron–Fonty, G., Kim, C.L.V., de Zamaroczy, M., Goursot, R. & Bernardi, B. (1984). A comparative study of the *ori* sequences from the mitochondrial genomes of twenty wild-type yeast strains. *Gene*, **32**, 459–73.

Genga, A., Bianchi, L. & Foury, F. (1986). A nuclear mutant of *Saccharomyces cerevisiae* deficient in mitochondrial DNA replication and polymerase activity. *Journal of Biological Chemistry*, **261**, 9328–32.

Gingold, E.B. (1981). Genetic analysis of the products of a cross involving a suppressive petite mutant of *S. cerevisiae*. *Current Genetics*, **3**, 213–20.

Gingold, E.B., Saunders, G.W., Lukins, H.B. & Linnane, A.W. (1969). Biogenesis of mitochondria X. Reassortment of the cytoplasmic genetic determinants for respiratory competence and erythromycin resistance in *Saccharomyces cerevisiae*. *Genetics*, **62**, 735–44.

Goldthwaite, C.D., Cryer, D.R. & Marmur, J. (1974) Effect of carbon source on the replication and transmission of yeast mitochondrial genomes. *Molecular and General Genetics*, **133**, 87–104.

Goursot, R., de Zamaroczy, M., Baldacci, B. & Bernardi, G. (1980). Supersuppressive 'petite' mutants of yeast. *Current Genetics*, **1**, 173–6.

Goursot, R., Mangin, M. & Bernardi, G. (1982). Surrogate origins of replication in the mitochondrial genomes of ori^0 petite mutants of yeast. *TheEMBO Journal*, **1**, 705–11.

Grimes, G.W., Mahler, H.R. & Perlman, P.S. (1974). Nuclear gene dosage effects on mitochondrial mass and DNA. *Journal of Cell Biology*, **61**, 565–74.

Grossman, L.I., Goldring, E.S. & Marmur, J. (1969). Preferential synthesis of yeast mitochondrial DNA in the absence of protein synthesis. *Journal of Molecular Biology*, **46**, 367–76.

Hall, R.M., Nagley, P. & Linnane, A.W. (1976). Biogenesis of mitochondria. XLII. Genetic analysis of the control of cellular mitochondrial DNA levels in *Saccharomyces cerevisiae*. *Molecular and General Genetics*, **145**, 169–75.

Hollenberg, C.P., Borst, P. & Van Bruggen, E.F.J. (1970). Mitochondrial DNA. V. A 25-micron closed circular duplex DNA molecule in wild-type yeast mitochondria. Structure and genetic complexity. *Biochimica et Biophysica Acta*, **209**, 1–15.

Hyman, B.C., Cramer, J.H. & Rownd, R.H. (1983). The mitochondrial genome of *Saccharomyces cerevisiae* contains numerous, densely spaced autonomously replicating sequences. *Gene*, **26**, 223–30.

Lazowska, J. & Slonimski, P.P. (1976). Electron microscopy analysis of the circular repetitive mitochondrial DNA molecules from genetically characterised *rho⁻* mutants of *Saccharomyces cerevisiae. Molecular and General Genetics*, **146**, 61–78.

Lee, E.H. & Johnson, B.F. (1977). Volume-related mitochondrial deoxyribonucleic acid synthesis in zygotes and vegetative cells of *Saccharomyces cerevisiae. Journal of Bacteriology*, **129**, 1066–71.

Linnane, A.W., Saunders, G.W., Gingold, E.B. & Lukins, H.B.(1968). The biogenesis of mitochondria. V. Cytoplasmic inheritance of erythromycin resistance in *Saccharomyces cerevisiae. Proceedings of the National Academy of Sciences, USA*, **59**, 903–10.

Michaelis, G., Petrochilo, E. & Slonimski, P.P. (1973). Mitochondrial genetics III. Recombined molecules of mitochondrial DNA obtained from crosses between cytoplasmic petite mutants of *Saccharomyces cerevisiae*: physical and genetic characterisation. *Molecular and General Genetics*, **123**, 51–65.

Molloy, P.L., Linnane, A.W. & Lukins, H.B. (1975). Biogenesis of mitochondria: analysis of deletion of mitochondrial antibiotic resistance markers in petite mutants of *Saccharomyces cerevisiae. Journal of Bacteriology*, **122**, 7–18.

Mounolou, J.C., Jakob, H. & Slonimski, P.P. (1966). Mitochondrial DNA from yeast 'petite' mutants: Specific changes of buoyant density coresponding to different cytoplasmic mutations. *Biochemical and Biophysical Research Communications*, **24**, 218–24.

Nagley, P. & Linnane, A.W. (1970). Mitochondrial DNA deficient petite mutants of yeast. *Biochemical and Biophysical Research Communications*, **39**, 989–96.

Nagley, P. & Linnane, A.W. (1972). Cellular regulation of mitochondrial DNA synthesis in *Saccharomyces cerevisiae. Cell Differentiation*, **1**, 143–8.

Nagley, P., Sriprakash, K.S. & Linnane, A.W. (1977). Structure, synthesis and genetics of yeast mitochondrial DNA. *Advances in Microbial Physiology*, **16**, 157–277.

Newlon, C.S. & Fangman, W.L. (1975). Mitochondrial DNA synthesis in cell cycle mutants of *Saccharomyces cerevisiae. Cell*, **5**, 423–8.

Newlon, C.S., Ludescher, R.D. & Walter, S.K. (1979). Production of petites by cell cycle mutants of *Saccharomyces cerevisiae* defective in mitochondrial DNA synthesis. *Molecular and General Genetics*, **169**, 189–94.

Perlman, P.S. & Birky, C.W. Jr. (1974). Mitochondrial genetics in baker's yeast: a molecular mechanism for recombinational polarity and suppressiveness. *Proceedings of the National Academy of Sciences, USA*, **71**, 4612–16.

Perlman, P.S. & Birky, C.W. Jr., Demko, C.A. & Strausberg, R.L. (1976). Confirmations and exceptions to the phage analogy model: Input bias, bud position effects, zygote heterogeneity and uniparental inheritance. In *Genetics and Biogenesis of Chloroplasts and Mitochondria*, ed. T.M. Bucher, W. Neupert, W. Sebald & S. Werner, pp. 405–14. Amsterdam: North Holland Publishing Company.

Prunell, A.H., Kopecka, H., Strauss, F. & Bernardi, G. (1977). The mitochondrial genome of wild-type yeast cells. V. Genome evolution. *Journal of Molecular Biology*, **110**, 17–52.

Sanders, J.P.M., Heyting, C., Verbeet, M.P., Keijlink, F.C.P.W. & Borst, P. (1977). The organisation of genes in yeast mitochondrial DNA. III. Comparison of the physical maps of the mitochondria DNAs from three wild-type *Saccharomyces* strains. *Molecular and General Genetics*, **157**, 239–61.

Saunders, G.W., Gingold, E.B., Trembath, M.K., Lukins, H.B. & Linnane, A.W. (1971). Mitochondrial genetics in yeast: Segregation of a cytoplasmic determinant in crosses and its loss or retention in the petite. In *Autonomy and Biogenesis of Mitochondria and Chloroplasts*, ed. N.K. Boardman, A.W. Linnane & R.M. Smillie, pp. 185–93. Amsterdam: North Holland Publishing Company.

Schatz, G., Haslbrunner, E. & Tuppy, H. (1964). Deoxyribonucleic acid associated with yeast mitochondria. *Biochemical and Biophysical Research Communications*, 15, 127–32.

Sena, E.P., Revet, B. & Moustacchi, E. (1986). *In vivo* homologous recombination intermediates of yeast mitochondrial DNA analyzed by electron microscopy. *Molecular and General Genetics*, 202, 421–8.

Sena, E.P., Welch, J. & Fogel, S. (1976). Nuclear and mitochondrial DNA replication during zygote formation and maturation in yeast. *Science*, 194, 433–5.

Sena, E.P., Welch, J.W., Halvorson, H.O. & Fogel, S. (1975). Nuclear and mitochondrial deoxyribonucleic acid replication during mitosis in *Saccharomyces cerevisiae*. *Journal of Bacteriology*, 133, 497–504.

Slonimski, P.P. (1953). *La formation des enzymes respiratoires chez la levure*. Paris: Masson.

Slonimski, P.P. & Ephrussi, B. (1949). Action de l'acriflavine sur les levures. V. Le système des cytochromes des mutants 'petite colonie'. *Annales de l'Institut Pasteur*, 77, 47–63.

Slonimski, P.P., Perrodin, G. & Croft, J.H. (1968). Ethidium bromide induced mutation of yeast mitochondria: complete transformation of cells into respiratory deficient non-chromosomal 'petites'. *Biochemical and Biophysical Research Communications*, 30, 232–9.

Sherman, F. (1963). Respiratory deficient mutants of yeast. I. Genetics. *Genetics*, 48, 375–85.

Sherman, F. & Ephrussi, B. (1962). The relationship between respiratory deficiency and suppressiveness in yeast as determined with segregational mutants. *Genetics*, 47, 695–700.

Strausberg, R.L. & Perlman, P.S. (1978). The effect of zygote bud positions on the transmission of mitochondrial genes in *Saccharomyces cerevisiae*. *Molecular and General Genetics*, 163, 131–44.

Tewari, K.K., Jayaraman, J. & Mahler, H.R. (1965). Separation and characterisation of mitochondrial DNA from yeast. *Biochemical and Biophysical Research Communications*, 21, 141–8.

Thomas, D.Y. & Wilkie, D. (1968). Inhibition of mitochondria synthesis in yeast by erythromycin: cytoplasmic and nuclear factors controlling resistance. *Genetic Research (Cambridge)*, 11, 33–41.

Thrailkill, K.M., Birky, C.W. Jr., Luckemann, G. & Wolf, K. (1980). Intracellular population genetics: evidence for random drift of mitochondrial allele frequencies in *Saccharomyces cerevisiae* and *Schizosaccharomyces pombe*. *Genetics*, 96, 237–62.

Tzagoloff, A., Akai, A. & Needleman, R.B. (1975). Assembly of the mitochondrial membrane system: characterisation of nuclear mutants of *Saccharomyces cerevisiae* with defects in mitochondrial ATPase and respiratory enzymes. *Journal of Biological Chemistry*, 250, 8228–35.

Waxman, M.F. & Birky, C.W. Jr. (1982). Partial pedigree analysis of the segregation of yeast mitochondrial genes during vegetative reproduction. *Current Genetics*, 5, 171–80.

Wilkie, D. (1963). The induction by monochromatic UV light of respiratory deficient mutants by aerobic and anaerobic cultures of yeast. *Journal of Molecular Biology*, **7**, 527–33.

Williamson, D.H. & Fennell, D. (1974). Apparent dispersive replication of yeast mitochondrial DNA as revealed by density labelling experiments. *Molecular and General Genetics*, **131**, 193–207.

Williamson, D.H., Johnston, L.H., Richmond, K.M.V. & Game, J.C. (1977). Mitochondrial DNA and the heritable unit of the yeast mitochondrial genome: a review. In *Mitochondria 1977. Genetics and Biogenesis of Mitochondria*, ed. W. Bandlow, R.J. Schweyen, K. Wolf & F. Kaudewitz, pp. 1–24. Berlin: Walter de Gruyter & Co.

Williamson, D.H. & Moustacchi, E. (1971). The synthesis of mitochondrial DNA during the cell cycle in the yeast *Saccharomyces cerevisiae*. *Biochemical and Biophysical Research Communications*, **42**, 195–201.

Williamson, D.H., Moustacchi, E. & Fennell, D.J. (1971). A procedure for rapidly extracting and estimating the nuclear and cytoplasmic DNA components of yeast cells. *Biochimica et Biophysica Acta*, **238**, 369–74.

Wintersberger, U. & Blutsch, H. (1976). DNA-dependent DNA polymerase from yeast mitochondria: dependence of enzyme activity on conditions of cell grwoth and properties of the highly purified polymerase. *European Journal of Biochemistry*, **68**, 199–207.

Wintersberger, U. & Wintersberger, E. (1970). Studies of DNA polymerase from yeast. 2. Partial purification and characterisation of mitochondrial DNA polymerase from wild-type and respiratory-deficient yeast cells. *European Journal of Biochemistry*, **13**, 20–7

R.J. ROSE

The role of membranes in the segregation of plastid DNA

The presence of DNA in plastids has been recognised for 25 years (Ris & Plaut, 1962). The importance of chloroplast DNA (cpDNA) to plastid biogenesis and photosynthesis is now well documented (e.g. Dyer, 1984), and the chloroplast genome has been completely sequenced in some species (Ohyama *et al.*, 1986; Shinozaki *et al.*, 1986). The continuity of plastid DNA is therefore central to the survival of green plants. However, in higher plants the segregation of plastid DNA at plastid division has received much less attention than other aspects of plastid DNA biology. The segregation of plastid DNA in *Chlamydomonas* has received more attention as it is amenable to genetic analysis (Sager, 1972; Gillham, 1978), but even in *Chlamydomonas* there is little understanding of the molecular cytology that facilitates segregation.

The segregation of plastid DNA at plastid division in the plastids of higher plants is considered here, focusing particularly on the chloroplast. Chloroplast division in enlarging cells, rather than the division of proplastids in the meristem, is responsible for most of the chloroplasts present in the green leaves of plants (Possingham & Lawrence, 1983). Proplastid division and DNA segregation is, however, very important in the transmission of plastid DNA to successive generations.

The case that the interaction of plastid DNA with membranes facilitates a relatively ordered segregation process will be developed. Furthermore, a role for membranes in the packaging, replication and transcription of the chloroplast nucleoid will be discussed. In this chapter the DNA in all plastid types, not only chloroplasts, will be abbreviated as cpDNA. The interconversion of the various plastid types has been extensively reviewed (see Possingham & Lawrence, 1983, for a list of these reviews).

Segregation of chloroplast DNA at chloroplast division

When a pulse of [3H]–thymidine is administered to chloroplasts undergoing division and DNA replication the radioactivity can be followed in succeeding generations by autoradiography. For chloroplasts of plant leaves this was carried out in cultured spinach discs (Rose, Cran & Possingham, 1974). [3H]–thymidine was pulsed for 24 h, and chased 4 days in unlabelled medium with 2.7 chloroplast division cycles occurring in the chase period. The experiment showed that cpDNA was segregated in fairly equal amounts to daughter chloroplasts and was consistent with semi-conservative replication. Similar data have been obtained in chloroplasts of the chrysophyte unicellular alga *Ochromonas* (Gibbs & Poole, 1973) and also with mitochondria in *Physarum polycephalum* (Kawano & Kuroiwa, 1979). More recently, Lawrence & Possingham (1986) have obtained micro-spectrofluorometric data in meristematic leaf cells of spinach which supports earlier proposals of equal segregation based on autoradiography (Rose, Cran & Possingham, 1974), rather than random segregation (Birky, 1983). Fig. 3 shows [^{3}H]–thymidine-labelled constricting chloroplasts partially synchronised by light treatments and clearly illustrates that both daughter chloroplasts receive DNA (Possingham & Rose, 1976a).

From the above experiments it might be assumed that a round of cpDNA replication occurs in each chloroplast division cycle. This does occur in cultured spinach discs (Possingham & Rose, 1976a; Tymms, Scott & Possingham, 1982) and in meristematic or very young leaf cells (Boffey et al., 1979; Boffey & Leech, 1982; Possingham & Lawrence, 1983; Lawrence & Possingham, 1986). It is, however, now well documented that cpDNA replication is largely confined to young leaf cells and little replication occurs in the subsequent phase of rapid cell enlargement where there is an associated large increase in chloroplast numbers (Lamppa, Elliot & Bendich, 1980; Scott & Possingham, 1980; Boffey & Leech, 1982). At some point in leaf development cpDNA synthesis is uncoupled from chloroplast division.

An important feature to consider in relation to chloroplast division and cpDNA segregation is that each chloroplast contains multiple copies of the chloroplast genome (Lamppa, Elliot & Bendich, 1980; Scott & Possingham, 1980; Boffey & Leech, 1982). In spinach leaves, numbers range from 57 to 353 (Scott & Possingham, 1980).

An idealised diagram that interprets the cpDNA segregation data from the study of Rose, Cran & Possingham (1974) is shown in Fig. 1. In this case there is duplication of DNA in each division cycle (Possingham & Rose, 1976a; Tymms, Scott & Possingham, 1982) and multiple copies of the genome in each plastid (Tymms, Scott & Possingham, 1982). The straight line to which the cpDNAs are attached represents a hypothetical segregation mechanism based on a membrane, from the ideas of Jacob, Brenner & Cuzin (1964). Is such a model tenable in plastids?

The location of cpDNA in chloroplasts
Autoradiographic and morphological studies
    A clear indication that cpDNA could be located in chloroplasts in a way that facilitated an ordered segregation came from the study of Bisalputra & Bisalputra (1969, 1970) with brown algae. Morphological studies with the electron microscope indicated that electron transparent areas (ETA) were organised in a ring around the perimeter of the chloroplast adjacent to the girdle thylakoids (Bisalputra & Bisalputra, 1969). Using the classification of Kuroiwa et al. (1981) who used the fluorochrome

Fig. 1. An idealised diagram that interprets the cpDNA segregation data from the study of Rose *et al.* (1974). A [^{3}H]–thymidine pulse labels the newly replicated DNA strands designated with a dashed line. The labelled strands are followed through in unlabelled medium as they are replicated and segregated. All daughter chloroplasts have similar amounts of label. The DNA molecules are associated with a hypothetical membrane. Further details of this model are available in the text.

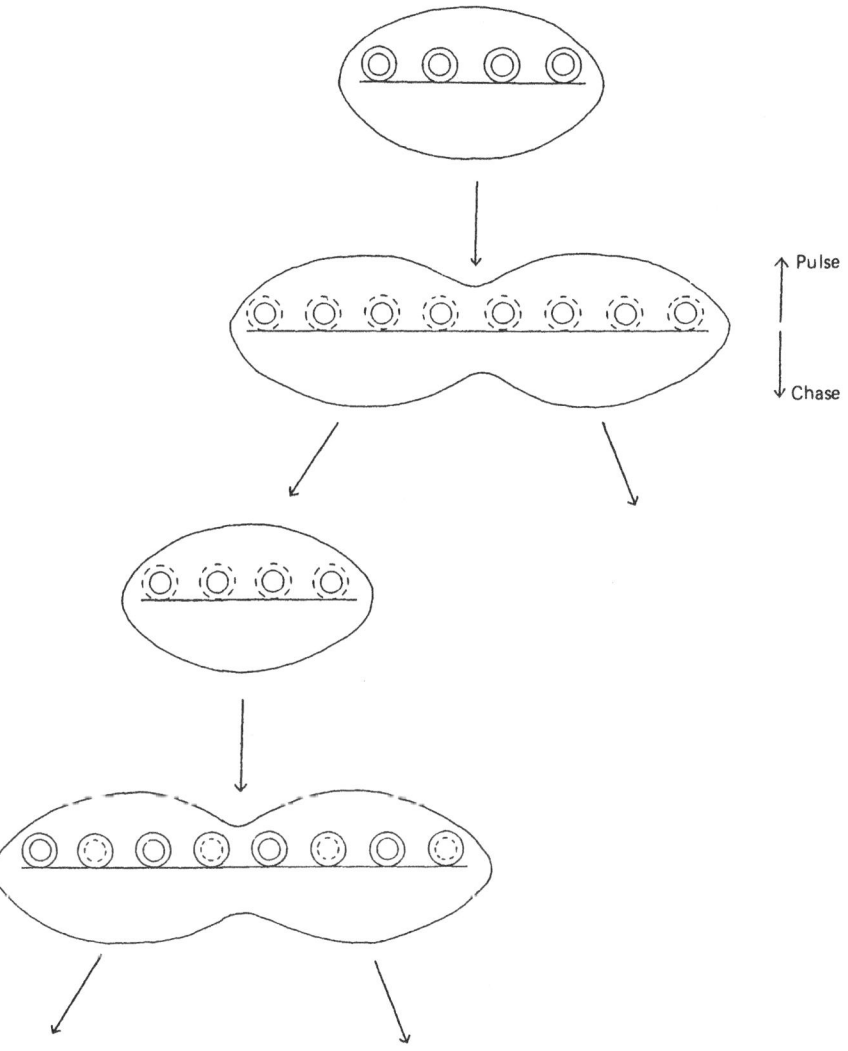

4',6–diamidino–2–phenylindole (DAPI) this is called a CL-type nucleoid distribution. Bisalputra & Bisalputra, (1970) showed in *Sphacelaria* sp. that the ring-like nucleoid segregated into two daughter loops, each of which was transmitted into a daughter chloroplast.

A similar type of system could have accounted for the segregation of cpDNA in the chloroplasts of green plants. However, no electron microscope studies of thin sections showed any comparable pattern of cpDNA location in the grana-containing chloroplasts of higher plants.

Electron microscope autoradiography of dividing chloroplasts of cultured spinach discs labelled with [3H]–thymidine shows that the silver grains are located predominantly over the thylakoids, particularly the grana (Fig. 2). Cultured spinach discs have chloroplasts that do not have a great deal of thylakoid stacking, but have high rates of chloroplast division and cpDNA replication (Possingham & Rose, 1976a; Tymms, Scott & Possingham, 1982). The limited thylakoid stacking allows for good resolution for distinguishing the label distribution amongst the different chloroplast membranes. The data (Rose & Possingham, 1976) are consistent with part of the cpDNA being attached to the membrane and the other part extending into the stroma. Morphological studies support this (Fig. 5).

Fig. 2. Histogram showing the location of silver grains in relation to the different chloroplast components Chloroplasts were labelled with [3H]–thymidine. 332 grains scored (from Rose & Possingham, 1976).

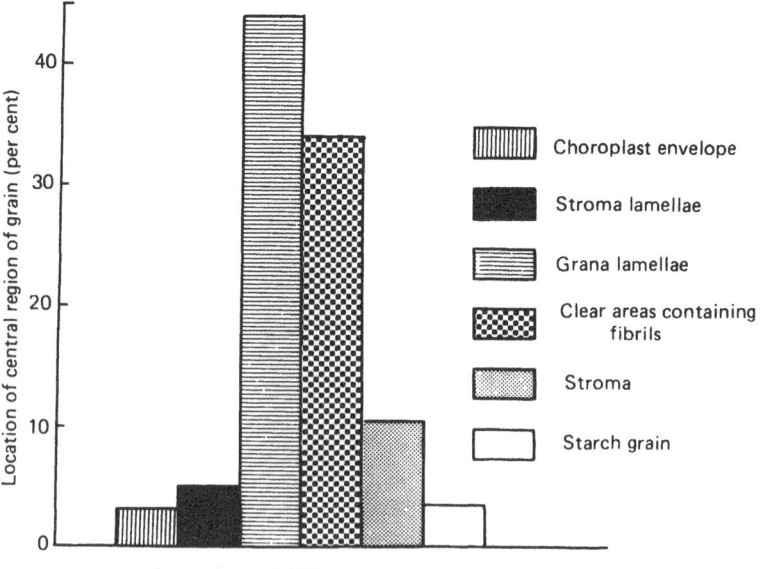

An analysis of 332 grain locations

Segregation could be explained (Fig. 1) if the cpDNA molecules were present along the length of a continuous thylakoid membrane system. The thylakoid system in a chloroplast is a continuum (Paolillo, 1970; Coombs & Greenwood, 1976). Two types of light microscope autoradiograms support the idea that cpDNA occurs along the membrane system. First, rare light autoradiographs show a spiral arrangement of grains (Possingham & Rose, 1976*b*) rather like the spiral arrangement of grana in chloroplasts sectioned in face view. Second, chloroplast vesicles free of outer envelope membranes (Fig. 6) allow the photosynthetic membrane system to be autoradiographed directly (Rose, 1979). Labelled DNA occurs along the membrane systems preferentially associated with the grana (Fig. 7; Rose, 1979).

Serial sectioning of chloroplasts to study the distribution of ETA has been carried out by two laboratories (Kowallik & Herrmann, 1972; Possingham *et al.*, 1983). These studies support the idea of DNA areas occurring scattered throughout the chloroplast. Interactions with the thylakoids occur, but it is not possible from the models presented to see if there is an ETA pattern in relation to the whole membrane system. The serial sectioning data depend on being able to assign ETA accurately in sections. The arrangement of the ETA does indicate considerable overlap of cpDNA molecules to give what amounts to a continuum (Possingham *et al.*, 1983), and is consistent with the autoradiography data discussed above. In considering ETA the question is raised about the packaging of the individual cpDNA molecules and transcriptional activity. This is considered later.

DAPI studies

The ideas formulated on the basis of morphological and autoradiographic studies in the light and electron microscope could be further tested with the advent of the fluorochrome DAPI to visualise cpDNA (Coleman, 1978; James & Jope, 1978).

The initial studies of cpDNA in higher plants with the fluorochrome DAPI supported earlier work, in that nucleoids were spread throughout the chloroplast except for the outer margins (James & Jope, 1978; Kuroiwa & Suzuki, 1980; Scott & Possingham, 1980). The binding of DNA to thylakoids would lead to such a distribution. DAPI studies, however, gave a new dimension to the concept of the nucleoid now that it could be visualised as a discrete highly fluorescent structure, reinforcing this as a structural entity first evident from the serial sections and light microscope autoradiography of Herrmann & Kowallik (Herrmann, 1970; Herrmann & Kowallik, 1970; Kowallik & Herrmann, 1972). The scattered nucleoid arrangement (Fig. 4) has been referred to as the SN-type arrangement by Kuroiwa *et al.* (1981) and is the terminology used here.

A survey of nucleoid arrangement in the chloroplasts of 65 species of higher and lower plants and algae by Kuroiwa and co-workers (1981), showed that 75 per cent had the SN-type nucleoid distribution. However, it soon became apparent that a number of important monocotyledon species had different nucleoid arrangements.

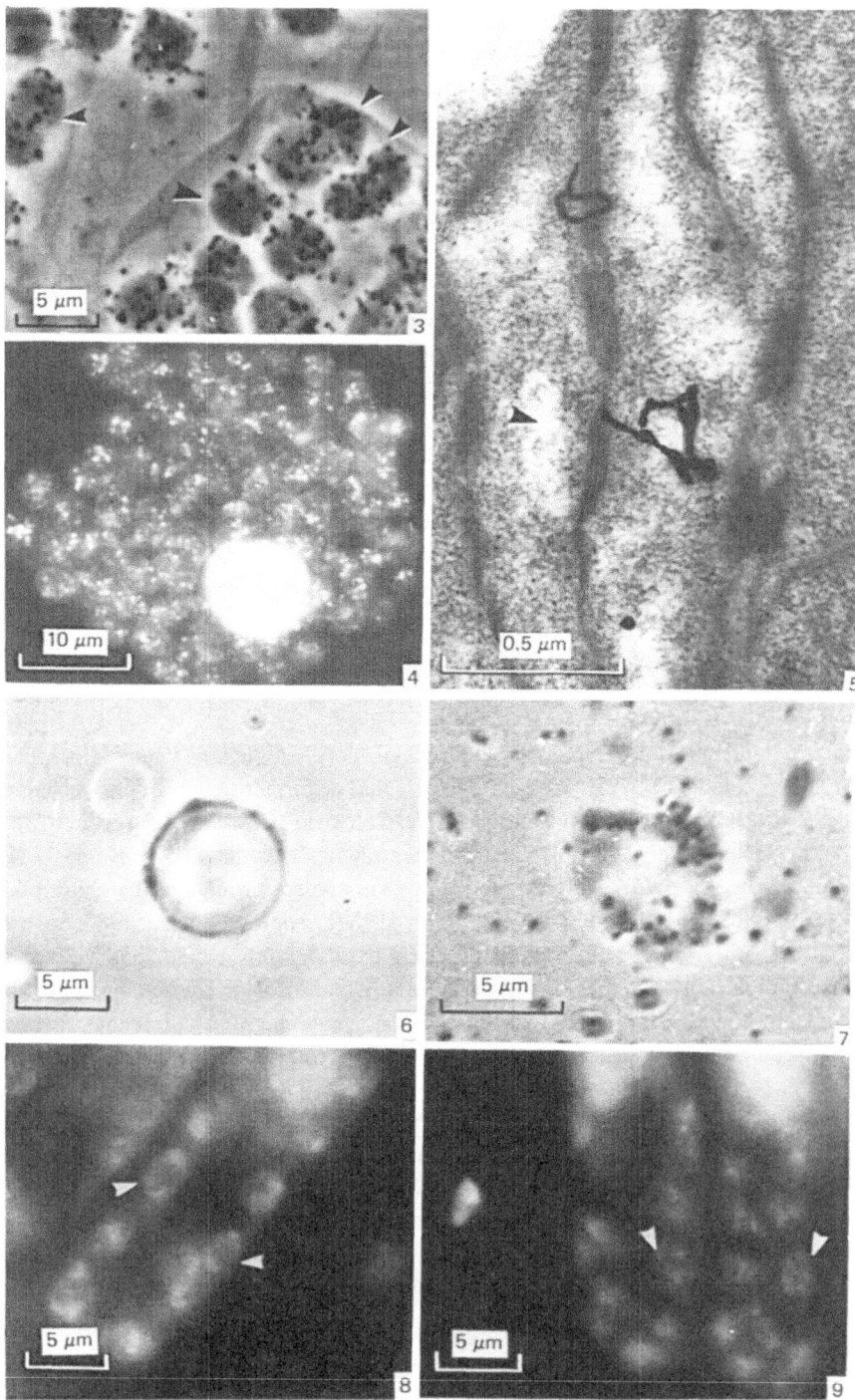

Fig. 3. Light microscope autoradiograph of constricting chloroplasts (small arrowheads) or recently divided chloroplasts (large arrowhead), labelled with [^{3}H]–thymidine. For further details see Possingham & Rose (1976*a*).

Fig. 4. DAPI-stained tobacco mesophyll protoplast (M.R. Thomas & R.J. Rose, unpublished).

Fig. 5. Electron microscope autoradiograph of chloroplast labelled with [^{3}H]–thymidine. Arrowhead indicates ETA containing 3 nm fibrils. For further details see Rose & Possingham (1976).

Fig. 6. Chloroplast vesicle, formed from chloroplasts treated with 3.5 mM Mg^{2+} (from Rose, 1979).

Fig. 7. Chloroplast vesicle labelled with [^{3}H]–thymidine (from Rose, 1979).

Fig. 8. Etiochloroplast formed from an etioplast illuminated for 5 h and stained with DAPI. Arrowheads indicate nucleoids (from Lindbeck *et al.*, 1987).

Fig. 9. Chloroplast formed from etioplast illuminated for 24 h and stained with DAPI. Arrowheads indicate nucleoids (from Lindbeck *et al.*, 1987).

Fig. 10. Chloroplast from mesophyll cells of maize stained with DAPI. Arrowheads indicate nucleoids (A.G.C. Lindbeck & R.J. Rose, unpublished).

Fig. 11. Chloroplast from bundle sheath cells of maize stained with DAPI. Arrowheads indicate nucleoids (A.G.C. Lindbeck & R.J. Rose, unpublished).

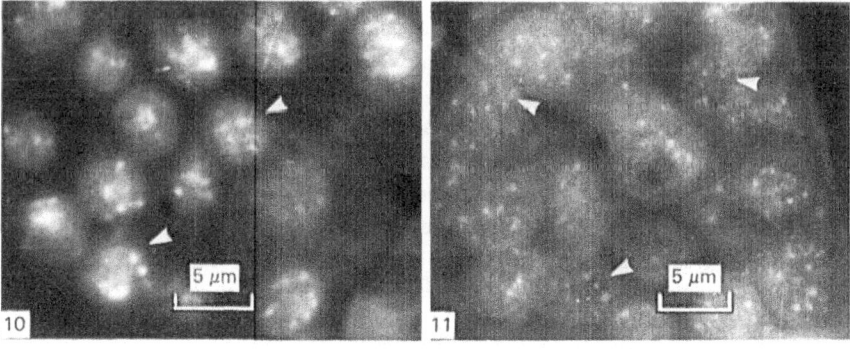

Selldén & Leech (1981) showed that, in wheat chloroplasts, discrete nucleoids were localised at the periphery of the plastid, frequently in a discrete band. Subsequently, a similar distribution was shown in the chloroplasts of barley (Scott, Cain & Possingham, 1982) and *Avena sativa* (Hashimoto, 1985). In the classification of Kuroiwa *et al.* (1981) this distribution has been called a PS-type distribution (i.e. peripheral scattered). It should be noted that this is not a characteristic of all members of the monocotyledon family Poaceae, as maize mesophyll (Fig. 10) and bundle sheath (Fig. 11) chloroplasts have an SN-type nucleoid distribution (Lindbeck, 1986).

Though it has not been shown conclusively, it seems likely that the nucleoids of PS-type chloroplasts are associated with the inner envelope, as they follow the perimeter of the envelope so closely. Both Selldén & Leech (1981) and Hashimoto (1985) have shown photographs of constricting chloroplasts, illustrating how the PS-type arrangement provides a convenient system for segregation.

The location of DNA in other plastids

The location of cpDNA has been studied in plastids other than chloroplasts, notably proplastids and etioplasts.

Proplastids of *Brassica juncea* contained one or two spherical or ovoid nucleoids (Kuroiwa *et al.*, 1981). Similarly in the albino mutant plastids of barley ('albostrians' mutant) which are undifferentiated and contain no plastid ribosomes (Borner, Schumann & Hageman, 1976) there is a single nucleoid (Scott, Cain & Possingham, 1982). A single nucleoid is also present in undifferentiated plastids of albino plantlets from cultured barley tissue (Thomas & Scott, 1985). Electron microscope autoradiography with [^{3}H]–thymidine by Possingham *et al.* (1983) shows that the cpDNA of spinach root proplastids is associated with the envelope and cpDNA also occurs in the central region. A similar arrangement occurs in the mutant albino plastids of *Pelargonium zonale* which are ribosome deficient (Knoth *et al.*, 1974). Presumably this reflects cpDNA that is attached at the membrane and extends into the stroma of the plastid. Herrmann, Kowallik & Bohnert (1974) have obtained electron micrographs which strongly support such an arrangement. Micrographs were obtained from serially sectioned proplastids whose matrix was partially digested with protease. Such a nucleoid arrangement readily provides for a membrane-based segregation mechanism. Electron micrographs of constricting proplastids show ample ETA in daughter proplastids (Chaly & Possingham, 1981).

The distribution of nucleoids as a diffuse sphere or shell of cpDNA surrounding the prolammelar body in etioplasts has been demonstrated by Sprey (1968), Kuroiwa *et al.* (1981), Hashimoto (1985) and Lindbeck *et al.* (1987), using electron microscopy and/or DAPI. It appears likely that this spherical nucleoid is composed of a number of smaller nucleoids. In the case of *Phaseolus vulgaris*, whose chloroplasts have an SN-nucleoid distribution, it appears that the cpDNA is attached to the prolamellar body and any associated prothylakoids, though ETA do impinge on the envelope in some sections. In the case of barley and *Avena sativa*, where chloroplasts have a PS-nucleoid distribution, there is clearly a lot of ETA associated with the envelope, as well as the prolamellar body. Sprey (1968) clearly showed that the ETA divided into two to be distributed to the two daughter etioplasts.

In chromoplasts of *Narcissus pseudonarcissus* (Hansmann *et al.*, 1985), 2–4 nucleoids are situated in close proximity to the envelope giving a PS-type distribution.

The membrane binding of plastid cpDNA: a central role in location and segregation

As has been discussed there is a strong cytological basis for the attachment of cpDNA to plastid membranes. This evidence, together with some additional evidence can be enumerated as follows.

(i) In ultrathin sections visualised in the electron microscope there is direct contact between DNA fibres and membranes. This has been perhaps most clearly shown in protease treated sections (Herrmann & Kowallik, 1970; Kowallik & Herrmann, 1972). Excellent examples also come from algae showing ring-shaped nucleoids associated with the girdle thylakoids (Bisalputra & Bisalputra, 1969; Gibbs, Cheng & Slankis, 1974).

(ii) Electron microscope autoradiography with [³H]–thymidine and thin sections have shown almost 50 per cent of the silver grains lie over the thylakoids, which only account for 20 per cent of the area of the chloroplast section (Rose & Possingham, 1976). Similar studies with spinach proplastids (Possingham *et al.*, 1983) show much of the cpDNA of proplastids associated with the envelope.

(iii) Photosynthetic membrane vesicles from spinach chloroplasts autoradiographed for light microscopy after [³H]–thymidine incorporation show preferential labelling of grana (Rose, 1979), which is consistent with a specific association.

(iv) Separation of [³H]–thymidine labelled spinach chloroplasts into thylakoid, envelope and supernatant fractions by the method of Douce, Holtz & Benson (1973) showed only 30 per cent of the label associated with soluble components (Rose & Possingham, 1976), and 61 per cent associated with the thylakoids. Photosynthetic membrane vesicles referred to in (iii) above retain 45 per cent of the label.

(v) Perhaps the most common argument in the literature for a membrane association of cpDNA comes from using cpDNA obtained from plastids by various disruption and isolation techniques followed by spreading (e.g. Woodcock & Fernandéz–Moran, 1968; Odinstova, Mikulska & Turischeva, 1970; Sprey & Gietz, 1972; Herrmann *et al.*, 1974; Knoth *et al.*, 1974;

Yoshida *et al.*, 1978; Hansmann *et al.*, 1985). Flower-like structures are usually formed with the cpDNA folded around a central proteinaceous body.

Flower structures can be induced in DNA preparations (Davis, Simon & Davidson, 1971) but the structures generated by the above authors have substantive and clearly defined cores which appear to have a structural role. Herrmann, Kowallik & Bohnert (1974) have addressed the question of artifacts by mixing mitochondrial DNA (mtDNA) with chloroplasts prior to lysis. Only 2 of 50 mtDNA molecules were found to be connected with proteinaceous material, but they did not form bouquets. Furthermore, the right degree of lysis is required to generate these folded structures; more highly purified cpDNA yields predominantly circular molecules. More recently Hansmann *et al.* (1985) have demonstrated a controlled formation of folded cpDNA from isolated nucleoids by using proteinase K of controlled ionic strength.

(vi) Using techniques developed by Miller and co-workers (1970) to visualise transcription, DNase sensitive transcriptional complexes have been shown attached to spinach chloroplast vesicles (Fig. 12; Rose & Lindbeck, 1982).

Fig. 12. A transcriptional complex released from a vesicle of the type shown in Fig. 6. The complex is still associated with membrane material (m). The predominant particle is about 24.5 nm (arrowheads) (from Lindbeck & Rose, 1987).

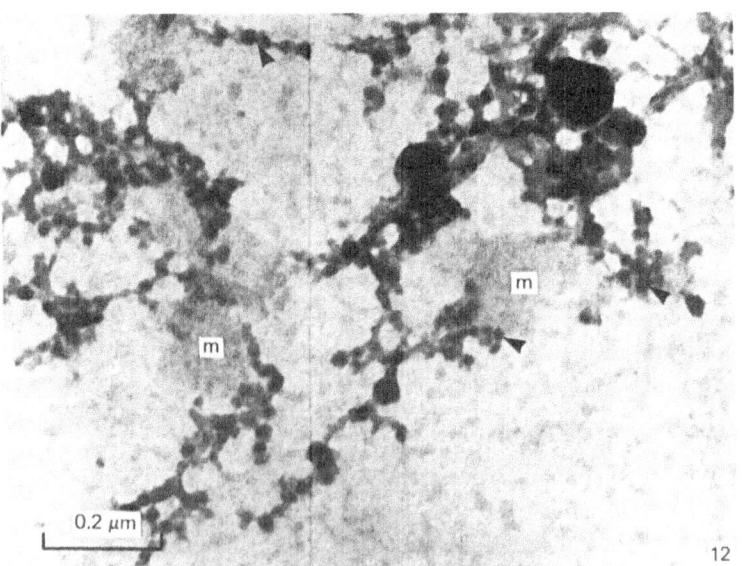

(vii) Nucleoid arrangements shown by the DAPI technique can be readily interpreted in terms of membrane attachment: SN-type chloroplasts show thylakoid attachment, PS-type chloroplasts show envelope attachment and CL-type chloroplasts show attachment to girdle thylakoids. As previously discussed, proplastid and etioplast nucleoids have a membrane association.

Perhaps the most important point that can be made about the evidence on cpDNA membrane binding is that we are not dealing with cpDNA simply adhering to membranous material. For a given plastid type in a particular species binding is to a specific membrane. Fig. 13 shows how the attachment of cpDNA nucleoids to membranes (or membrane precursors in the case of etioplasts) could lead to the segregation of cpDNA at plastid division, in line with Fig. 1.

Separation of nucleoids by membrane growth

A membrane-based segregation mechanism requires that membrane growth separates the newly replicated cpDNA molecules (Jacob, Brenner & Cuzin, 1964).

Fig. 13. Idealised diagram showing nucleoid segregation in dividing chloroplasts, etioplasts and proplastids; based on information in the text. A, SN-type chloroplast; B, etioplast from plant with SN-type chloroplasts; C, PS-type chloroplast; D, etioplast from plant with PS-type chloroplasts; E, proplastid from plant with SN-type or PS-type chloroplasts. The plastid envelope is shown as a single line. N, nucleoid; P, prolamellar body; G, granum.

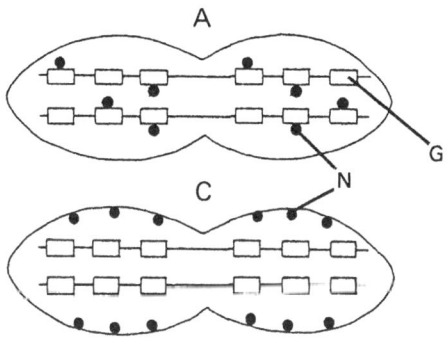
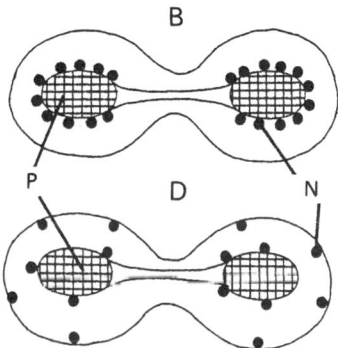
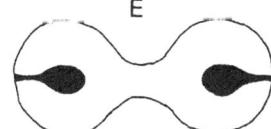

This has been well illustrated by a timecourse series of electron microscope pictures showing nucleoid separation in the enlarging bacterial cell (Ryter, 1968). Such a series is not feasible for the more complex system of higher plant plastids. However, in greening etioplasts there is an opportunity to study nucleoid location as the membrane system of the chloroplast develops.

In *Phaseolus vulgaris* etioplasts there is initially a diffuse sphere of cpDNA surrounding most of the prolamellar body and it appears to be made up of a number of smaller nucleoids (Lindbeck *et al.*, 1987). The distribution of these nucleoids which form a complete ring of fluorescence shortly after illumination (Fig. 8) has been followed by DAPI and electron microscopy as chloroplast development occurs during continuous illumination (Lindbeck *et al.*, 1987). The nucleoids become located between the developing prothylakoids (Fig. 14) and, as the prolamellar body disappears and the thylakoids develop, the nucleoids locate along the thylakoids giving an SN-type distribution (Fig. 9) with evidence of thylakoid attachment. These changes in nucleoid location occur without changes in DNA levels per plastid, and without plastid replication. It seems clear that the growing membranes have separated and arranged the existing nucleoids throughout the chloroplast.

Hashimoto (1985) has studied etioplasts in barley from different regions of the leaf. Barley has a PS-type nucleoid arrangement in the chloroplast. Etioplasts with well developed prolamellar bodies appear to have peripherally located nucleoids (PS-

Fig. 14. Electron micrograph of etiochloroplast formed from etioplast after 5 h illumination. ETA, electron transparent area (from Lindbeck *et al.*, 1987).

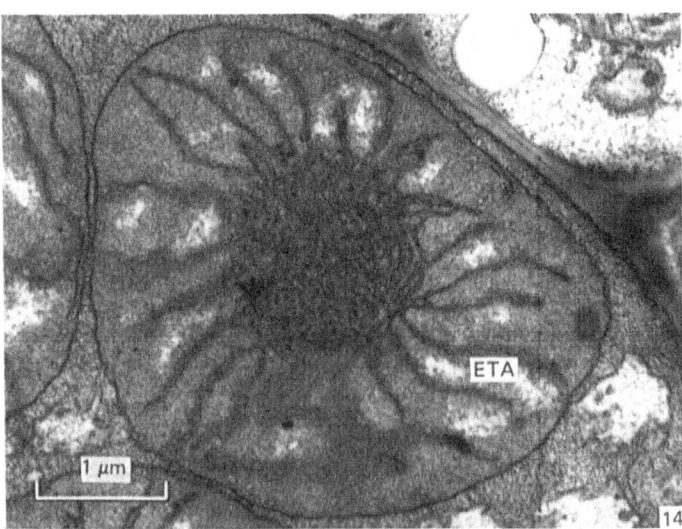

type), but there is certainly ETA and prolamellar body associations. Etioplasts with more prothylakoids have a more scattered nucleoid distribution. Hashimoto (1985) and Lindbeck *et al.* (1987) argue for a spatial nucleoid arrangement due to membrane biogenesis, because of nucleoid-membrane associations. (See also Kowallik & Herrmann, 1974; Kuroiwa *et al.*, 1981.)

Tobacco mesophyll protoplasts when stimulated to divide dedifferentiate and the chloroplasts convert to proplastids (Thomas & Rose, 1983). The loss of the thylakoid system is occurring at the time of the earliest plastid divisions, and it appears that the segregation of nucleoids is asymmetric in this situation where the SN-type chloroplasts (Fig. 4) progressively lose their thylakoid system (Thomas & Rose, 1983). This provides some support for the idea that an ordered thylakoid system facilitates ordered segregation in dividing chloroplasts. Asymmetric distribution of nucleoids occurs in *Acetabularia* and leads to asymmetric segregation with some daughter plastids even devoid of DNA (Woodcock & Bogorad, 1970; Luttke & Bonotto, 1981). More detailed plastid developmental studies may help to resolve this unusual situation.

Why is there variable nucleoid location in plastids?

Nucleoid location in chloroplasts appears to have a phylogenetic basis, e.g. CL-type, SN-type, PS-type, although each arrangement is related to binding to specific membranes. The PS-types so far reported are monocotyledons, though not all monocotyledons have a PS-arrangement (Figs. 10 & 11; Lindbeck, 1986). Plants having SN-type chloroplasts with thylakoid-bound nucleoids may have envelope bound nucleoids in proplastids (Herrmann, Kowallik & Bohnert, 1974; Knoth *et al.*, 1974; Possingham *et al.*, 1983) and a PS-type arrangement in chromoplasts (Hansmann *et al.*, 1985). In the case of *Avena sativa* (PS-type chloroplasts) internally scattered nucleoids have been observed in etioplasts with well developed prothylakoids and in fully developed chloroplasts with highly stacked grana (Hashimoto, 1985).

It is possible that many plants can bind cpDNA at envelope or thylakoid sites and in either case this can serve to segregate cpDNA at chloroplast division, but it is a dynamic rather than static situation depending on plant type and plastid development stage. However, it is important to note that a plastid of a particular species at a particular stage of development will have a characteristic nucleoid arrangement. Clearly, a lot more needs to be known about the mechanism of cpDNA–membrane binding.

The mechanism of cpDNA-membrane binding

Until the molecular basis of the cpDNA–membrane attachment is understood there will remain some doubts about its proposed biological significance. It is encouraging that similar types of structural associations between the cell membrane of bacteria and DNA (Ryter, 1968; Kusano *et al.*, 1984) and the nuclear scaffold of eukaryotes and DNA (Valenzuela, Mueller & Dasgupta, 1983; Zehnbauer & Vogelstein, 1985) is receiving increasing attention. There is no reason why organelle studies should not make useful contributions to this area, and it seems likely that fruitful insights into DNA biology will emerge (Gasser & Laemmli, 1986*a,b*).

There are indications that membrane proteins are involved in cpDNA membrane-binding. Hansmann *et al.* (1985) found that isolated nucleoids from daffodil had several proteins in common with integral envelope membrane proteins and suggested their involvement in a DNA–protein–membrane complex. Thylakoid proteins involved in DNA binding may have escaped detection, it was suggested, due to the large quantities of thylakoid proteins involved in photosynthesis. Daffodil chloroplasts have an SN-type nucleoid arrangement. Interestingly, it was argued, based on literature comparisons of inner and outer envelope proteins, that the DNA-binding proteins were mainly associated with the outer membrane of the envelope.

Working with spinach thylakoid membranes, Lindbeck & Rose (1987) have found [^{3}H]–thymidine labelled DNA fragments bound to thylakoid proteins. Heating the complex releases the cpDNA in an analogous way to the release of chlorophyll when chlorophyll–protein complexes are heated, suggesting non-covalent interactions are involved.

Other studies also suggest membrane proteins are involved in cpDNA membrane-binding. Incomplete lysis of chloroplasts yields folded chromosomes attached to a core. These latter structures are converted to circular DNA with phenol treatment (Herrmann *et al.*, 1974). Similar conclusions have been drawn by Yoshida *et al.* (1978), though drastic proteinase treatments were required.

In the plastids of higher plants both envelope membranes and thylakoid membranes can be involved with the association, yet at any one time depending on species and development stage there is membrane specificity. Plastid DNA-binding proteins may exist in both envelope and thylakoid membranes but compartmentalisation in the plastid could restrict access to the DNA. Alternatively there may be more complex control of protein location in relation to plastid development.

Given the likelihood that membrane proteins that bind DNA exist it is not clear whether such proteins would interact directly with cpDNA or with protein(s) associated with cpDNA. This is an area that has received little attention. In bacteria structural DNA-binding proteins known as histone-like proteins (HLPs) are present (Varshavsky *et al.*, 1977; Pettijohn, 1982) and may have a similar role to eukaryote histones (Rouvière–Yaniv & Gros, 1975). Histone-like proteins have been found in mitochondria (Caron, Jacq & Rouvière–Yaniv, 1979). Basic proteins with a molecular weight of less than 23 kDa have been extracted from a transcriptionally active DNA-protein complex from spinach chloroplasts (Briat *et al.*, 1982). Similar gel patterns have been obtained from nucleoid proteins of daffodils (Hansmann *et al.*, 1985), but it is not known whether these proteins can be considered histone-like.

The other aspect of DNA-membrane binding is the DNA itself. In an attempt to examine if specific regions of the cpDNA were involved in interacting with spinach thylakoids the thylakoid-bound DNA associated with chloroplast vesicles (Fig. 7) was compared with total chloroplast DNA using restriction endonucleases (Lindbeck & Rose, 1987). All cpDNA sequences are represented in the vesicle-bound cpDNA, as cpDNA from vesicles and chloroplasts have similar fragmentation patterns (Fig. 15). The simplest explanation for these data is that random DNA sequences are responsible for the DNA–membrane interactions. However, many specific interaction sites scattered around the chromosome would give the same restriction data. Furthermore, membrane–DNA interactions may be dynamic or static. However, it is clear that a single binding site with DNA extending into the stroma is inconsistent with the data. A folded chromosome attached to the membrane at a number of sites might explain these data and this is the model that fits morphological observations (Yoshida *et al.*, 1978; Briat *et al.*, 1982; Hansmann *et al.*, 1985).

It is important to note that in bacteria where similar restriction data have been obtained (Drlica, Burgi & Worcel, 1978) special regions of the DNA have been found associated with the cell membrane: the replication fork (Valenzuela & Inman, 1981) and the origin and terminus (Kusano *et al.*, 1984; Sargent & Bennett, 1985). We have recently obtained evidence for an association of replication forks with spinach thylakoids (Lindbeck & Rose, 1986). It still seems feasible that certain special regions of the chromosome are involved in membrane interaction. In mitochondria, Kawano & Kuroiwa (1985) have obtained a membrane–mtDNA complex which is AT rich.

Other roles for membrane bound cpDNA

There is a good case for the membrane binding of cpDNA being a key factor in the segregation of cpDNA in the chloroplast division cycle. However, it is likely that there are other roles of the cpDNA-membrane association that are of significance for the *in vivo* functioning of cpDNA. These other putative roles are discussed below.

DNA replication

A role for the bacterial membrane in DNA replication linked to the segregation mechanism was postulated by Jacob, Brenner & Cuzin (1964). The bacterial evidence for a membrane role in the DNA replicative cycle centres on key chromosome regions involved in replication being membrane-associated. These regions are the origin and terminus (Kusano *et al.*, 1984; Sargent & Bennett, 1985) and the replication fork (Valenzuela & Inman, 1981). Such an association is not important for *in vitro* replication systems, and may in part explain the relative scarcity of understanding in this area (Kornberg, 1984). Presumably, the membrane

Fig. 15. Restriction endonuclease fragmentation patterns of cpDNA from chloroplast vesicles (A) and intact chloroplasts (B) digested with *Eco*R1 and separated by electrophoresis on a 0.85 per cent agarose gel. Arrowheads correspond to size marker fragments produced by lambda DNA with *Hind* III (23,1, 9.4, 6.7, 4.4, 2.3, 2.0 and 0.5 kb) (from Lindbeck & Rose, 1987).

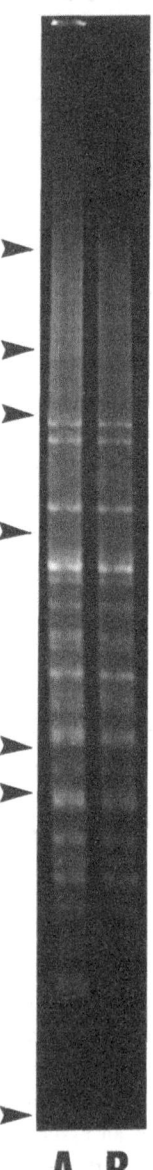

associations are important *in vivo* to provide orderly replication in concert with the growth and division of the cell (Sargent & Bennett, 1985).

Recently, using a similar approach to that of Valenzuela & Inman (1981) membrane bound cpDNA was examined for replication forks (Lindbeck, 1986; Lindbeck & Rose, 1986). Chloroplast vesicles (Fig. 6) were isolated, digested with *Eco* RI to remove DNA not tightly associated with the membrane, the DNA isolated and then examined for replication forks after standard spreading techniques. When the frequency of the replication forks from this tightly bound fraction was compared with the frequency of replication forks in total DNA digested with *Eco* RI, the membrane-bound fraction was enriched for replication forks. Furthermore, the branched fragments from the membrane-bound cpDNA frequently contained branches of unequal length, probably caused by the blockage of endonuclease cleavage sites, supporting the presence of membrane sites interacting with the replication system.

Support for a connection between membrane-binding and cpDNA replication also comes from the work of Herrmann, Kowallik & Bohnert (1974). The membrane associated complexes they isolated, which appear as several folded chromosomes attached to a proteinaceous core, were capable of converting deoxyribonucleoside triphosphates into acid-insoluble, DNase sensitive products. The large replication complex or replisome (Kornberg, 1984) consisting of polymerase, primases and helicases would seem a good candidate for membrane-association.

A linkage between cpDNA replication and segregation makes a great deal of functional sense, helping to regulate the division cycle. In a division cycle there has to be appropriate timing for replication, segregation and organelle growth. Due to multiple genome copies in plastids, the coupling between these processes can be more flexible than in bacteria.

Future work to investigate the membrane-bound cpDNA will require more detailed understanding of cpDNA replication and probes based on such regions of the genome as the origin of replication.

Kolodner & Tewari (1975) have studied cpDNA replication in pea and corn. They conclude that cpDNA replication is initiated by the formation of two displacement loops which expand towards each other to form a Cairns replicative forked structure. A Cairns round of replication may be used to initiate a rolling circle round of replication. It was suggested by Kolodner & Tewari (1975) that the rolling circle mechanism might be used for the rapid synthesis of cpDNA while the Cairns mechanism might be used for normal duplication of cpDNA. This is an interesting comment in view of more recent understanding of the pattern of cpDNA synthesis in leaf development (Lamppa, Elliott & Bendich, 1980; Scott & Possingham, 1980; Boffey & Leech, 1982) where cpDNA synthesis in meristematic tissue keeps pace with chloroplast division but cpDNA is amplified in the early phase of cell enlargement (Lawrence & Possingham, 1986).

The sites of the replication origins of corn and pea are not known. Autonomous replication sequences (ars) capable of acting as origins of replication for chimaeric plasmids in yeast are known for tobacco (Ohtani *et al.*, 1984) and *Petunia* (Overbeeke *et al.*, 1984); but their *in vivo* role in cpDNA replication is not as yet clear.

Transcription

Studies based on morphology, nuclease digestion, labelling and electrophoresis of thylakoid-bound cpDNA from chloroplast vesicles have shown that there is a coupling of transcription and translation (Fig. 12; Rose & Lindbeck, 1982; Lindbeck & Rose, 1987). This is consistent with the prokaryotic nature of this DNA as both bacterial and mitochondrial DNA show coupling of transcription and translation (Miller, Hamkalo & Thomas, 1970; Laird *et al.*, 1973). However, it does indicate that the membrane-bound cpDNA is transcriptionally active. Similar conclusions have been reached by Filippovich *et al.*, (1975) using grana obtained from sucrose gradient fractionation of detergent-treated chloroplasts.

It has been reported in a number of studies that membrane-bound chloroplast polyribosomes are bound to the chloroplast thylakoids by nascent proteins (Chua *et al.*, 1973; Margulies & Michaels, 1975; Yamamoto *et al.*, 1981) and play a significant role in membrane biogenesis (e.g. Herrin & Michaels, 1985). In chloroplasts containing thylakoid-bound DNA as in spinach chloroplasts (SN-type) a proportion of the ribosomes bound to the thylakoids would be associated with nascent transcripts.

DNA packaging

Like other DNAs cpDNA has to be appropriately packaged. Chloroplast DNA varies in size from 120–150 kb (40–50 μm) depending on species (Dyer, 1984). Nucleoids in SN-type chloroplasts measured by Kuroiwa *et al.* (1981) had a diameter of 0.2 to 0.4 μm. It is clear from observations of published micrographs that there is substantive variation in nucleoid size. This would most likely be due to variation in the numbers of genomes per nucleoid (Herrmann, Kowallik & Bohnert, 1974).

Different conclusions in the literature on the number of cpDNA copies per nucleoid may simply reflect the potential for variation, although different measurement techniques have been used. Kuroiwa *et al.* (1981) suggested one cpDNA molecule per nucleoid based on visual comparisons of the fluorescence of DAPI-stained T4 phage particles. Scott & Possingham (1980) using reassociation kinetics showed that 10–15 plastid nucleoids in mature spinach chloroplasts contained 2–3 copies of cpDNA per nucleoid. Using diphenylamine, Boffey & Leech (1982) found mature wheat chloroplasts contained 11–12 nucleoids with 24 genome copies per nucleoid. Much earlier studies by Herrmann, Kowallik & Bohnert (1974) concluded that there were 4–8 cpDNA genomes per nucleoid, with nucleoid numbers determined by serial sectioning. Isolated cpDNA complexes showed multiple cpDNA copies bound to a protein core.

Some evidence has been obtained to support the idea that the ease with which segregation of newly replicated cpDNAs occurs influences the size of nucleoids and their genome copies. In maize both bundle sheath chloroplasts and mesophyll chloroplasts have an SN-type nucleoid distribution. Mesophyll chloroplasts have about 12 nucleoids with a mean size of 0.38 μm^2 (Fig. 10) while bundle sheath chloroplasts have about 27 nucleoides with a mean size of 0.2 μm^2 (Fig. 11), yet both chloroplasts have similar amounts of DNA (Lindbeck, 1986). Bundle sheath cells do not have grana and it is feasible that the thylakoid-bound cpDNA is segregated more readily leading to fewer genome copies per nucleoid. The presence of grana, where some of the cpDNA is bound (Lindbeck, 1986), probably makes segregation more complex, though it clearly can occur as judged by the nucleoid distribution.

Hansmann & co-workers (1985) have been able to isolate nucleoids from daffodil and unfold the DNA by a proteinase or high salt treatment, clearly showing how cpDNA folding is mediated by proteins. Because some of these proteins are common to membranes, Hansmann *et al.* (1985) have suggested that they assist binding of nucleoids to the membrane. These studies provide support for the presence of structures observed by cytochrome *c* spreading which have a central protein-containing region and loops of naked DNA (Herrmann, Kowallik & Bohnert, 1974; Yoshida *et al.*, 1978). As outlined earlier, Herrmann, Kowallik & Bohnert (1974) have argued that these are membrane-bound structures with a core of membrane origin.

In considering the chloroplast information it is interesting to note the current focus on the proteins of the nuclear skeleton (Gasser & Laemmli, 1986*a,b*). Not only is it apparent that these latter interactions provide a structural basis for the organisation of the nucleus, but by appropriate chromatin unfolding are important in providing transcription sites (Zehnbauer & Vogelstein, 1985). These nuclear skeleton sites are also important in replication (Valenzuela *et al.*, 1983; Zehnbauer & Vogelstein, 1985).

The plastid nucleoid

A large amount of morphological data and more limited molecular data supports the notion that membrane-binding of plastid DNA is central to the overall organisation of the plastid nucleoid.

The nucleoid as visualised in the light microscope by DAPI is of variable size and shape, but is frequently spherical (Figs. 4, 9, 10 & 11); and Kuroiwa *et al.*, 1981). It consists of perhaps several genome copies (Possingham & Lawrence, 1983). These copies are linked to a membrane–protein complex (Herrmann, Kowallik & Bohnert, 1974; Hansmann *et al.*, 1985). Specific membrane proteins (Hansmann *et al.*, 1985) and a number of regions of the genome are involved (Lindbeck & Rose, 1987). Whether specific DNA sites are involved is an open question.

Clearly the naked DNA loops visualised in the studies of Herrmann, Kowallik & Bohnert, (1974) and Yoshida *et al.* (1978) have to be packaged in the nucleoid, membrane attachment alone being insufficient. It seems likely that basic DNA-binding proteins described by Briat *et al.* (1982) and Hansmann *et al.* (1985) are involved, together with supercoiling. Higher plant cpDNA is negatively supercoiled (Stirdivant, Crossland & Bogorad, 1985) and the basic proteins may play a role in maintaining the supercoiling (Briat *et al.*, 1982). It is interesting that an irregularly beaded substructure has been isolated from chloroplast nucleoids of algae (Chiang *et al.*, 1981) and spinach (Briat *et al.*, 1982).

The amounts and composition of nucleoid protein are influenced by the amount of transcriptional activity. The plastid genome is inactive in daffodil chromoplasts and has more protein and different types of protein than daffodil chloroplasts (Hansmann *et al.*, 1985). Reiss & Link (1985) have studied transcriptionally active genomes (Rushlow & Hallick, 1982) from mustard etioplasts and chloroplasts. The transcriptionally active genomes represent part or all of the nucleoid and have been visualised by Briat *et al.* (1982). The transcriptionally active genomes of etioplasts and chloroplasts showed qualitative and quantitative differences in proteins, but contained a group of comparable proteins in similar amounts that may be involved in establishing the higher-order structural organisation and membrane attachment (Reiss & Link, 1985).

In considering what is known of nucleoids, one question that arises is how the organisation discussed above relates to the ETA seen in thin sections. Is the ETA part or all of the nucleoid? The ETA lack ribosomes and because of the coupling of transcription and translation (Rose & Lindbeck, 1982; Lindbeck & Rose, 1987) these areas may be transcriptionally inactive or the sites of rRNA synthesis. How a nucleoid operates *in vivo* is not yet known.

Acknowledgements
I wish to thank Dr John Possingham and Dr Graeme Lindbeck for helpful discussions on this topic.

References
Birky, C.W. Jr. (1983). The partitioning of cytoplasmic organelles at cell division. In *Aspects of Cell Regulation, Supplement 15, International Review of Cytology*, ed. J.F. Danielli, pp. 49–89. New York: Academic Press.

Bisalputra, T. & Bisalputra, A.A. (1969). The ultrastructure of chloroplast of a brown alga. *Sphacelaria* sp.1. Plastid DNA configuration – the chloroplast genophore. *Journal of Ultrastructure Research*, **29**, 151–70.

Bisalputra, T. & Bisalputra, A.A. (1970). The ultrastructure of chloroplast of a brown alga. *Sphacelaria* sp.111. *Journal of Ultrastructure Research*, **32**, 417–29.

Boffey, S.A., Ellis, J.R., Selldén, G. & Leech, R.M. (1979). Chloroplast division and DNA synthesis in light-grown wheat leaves. *Plant Physiology*, 64, 502–5.

Boffey, S.A. & Leech, R.M. (1982). Chloroplast DNA levels and the control of chloroplast division in light-grown wheat leaves. *Plant Physiology*, 69, 1387–91.

Borner, T., Schumann, B. & Hageman, R. (1976). Biochemical studies on a plastid ribosome-deficient mutant of *Hordeum vulgare*. In *Genetics and Biogenesis of Chloroplasts and Mitochondria*, eds Th. Bucher, W. Neupert, W. Sebald and S. Werner, pp. 41–8. Amsterdam: Elsevier/North Holland Biomedical Press.

Briat, J.F., Gigot, C., Laulhère, J.P. & Mache, R. (1982). Visualization of a spinach plastid transcriptionally active DNA–protein complex in a highly condensed structure. *Plant Physiology*, 69, 1205–11.

Caron, F., Jacq, C. & Rouvière-Yaniv, J. (1979). Characterization of a histone-like protein extracted from yeast mitochondria. *Proceedings of the National Academy of Sciences, USA*, 76, 4265–9.

Chaly, N. & Possingham, J.V. (1981). Structure of constricted proplastids in meristematic plant tissues. *Biology of the Cell*, 41, 203–10.

Chiang, K.S., Friedman, E., Malavasic, M.J., Feng, M.L., Eves, E.M., Feng, T–Y. & Swinton, D.C. (1981). On the folding and organization of chloroplast DNA in *Chlamydomonas reinhardtii*. *Annals of the New York Academy of Sciences*, 361, 219–47.

Chua, N–H., Blobel, G., Siekevitz, P. & Palade, G.E. (1973). Attachment of chloroplast polysomes to thylakoid membranes in *Chlamydomonas reinhardtii*. *Proceedings of the National Academy of Sciences, USA*, 70, 1554–8.

Coleman, A.W. (1978). Visualization of chloroplast DNA with two fluorochromes. *Experimental Cell Research*, 114, 95–100.

Coombs, J. & Greenwood, A.D. (1976). Compartmentation of the photosynthetic apparatus. In *The Intact Chloroplast*, ed A. Barber. pp. 1–51. Amsterdam: Elsevier/North Holland.

Davis, R.W., Simon, M. & Davidson, N. (1971). Electron microscope heteroduplex methods for mapping regions of base sequence homology in nucleic acids. In *Methods in Enzymology, Vol 21, Nucleic Acids, Part D*. eds L. Grossman & K. Moldave. pp. 413–28. New York: Academic Press.

Douce, R., Holtz, R.B. & Benson, A.A. (1973). Isolation and properties of the envelope of spinach chloroplasts. *The Journal of Biological Chemistry*, 248, 7215–22.

Drlica, K., Burgi, E. & Worcel, A. (1978). Association of the folded chromosome with the cell envelope of *Escherichia coli*: Nature of the membrane-associated DNA. *Journal of Bacteriology*, 134, 1108–6.

Dyer, T.A. (1984). The chloroplast genome: Its nature and role in development. In *Topics in Photosynthesis – Vol. 5. Chloroplast Biogenesis*. eds. N.R. Baker & J. Barber. pp. 24–69. Amsterdam: Elsevier.

Filippovich, I.I., Alina, B.A. Bezsmertnaya, I.N., Tongur, A.M. & Oparin, A.I. (1975). Relationship between the protein-synthesizing system and the structure of chloroplasts. In *Genetic Aspects of Photosynthesis*. eds. Y.S. Nasyrov & Z. Sestak. pp. 105–13. The Hague: Dr. W. Junk b.v.

Gasser, S.M. & Laemmli, U.K. (1986a). The organization of chromatin loops: characterization of a scaffold attachment site. *The EMBO Journal*, 5, 511–18.

Gasser, S.M. & Laemmli, U.K. (1986b). Cohabitation of scaffold binding regions with upstream/enhancer elements of three developmentally regulated genes of *D. melanogaster*. *Cell*, 46, 521–30.

Gibbs, S.P., Cheng, D. & Slankis, T. (1974). The Chloroplast Nucleoid in *Ochromonas danica* I. Three-dimensional morphology in light- and dark-grown cells. *Journal of Cell Science*, 16, 557–77.

Gibbs, S.P. & Poole, R.J. (1973). Autoradiographic evidence for many segregating DNA molecules in the chloroplast of *Ochromonas danica*. *Journal of Cell Biology*, 59, 318–28.

Gillham, N.W. (1978). *Organelle Heredity*. New York: Raven Press.

Hansmann, P., Falk, H., Ronai, K. & Sitte, P. (1985). Structure, composition, and distribution of plastid nucleoids in *Narcissus pseudonarcissus*. *Planta*, 164, 459–72.

Hashimoto, H. (1985). Changes in distribution of nucleoids in developing and dividing chloroplasts and etioplasts of *Avena sativa*. *Protoplasma*, 127, 119–27.

Herrin, D. & Michaels, A. (1985). In vitro synthesis and assembly of the peripheral subunits of coupling factor CF_1 (α and β) by thylakoid-bound ribosomes. *Arch. Biochem. Biophys.*, 237, 224–36.

Herrmann, R.G. (1970). Multiple amounts of DNA related to the size of chloroplasts. I. An autoradiographic study. *Planta*, 90, 80–96.

Herrmann, R.G. & Kowallik, K.V. (1970). Multiple amounts of DNA related to the size of chloroplasts. II. Comparison of electron-microscope and autoradiographic data. *Protoplasma*, 69, 365–72.

Herrmann, R.G., Kowallik, K.V. & Bohnert, H.J. (1974). Structural and functional aspects of the plastome. 1 – The organisation of the plastome. *Portugaliae Acta Biologica Serie A.*, 14, 91–110.

Jacob, F., Brenner, S. & Cuzin, F. (1964). On the regulation of DNA replication in bacteria. *Cold Spring Harbor Symposium of Quantitative Biology*, 28, 329–48.

James, T.W. & Jope, C. (1978). Visualization by fluorescence of chloroplast DNA in higher plants by means of the DNA-specific probe 4'6–diamidino–2–phenylindole. *Journal of Cell Biology*, 79, 623–30.

Kawano, S. & Kuroiwa, T. (1979). Studies on mitochondrial structure and function in *Physarum polycephalum*. VIII. Distribution of mitochondrial DNA molecules during mitochondrial division. *Cell Structure and Function*, 4, 99–108.

Kawano, S. & Kuroiwa, T. (1985). Isolation of a membrane–DNA complex in the mitochondria of *Physarum polycephalum*. *Experimental Cell Research*, 161, 460–72.

Knoth, R., Herrmann, F.H., Bottger, M. & Borner, T. (1974). Struktur und Funktion der Genetischen Information in den Plastiden. XI. DNA in normalen und mutierten Plastiden der Sorte 'Mrs. Parker' von *Perlargonium zonale*. *Biochemie und Physiologie Pflanzen*, 166, 129–48.

Kornberg, A. (1984). DNA replication. *Trends in Biochemical Sciences*, 9, 122–4.

Kolodner, R.D. & Tewari, K.K. (1975). Chloroplast DNA from higher plants replicates by both the Cairns and the rolling circle mechanism. *Nature*, 256, 708–11.

Kowallik, K.V. & Herrmann, R.G. (1972). Variable amounts of DNA related to the size of chloroplasts. IV. Three-dimensional arrangement of DNA in fully differentiated chloroplasts of *Beta vulgaris* L. *Journal of Cell Science*, 11, 357–77.

Kowallik, K.V. & Herrmann, R.G. (1974). Structural and functional aspects of the plastome. *Portugaliae Acta Biologica Serie A*, 14, 111–26.

Kuroiwa, T. & Suzuki, T. (1980). An improved method for the demonstration of the *in situ* chloroplast nuclei in higher plants. *Cell Structure and Function*, 5, 195–7.

Kuroiwa, T., Suzuki, T., Ogawa, K. & Kawano, S. (1981). The chloroplast nucleus: Distribution, number, size and shape, and a model for the multiplication of the chloroplast genome during chloroplast development. *Plant and Cell Physiology*, **22**, 381–96.

Kusano, T., Steinmetz, D., Hendrickson, W.G., Murchie, J., King, M., Benson, A. & Schaechter, M. (1984). Direct evidence for specific binding of the replicative origin of the *Escherichia coli* chromosome to the membrane. *Journal of Bacteriology*, **158**, 313–16.

Laird, C.D., Chooi, W.Y., Cohen, E.H., Dickson, E., Hutchinson, N. & Turner, S.H. (1973). Organisation and transcription of DNA in chromosomes and mitochondria of *Drosophila*. *Cold Spring Harbor Symposium of Quantitative Biology*, **38**, 311–27.

Lamppa, G.K., Elliot, L.V. & Bendich, A.J. (1980). Changes in chloroplast number, during pea leaf development. An analysis of a protoplast population. *Planta*, **148**, 437–43.

Lawrence, M.E. & Possingham, J.V. (1986). Microspectrofluorometric measurement of chloroplast DNA in dividing and expanding leaf cells of *Spinacia oleracea*. *Plant Physiology*, **81**, 708–10.

Lindbeck, A.G.C. (1986). The relationship of chloroplast membranes to the distribution, transcription and replication of chloroplast nucleoids. Ph.D. Thesis. The University of Newcastle, N.S.W. Australia.

Lindbeck, A.G.C. & Rose, R.J. (1986). Thylakoid-bound DNA of spinach chloroplasts is enriched for replication forks. *Proceedings of the Australian Biochemical Society*, **18**, 50.

Lindbeck, A.G.C. & Rose, R.J. (1987). Is DNA associated with spinach chloroplast vesicles at specific sites? *Journal of Plant Physiology*, **129**, 425–34.

Lindbeck, A.G.C. & Rose, R.J., Lawrence, M.E. & Possingham, J.V. (1987). The role of chloroplast membranes in the location of chloroplast DNA during the greening of *Phaseolus vulgaris* etioplasts. *Protoplasma*, **139**, 92–9.

Luttke, A. & Bonotto, S. (1981). Chloroplast DNA of *Acetabularia mediterranea*: Cell cycle related changes in distribution. *Planta*, **153**, 536–42.

Margulies, M.M. & Michaels, A. (1975). Free and membrane-bound chloroplast polyribosomes in *Chlamydomonas reinhardtii*. *Biochimica et Biosphysica Acta*, **402**, 297–308.

Miller, O.L. Jr., Hamkalo, B.A. & Thomas, C.A. Jr. (1970). Visualization of bacterial genes in action. *Science*, **169**, 392–5.

Odinstova, M.S., Mikulska, E. & Turischeva, M.S. (1970). Electron microscopy of DNA in pea chloroplasts. *Experimental Cell Research*, **61**, 423–32.

Ohtani, T., Uchimiya, H., Kato, A., Harada, H., Sugita, M. & Sugiura, M. (1984). Location and nucleotide sequence of a tobacco chloroplast DNA segment capable of replication in yeast. *Molecular and General Genetics*, **195**, 1–4.

Ohyama, K., Fukuzawa, H., Kohchi, T., Shirai, H., Sano, T., Sano, S., Umesowo, K., Shiki, Y., Takeuchi, M., Chang, Z., Aota, S I., Inokuchi, H. & Ozeki, H. (1986). Chloroplast gene organization deduced from complete sequence of liverwort *Marchantia polymorpha* chloroplast DNA. *Nature*, **322**, 572–4.

Overbeeke, N., Haring, M.A., John, H., Nijkamp, J. & Kool, A.J. (1984). Cloning of *Petunia hybrida* chloroplast DNA sequences capable of autonomous replication in yeast. *Plant Molecular Biology*, **3**, 235–41.

Paolillo, D.J. Jr. (1970). The three dimensional arrangement of intergranal lamellae in chloroplasts. *Journal of Cell Science*, **6**, 243–5.

Pettijohn, D.F. (1982). Structure and properties of the bacterial nucleoid. *Cell*, **30**, 667–9.

Possingham, J., Chaly, N., Robertson, M. & Cain, P. (1983). Studies of the distribution of DNA within spinach plastids. *Biology of the Cell*, **47**, 205–12.

Possingham, J.V. & Lawrence, M.E. (1983). Controls to plastid division. *International Review of Cytology*, **84**, 1–56.

Possingham, J.V. & Rose, R.J. (1976*a*). Chloroplast replication and chloroplast DNA synthesis in spinach leaves. *Proceedings of the Royal Society of London, Series B*, **193**, 295–305.

Possingham, J.V. & Rose, R.J. (1976*b*). Studies of the growth and replication of spinach chloroplasts and of the location and segregation of their DNA. In *Genetics and Biogenesis of Chloroplasts and Mitochondria*, eds. Th. Bucher, W. Neupert, W. Sebald & S. Werner, pp. 387–90. Amsterdam: Elsevier/North-Holland Biomedical Press.

Reiss, T. & Link, G. (1985). Characterization of transcriptionally active DNA–protein complexes from chloroplasts and etioplasts of mustard (*Sinapis alba* L.). *European Journal of Biochemistry*, **148**, 207–12.

Ris, H. & Plaut, W. (1962). Ultrastructure of DNA-containing areas in the chloroplast of *Chlamydomonas*. *Journal of Cell Biology*, **13**, 383–91.

Rose, R.J. (1979). The association of chloroplast DNA with photosynthetic membrane vesicles from spinach chloroplasts. *Journal of Cell Science*, **36**, 169–83.

Rose, R.J., Cran, D.G. & Possingham, J.V. (1974). Distribution of DNA in dividing spinach chloroplasts. *Nature*, **251**, 641–2.

Rose, R.J. & Lindbeck, A.G.C. (1982). Morphological studies on the transcription of spinach chloroplast DNA. *Zeitschrift für Pflanzenphysiologie*, **106**, 129–37.

Rose, R.J. & Possingham, J.V. (1976). The localization of [^{3}H]–thymidine incorporation in the DNA of replicating spinach chloroplasts by electron-microscope autoradiography. *Journal of Cell Science*, **20**, 341–55.

Rouvière–Yaniv, J. & Gros, F. (1975). Characterization of novel, low-molecular-weight DNA-binding protein from *Escherichia coli*. *Proceedings of the National Academy of Sciences, USA*, **72**, 3428–32.

Rushlow, K.E. & Hallick, R.B. (1982). The isolation and purification of a transcriptionally active chromosome from chloroplasts of *Euglena gracilis*. In *Methods in Chloroplast Molecular Biology* eds. M. Edelman, N–H. Chua & R. Hallick. pp. 543–50. Amsterdam: Elsevier Biomedical Press.

Ryter, A. (1968). Association of the nucleus and the membrane of bacteria: a morphological study. *Bacteriological Reviews*, **29**, 277–93.

Sager, R. (1972). *Cytoplasmic Genes and Organelles*. New York: Academic Press.

Sargent, M.A. & Bennett, M.F. (1985). Amplification of a major membrane-bound DNA sequence of *Bacillus subtilis*. *Journal of Bacteriology*, **161**, 589–95.

Scott, N.S., Cain, P. & Possingham, J.V. (1982). Plastid DNA levels in albino and green leaves of the 'albostrians' mutant of *Hordeum vulgare*. *Zeitschrift für Pflanzenphysiologie*, **108**, 187–91.

Scott, N.S. & Possingham, J.V. (1980). Chloroplast DNA in expanding spinach leaves. *Journal of Experimental Botany*, **31**, 1081–92.

Selldén, G. & Leech, R.M. (1981). Localization of DNA in mature and young wheat chloroplasts using the fluorescent probe 4'–6–diamidino–2–phenylindole. *Plant Physiology*, **68**, 731–4.

Shinozaki, K., Ohme, M., Tanaka, M., Wakasugi, T., Hayashida, N., Matsubayashi, T., Zaita, N., Chunwongse, J., Obokata, J., Yamaguchi–Shinozaki, K., Ohto, C., Torazawa, K., Meng, B.Y., Sugita, M., Deno, H., Kamogashira, T., Yamada, K., Kusuda, J., Takaiwa, F., Kato, A., Tohdoh, N., Shimada, H. & Sugiura, M. (1986). The complete nucleotide sequence of the tobacco chloroplast genome: its gene organisation and expression. *The EMBO Journal*, **5**, 2043–9.

Sprey, B. (1968). Zum verhalten DNS-haltiger Areale des Plastidenstromas bei der Plastidenteilung. *Planta*, **78**, 115–33.

Sprey, B. & Gietz, N. (1972). Isoleirung von etioplasten und elektronen-mikroskopische Abbildung membranassoziierter Etioplasten–DNA. *Zeitschrift für Pflanzenphysiologie*, **68**, 397–414.

Stirdivant, S.M., Crossland, L.D. & Bogorad, L. (1985). DNA supercoiling effects *in vitro* transcription of two maize chloroplast genes differently. *Proceedings of the National Academy of Sciences, USA*, **82**, 4886–9.

Thomas, M.R. & Rose, R.J. (1983). Plastid number and plastid structural changes associated with tobacco mesophyll protoplast culture and plant regeneration. *Planta*, **158**, 329–38.

Thomas, M.R. & Scott, K.J. (1985). Plant regeneration by somatic embryogenesis from callus initiated from immature embryos and immature inflorescences of *Hordeum vulgare*. *Journal of Plant Physiology*, **121**, 159–69.

Tymms, M.J., Scott, N.S. & Possingham, J.V. (1982). Chloroplast and nuclear DNA content of cultured spinach leaf discs. *Journal of Experimental Botany*, **33**, 831–7.

Valenzuela, M.S. & Inman, R.B. (1978). Restriction enzyme cleavage of DNA resulting from gently lysed *Escherichia coli*. *Molecular and General Genetics*, **166**, 245–9.

Valenzuela, M.S., Mueller, G.C. & Dasgupta, S. (1983). Nuclear matrix–DNA complex resulting from *Eco* RI digestion of Hela nucleoids is enriched for DNA replicating forks. *Nucleic Acids Research*, **11**, 2155–64.

Varshavsky, A.J., Nedospasov, S.A., Bakayev, V.V., Bakayeva, T.G. & Georgiev, G.P. (1977). Histone-like proteins in purified *Escherichia coli* deoxyribonucleoprotein. *Nucleic Acids Research*, **4**, 2725–45.

Woodcock, C.L.F. & Bogorad, L. (1970). Evidence for variation in the quantity of DNA among plastids of *Acetabularia*. *Journal of Cell Biology*, **44**, 261–75.

Woodcock, C.L.F. & Fernandéz–Moran, H. (1968). Electron microscopy of DNA conformations in spinach chloroplasts. *Journal of Molecular Biology*, **31**, 627–31.

Yamamoto, T., Burk, J., Autz, G. & Jagendorf, A.T. (1981). Bound ribosomes of pea chloroplast thylakoid membranes: location and release *in vitro* by high salt, puromycin and RNase. *Plant Physiology*, **67**, 940–9.

Yoshida, Y., Laulhère, J–P., Rozier, C. & Mache, R. (1978). Visualization of folded chloroplast DNA from spinach. *Biologie Cellulaire*, **32**, 187–90.

Zehnbauer, B.A. & Vogelstein, B. (1985). Supercoiled loops and the organization of replication and transcription in eukaryotes. *Bio Essays*, **2**, 52–4.

R.F.ROSENBERGER

Homologies between the nuclear and mitochondrial genomes of mammalian cells

With the development of techniques for restriction enzyme analysis, DNA cloning and DNA sequencing, it has become clear that genomic DNA is remarkably mobile. Within a prokaryotic genome or the nuclear genome of eukaryotes, a variety of rearrangements can be catalysed by transposons (Kleckner, 1981), or by retroviral-like elements (Jelinek & Schmid, 1982). Once gene duplications have appeared, homologous recombination will introduce further changes. However, in the excitement of explaining the presence of repetitive genes and the mobilisation of oncogenes by retroviral elements, the movement of DNA between the multiple genomes in eukaryotic cells has been largely neglected. The present paper will attempt to summarise data on the transposition of DNA between the mitochondrion and the nucleus in mammalian cells and to discuss some of the implications of the interchanges.

The movement of DNA from organelles to the nucleus is, in fact, implied in a widely accepted view of organelle origins. This view holds that mitochondria and chloroplasts descended from free-living organisms endocytosed by the progenitors of eukaryotes (Wallace, 1982). Organelle genomes have since lost the information coding for the vast majority of proteins needed to replicate and perform their biochemical functions, these proteins now being specified by nuclear genes. One cannot eliminate the possibility that such genes were present in the ancestral eukaryotes, but it seems simple and natural to suggest that they moved from the organelles during evolution.

Fortunately, the movement of DNA from mitochondria to the nucleus has not been limited to the early days of evolution, where it would be out of our reach. Homologous duplications occur in the two genomes of present-day organisms. The number of detailed studies published on this subject has been relatively small and the majority of papers have dealt with lower eukaryotes (e.g. Farrelly & Butow, 1983; Wright & Cummings, 1983; Jacobs et al., 1983; Gellissen et al., 1983). Studies on homologies in mammalian cells can easily be counted in single figures (Hadler, Dimitrijevic & Mahalingham, 1983; Tsuzuki et al., 1983; Nomiyama et al., 1984; Wakasugi et al., 1985; Kristensen & Prydz, 1986). However, the existence of such homologies poses challenging opportunities and questions and some of these will be discussed below.

In theory, at least, the presence of duplicated segments in the two genomes should permit an estimate of the times when the transpositions occurred. The frequencies of silent mutations in the mitochondrial (*mt*) and nuclear genomes have been determined over a range of species and genes (Miyata *et al.*, 1982; Vawter & Brown, 1986). Such mutations specify changes in the third codon base, do not produce alterations in the amino acid sequence and occur at a high and reasonably constant rate in mitochondria. In the nuclei of vertebrates, the frequency of silent mutations is 6–10 times less, although in lower eukaryotes it can increase to match that in the organelle (Vawter & Brown, 1986). Where the base sequences in the nuclear fragment and the corresponding gene in the mitochondrion have been determined, it should be possible to calculate the timing of the event. The very limited data available (Farrelly & Butow, 1983; Jacobs *et al.*, 1983; Wakasugi *et al.*, 1985) indicate that at least some of the transpositions occurred several million years ago.

Sequencing data on the regions where the *mt* insert joins the nuclear DNA could also indicate what mechanisms were responsible for the transpositions. Movements via RNA intermediates or through the other major class of transposons should result in characteristic DNA repeats, both direct and inverted, at the insertion sites (Majors *et al.*, 1981). The lack of any repeats would focus attention on the possibility of a direct uptake of DNA from damaged or disintegrated mitochondria. Exogenous DNA fragments are well known to readily enter the eukaryotic nucleus and in mammals, at least, the mitochondria of the sperm appear to undergo disintegration shortly after fertilisation (Zamboni *et al.*, 1966). So far, the sequencing data of the insertion sites are insufficient to provide any answers.

Turning to a different aspect, one may ask if the transposed organelle DNA can have any physiological or biochemical effects on the cells. A strong impetus to posing this question comes from studies on the ascomycete *Podospora anserina*. The vegetative hyphae of this fungus regularly undergo a time-dependent deterioration of growth and metabolism. This shares two properties with senescence in cultured mammalian cells; it invariably ends in cell death and is dominant in hybrids between young and old cells (Esser *et al.*, 1980; Marcou, 1961; Osiewacz & Esser, 1984). In *Podospora*, senescence is caused by the excision and uncontrolled replication of an intron in the cytochrome oxidase subunit I (COI) gene of the mitochondrion (Wright & Cummings, 1933; Kuck *et al.*, 1985; Belcour & Vierney, 1986). Occasionally other *mt*DNA changes accompany senescence, but the amplification of the intron, also called *sen*DNA , appears to be the primary cause. The physiological changes do not appear to be due to the loss of the *sen*DNA from the *mt* genome and the subsequent lack of cytochrome oxidase activity in the hyphae. This is shown by the isolation of non-senescing mutants which have that very bit of *mt*DNA deleted (Belcour & Vierny, 1986). Infection of young hyphae by purified *sen*DNA induces all the senescent changes, presumably because of its extensive replication in the mycelium. The replication of *sen*DNA as a plasmid is not surprising, since, like a number of other organelle DNAs, it contains sequences which allow autonomous plasmid replication in fungal nuclei (Wallace, 1982; Osiewacz & Esser, 1984).

To summarise, the amplification of a *mt*DNA fragment induces striking physiological changes in *Podospora* and it is most likely that the *mt* fragment replicates in the nucleus. However, there is still discussion if *sen*DNA actually integrates into the nuclear genome (Koll, 1986), and how it influences cell metabolism is not yet clear on the molecular level (Belcour & Vierny, 1986; Michel & Lang, 1985). It is difficult to escape the conclusion that *sen*DNA must at least be transcribed to produce its effects. It has been suggested that *sen*DNA has significant homology to genes encoding reverse transcription (Michel & Lang, 1985) and such an activity could certainly bring about metabolic alterations, but the translation of messages from *mt*DNA must be interpreted in the light of the slightly different codes and ribosome recognition sites used for protein translation by the cytosol and by mitochondria (Wallace, 1982). Apart from the *Podospora* studies discussed above, there are speculations (e.g. Jacobs *et al.*, 1983), but no clear data which could answer the questions if (a) transposed organelle DNA is transcribed or not and (b) if the putative messages are translated as normal polypeptides, as truncated ones, or not at all.

To clarify the frequency of *mt*DNA transpositions, the mechanisms responsible for their movement and their impact on cell metabolism, much more data on the detailed nature of the duplications are required. The next sections will describe some experimental approaches and problems in studying homologies in mammalian cells and our attempts to obtain more information about the duplications.

Experimental approaches to detecting homologies in mammalian cells

The first essential in detecting homologies is to obtain both pure *mt*DNA and pure nuclear DNA. The purification of *mt*DNA can be achieved quite readily. It is a relatively small, circular molecule which can be separated from the main genome by isolation of mitochondria and conventional centrifugations of the extracted DNA. Purification can then be completed by cloning sections in suitable vectors. Such clones representing the *mt*DNA molecule have been obtained from a number of mammalian species and have further been sequenced to give the complete base order in the organelle genomes (Anderson *et al.*, 1981; Anderson *et al.*, 1982).

The purification of nuclear DNA presents much greater difficulties. Mammalian cells can be broken carefully, the mitochondria and nuclei separated by centrifugation steps and the nuclei purified until no traces of mitochondrial enzymes can be found. If any *mt*DNA is now detected, one faces the problem of eliminating all possibilities of artifactual contamination. The reality of this difficulty is clearly illustrated by the studies of Prydz and his colleagues (Krokan, Bjorklid & Prydz, 1975; Kristensen & Prydz, 1986). They purified nuclei as described above from cultured HeLa cells, a permanent human fibroblast line. The nuclei were used for studies on *in vitro* DNA synthesis and newly synthesised DNA was indeed obtained. However, analysis showed that all of this was full-length *mt*DNA and that the nuclei must have contained monomer length *mt* genomes which had acted as templates. Prydz and his colleagues went a long way to eliminating contamination as an explanation. They compared their normal system to others where exogenous *mt*DNA was added during breakage or where cells depleted of *mt*DNA by growth with a dye were used. The similarity of the results indicated that contamination was not involved. Clearly, though, even the most capable purification of mammalian nuclei needs to be backed up by additional criteria in order to establish homologies.

One relatively simple system for demonstrating that *mt*DNA sequences are actually integrated in the purified nuclear DNA, uses restriction enzyme digests, agarose gel electrophoresis and Southern blotting. The restriction enzyme or enzymes have to be chosen so that they either do not cut in the *mt* genome or produce only a small number of fragments from the *mt* genome. The purified nuclear DNA is restricted, electrophoresed, blotted and hybridised with a radioactive *mt* probe. One then looks for bands whose sizes are different from those predicted by the *mt* base sequences. Such fragments would originate either from two cuts in flanking genomic DNA or from one cut in the *mt* insert and a second in the flanking non-*mt* DNA.

Hadler, Dimitrijevic & Mahalingam (1983) used this approach to study *mt*DNA transposed to the nucleus of rat liver cells. They found a band which could not be accounted for by digestion of contaminating *mt*DNA and by hybridisations showed that it contained part of the *mt* D-loop and rRNA genes. While simple and effective, this method may have limitations in the range of transpositions that can be detected.

At first sight, the best way to purify nuclear DNA would appear to be through cloning random fragments in a suitable phage vector and then to screen the plaques of the genomic library with *mt* probes. Contamination would be completely eliminated and isolation of plaques containing *mt* inserts would allow access to sequencing and the detailed structural information needed.

Because of these attractive features, the screening of genomic libraries with *mt* probes has been widely used to search for homologies. Farrely & Butow (1983) and Jacobs *et al.* (1983) were among workers using genomic libraries in lower eukaryotes and Shimada's group at Kumamoto University has made an extensive study of human genomic libraries (Tsuzuki *et al.*, 1983; Nomiyama *et al.*, 1984; Wakasugi *et al.*, 1985). However, with human cells an unpredicted problem has arisen. Human *mt*DNA is toxic to *Escherichia coli*, the vector used in all these studies, and repeated attempts to clone the whole human *mt* genome in *E.coli* have failed (Tapper, Van Etten & Clayton, 1983). It can be cloned in quite a large number of separate fragments (Drouin, 1980) and, while the actual toxic sequences are not known, they may be quite widely distributed throughout the genome. The human *mt*DNA could be a singular example of toxicity, since the whole mouse *mt* genome can be cloned and maintained in *E.coli* (Tapper *et al.*, 1983). But even with mouse *mt*DNA, plasmids carrying certain sections can interfere with *E.coli* growth and this also appears to be the case with rat *mt*DNA (Kearsey, Flanagan & Craig, 1980).

The group at Kumamoto screened 3.5×10^5 phage plaques from a human genomic library and found 14 which hybridised to human *mt*DNA. Some of these inserts were analysed in detail and found to contain DNA homologous to *mt* 16S rRNA, cytochrome oxidase subunits I and II, Unidentified Reading Frame (*URF*)2, *URF*4 and *URF*5 genes. Some inserts also had a human genomic repetitive sequence *Kpn*I, in the nuclear *mt* genes. The work of the Japanese group represents the most extensive studies on transposed mammalian *mt*DNA published to date. But, in view of the toxicity of human *mt*DNA to *E .coli*, these investigations are very likely to have underestimated both the frequency and size of organelle transpositions in human cells. Until more is known about the effects of cosmids carrying *mt* inserts on the vectors in which they have been cloned, the screening of gene libraries must be interpreted with caution.

My own work on *mt*DNA movements in mammalian cells originated from our department's long-standing interest in the senescence of human cells (Holliday, 1984; Holliday, 1986; Kirkwood & Holliday, 1979; Kirkwood, Holliday & Rosenberger, 1984). *P. anserina* mycelium senesces because of the excision and uncontrolled replication of a section of *mt*DNA, as described in a previous section. Although the differences between the *mt* genomes of lower eukaryotes and mammals are marked (Wallace, 1982) it seemed reasonable to investigate possible changes in the *mt*DNA of young and old human cells. These experiments and others suggested by them are described in the following sections.

Mitochondrial DNA in young and old human fibroblasts

Many studies of senescence in mammalian cells have used the changes in normal cells during laboratory subcultures as a model system (for a review and discussion of cellular ageing, see Holliday, 1984). Briefly, the main features of cellular ageing are as follows. When cells are isolated from healthy mammalian tissues, they will only grow and replicate for a limited number of generations (Hayflick, 1965; Holliday & Tarrant, 1972). The total number of divisions that are attained in cultures is related to the age of the donors within a species (Bierman, 1978) and to the lifespans characteristic of the animals in comparisons between species (Rohme, 1981). Some cell types will spontaneously produce infrequent variants which then develop into permanent, immortal lines (e.g. Todaro & Green, 1963). Others can be induced to form immortal lines by treatments with mutagens, carcinogens or oncogenic viruses (Newbold, Overell & Connell, 1982; Huschtscha & Holliday, 1983). But in untreated cells from healthy human tissues, the progression from isolation to loss of growth potential appears to mirror the changes in ageing populations. For studying human cells, cellular senescence is further the only convenient experimental approach available.

Our experiments were carried out with MRC5 cells, a strain of human foetal lung fibroblasts (Jacobs, Jones & Baillie, 1970). MRC5 cells have been used extensively for studies on cellular ageing; they grow rapidly for 40–45 population doublings (PD), more slowly for the next 10–15 PDs and growth finally ceases around 60 PDs (Holliday & Tarrant, 1972; Kirkwood & Holliday, 1975). We harvested cells at different points of their cultured lifespans and extracted the total DNA, i.e. both the nuclear and *mt* genomes, by conventional methods (Maniatis, Fritsch & Sambrook, 1982). The DNA was analysed, both without and after restriction enzyme digestions, by agarose gel electrophoresis and Southern blotting (Maniatis *et al.*, 1982). To visualise the *mt*DNA on Southern blots, we hybridised with radiolabelled probes of cloned human *mt*DNA, the clones being a kind gift of Dr Anderson, MRC, Cambridge.

DNA undigested by restriction enzymes contained normal length *mt*DNA and a minor proportion of larger segments hybridising with the *mt* probes. These latter bands could have been multimeric *mt* genomes (Clayton, 1982) or *mt* sequences in large fragments of chromosomal DNA. We obtained this result for all the cell ages tested. No clear bands of *mt*DNA smaller than the canonical genome size were found, even in cells at the very end of their lifespan.

After digestion with a variety of restriction enzymes, the *mt* fragment patterns were very similar for young and old cells. The results for *Xba*1, *Sau*3A and *Eco*R1 are shown in Fig. 1, and closely corresponding findings were obtained after digestion with *Hha*I and *Hae*III (results not shown). We thus did not find an amplification of discrete *mt* DNA segments nor a significant accumulation of altered *mt* genomes. These, however, were the major changes detected in senescing *P. anserina* mycelia. If *mt*DNA is involved in the senescence of human cells, its actions must be quite different from that in the ascomycete. A similar conclusion was reached independently by Shmookler–Reis & Goldstein (1983), who used human *mt*DNA probes purified by centrifugation procedures, but not by cloning.

While no age-related changes in gross *mt*DNA structures appeared, there were indications of *mt*DNA rearrangements. *EcoRI* has three cutting sites in human *mt*DNA and gives fragments of 1.1, 7.4 and 8.05 kb (Anderson *et al.*, 1981). The bands at 7–8 kb could be clearly seen, but fainter bands of approximately 6.5 and 3.5 kb were also present after *EcoRI* digestion (Fig. 1, lanes 5 and 6). These could not have been due to partial digestion nor could they be explained by the canonical base sequence. They could have originated from rare polymorphisms in *mt*DNA (Giles *et al.*, 1980), from *mt*DNA integrated in the nuclear genome or some other, unsuspected, rearrangements.

To clarify the position, we homogenised cells of MRC5 and of an immortal derivative of MRC5, VI (Huschtscha & Holliday, 1983). The nuclear and mitochondrial fractions were separated by differential centrifugation (Pederson *et al.*, 1978) and the nuclei purified either by the method of Schibler & Weber (1974), or by following the protocol of Kristensen & Prydz (1986). The DNA was extracted and purified from both fractions by conventional methods (Maniatis *et al.*, 1982). After restriction enzyme digestion, agarose gel electrophoresis, Southern blotting and hybridisation with probes, the DNA from the *mt* fraction behaved as described in the literature. The results with DNA from the nuclear fraction can be summarised as follows. When not digested with restriction enzymes, a single rather faint band around 16.5 kb was detected (Fig. 2A, lane 1). *Eco*R1 digestion produced much stronger bands between 7–8 kb Fig. 2A, lane 2) and treatment with *Bam* H1, which has a single recognition site in human *mt*DNA, gave a marked band at 16.5 kb (results not shown). In lanes 1 and 2 of Fig. 2A, equal amounts of uncut and cut DNA were loaded. *Eco*R1 also gave weak bands around 6.5 and 2.5 kb, (Fig. 2A, lane 2) which correspond to those seen in total DNA extracted from cells.

Fig. 1. Comparisons of the patterns of *mt*DNA restriction enzyme fragments in young and old MRC5 fibroblasts. Young (23 PDs) and senescing (53 PDs) cells were harvested and the total DNA extracted and purified. After digestions with various restriction enzymes, the DNA was electrophoresed on 0.7 per cent agarose gels and blotted onto nitrocellulose filters (Maniatis *et al.*, 1982). The filters were hybridised with a mixture of cloned *mt* fragments covering about 80 per cent of the *mt* genome, the probe being radiolabelled by nick translation. Lanes 1 and 2: *Xba*1 digestion, 1 young and 2 senescing cells. Lanes 3 and 4: *Sau*3a digestion, 3 young and 4 senescing cells. For both these enzymes, the digestion was deliberately incomplete to show that partially-cut intermediates also correspond in size. Lanes 5 and 6: *Eco*R1 digestion, 5 young and 6 senescing cells. The arrows indicate the faint 6.5 and 2.5 kb fragments. The size markers were a 1 kb ladder and λ *Hind*III fragments from Bethesda Research Laboratories.

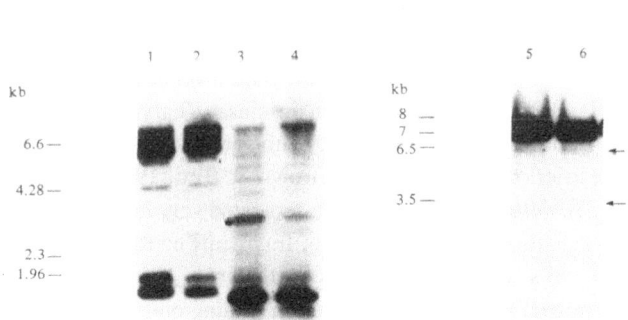

These data are insufficient for firm conclusions to be drawn, but they suggest the following working hypothesis. Unit length *mt*DNA had entered the nuclei, as shown by the band at 16.5 kb without restriction enzyme treatment and the strong bands at 7–8 kb after *Eco*R1 digestion. Here we agree with the conclusions of Kristensen & Prydz (1986) that, even after careful purification, human nuclei contain full-length *mt*DNA. The amount of DNA hybridising in the uncut material appeared to be much less than that after restriction enzyme digestions, and this suggests that only a fraction of the *mt*DNA is present as free unit-length genomes. A major part would need to be in a form which did not enter the gel, either because it is integrated in large chromosomal fragments or else because of some unusual topological configurations. The fainter bands at 6.5 and 3.5 kb could represent fragments of *mt*DNA integrated separately or sections cut from the joins of the integration sites.

To obtain more incisive information on the *mt*DNA in the nuclei we needed an approach rather different from the above. Probing human gene banks in phage vectors seemed unlikely to be helpful because of the previously discussed toxicity of large *mt* inserts. We therefore turned to the electrophoretic analysis of large DNA fragments. The advantages here are that free *mt* genomes and their concatamers should be separated from large chromosomal fragments if the latter were greater than say 100 kb. Recent technological advances have shown how such a system can work. Whole cells are embedded in agarose blocks and prepared for DNA electrophoresis while in the agarose (Schwartz & Cantor, 1984; Brown & Bird, 1986). Mammalian cells are lysed *in situ* with detergents and the proteins stripped off the chromosomes by detergent and protease action. The gel protects the fragile DNA from mechanical damage and very large pieces of DNA can be prepared with restriction enzymes having only a few cutting sites. Fragments up to at least 1000 kb can then be separated in non-uniform electrophoresis fields and blotted and hybridised as usual (Brown & Bird, 1986; Carle, Frank & Olson, 1986).

Non-uniform gel electrophoresis runs were made in collaboration with my colleague, Dr D.H. Williamson of the Laboratory of Cell Propagation, NIMR. Our experiments have so far concentrated on using the field inversion system of Carle *et al.* (1986). MRC5 and VI cells were grown in flasks, harvested, washed and embedded in agarose blocks. They were lysed with sarcosyl and proteinase K and digested with the restriction enzyme *Sma*1 (Brown & Bird, 1986). Two different field inversion regimes were used, one of which separated very large fragments (Fig. 2C) and a second which allows differentiation of fragments up to about 200 kb (Fig. 2B). Both types of cells showed fragments well above 100 kb in size which hybridised with *mt* probes. We are now engaged in finding out what parts of the *mt* genome are contained in these fragments by using clones with different *mt* inserts for hybridisations.

Fig. 2. DNAs from purified nuclei and from agarose embedded cells and their hybridisation with *mt* probes. 2A: DNA from purified nuclei of MRC5 was electrophoresed in 0.7 per cent agarose and blotted and hybridised as in Fig. 1. Lane 1 is DNA without digestion and lane 2 after *Eco*R1 digestion. Both lanes were loaded with equal amounts of DNA and the size markers were as in Fig. 1. 2B: Cultured cells were embedded in agarose and digested with *Sma*1. The gel was run in 1 per cent agarose, 10 V cm^{-1} for 18 h and under a field inversion regime of 3 second forward and 1 second back. The gels were treated with

0.25 M HCl to break up large fragments before Southern blotting. Size markers were T4 DNA and a lambda ladder, it should be pointed out that under these conditions fragment sizing tends to be approximate rather than exact (Carle *et al.*, 1986). Lane 1 MRC5 cells, lane 2 MRC5 undigested, lane 3 V1 cells. 2C: As 2B, but with a field switching of 9–60 seconds forward and 3–20 seconds reverse over a run of 16 h. Size markers as in 2B and with the same reservations. Lane 1 MRC5, Lane 2 V1.

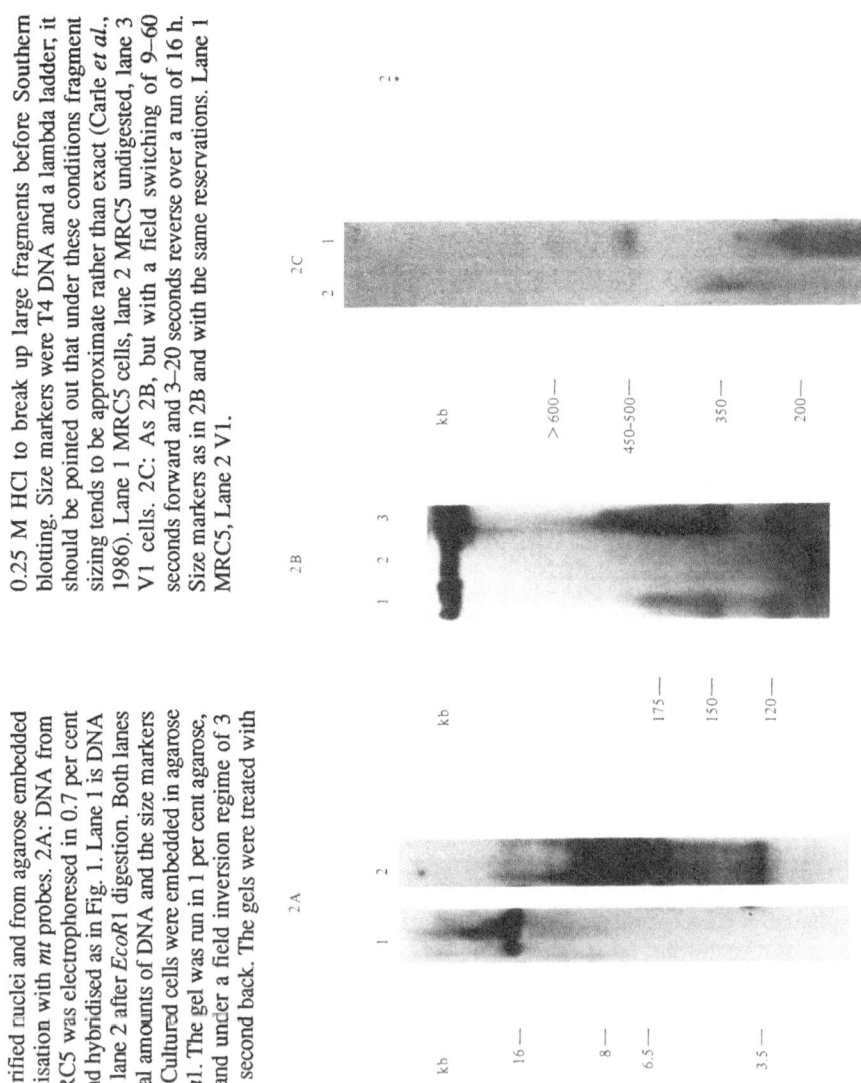

Concluding remarks

The homologies between nuclear and mtDNA will eventually allow conclusions to be drawn about the timing, frequencies and mechanisms of the transposition events. What is missing at present is sufficient data about the base sequences of the homologous stretches and on the sequences of the joins at the insertion sites. One difficulty in obtaining these data is the isolation of the organelle DNA transposed to the nuclear genome. Although the screening of genomic banks would appear to be an obvious approach, it could be quite misleading because of the toxic effects of mt inserts on vectors. A second problem is the weight of work involved in sequencing a number of quite long stretches of DNA. However, it seems likely that such data will become available in the quite near future.

A more speculative, but exciting aspect of the homologies is that the out-of-place organelle DNA may have physiological effects. The only well-documented case of this is senescence in *Podospora*. But here there is not only movement of mtDNA from the organelle, but also a marked amplification of the senDNA. This naturally leads to the question if transposed sections of mtDNA could have any significant effects unless they are greatly amplified. A mammalian cell contains about 5000 mt genomes (Clayton, 1982) and one may question if the movement of a single or a few mt genes or genomes to the nucleus could induce any significant metabolic alteration. Surprisingly, the answer may be yes. For example, the mt respiratory chain contains the mt-encoded proteins cytochrome b and cytochrome oxidase and the nuclear-encoded protein cytochrome c. While the gene dosage differs by three orders of magnitude, the organelle and nuclear-encoded proteins are present in virtually equimolar amounts (Wainio, 1970). We know very little about the factors that control the output of organelle genes irrespective of the position they are in. This is highlighted by the markedly different levels in mtRNAs found when normal and transformed mammalian cells are compared and for which there is as yet no explanation (Glaichennaus, Leopold & Cuzin, 1986). If nothing else, studying the physiological impacts of transpositions may clarify these problems.

Acknowledgements

The author would like to express his thanks to Drs R. Holliday and T.B.L. Kirkwood for helpful discussions, to Dr D.H. Williamson for his collaboration in non-uniform field gel electrophoresis, to Mrs S. Davies for her excellent technical assistance and to Ms S. Moore for her help in the preparation of the manuscript.

References

Anderson, S., Bankier, A.T., Barrell, B.G., De Bruijn, N.H.L., Coulson, A.R., Drouin, J., Eperon, I.C., Nierlich, D.P., Roe, B.A., Sanger, F., Schreir, P.H., Smith, A.J.H., Staden, R. & Young, I.G. (1981). Sequence and organisation of the human mitochondrial genome. *Nature*, **290**, 457–65.

Anderson, S., de Bruijn, M.H.L., Coulson, A.R., Eperon, I.C., Sanger, F. & Young, I.G. (1982). Complete sequences of bovine mitochondrial DNA. *Journal of Molecular Biology*, **156**, 683–717.

Belcour, L. & Vierny, C. (1986). Variable DNA splicing sites of a mitochondrial intron: relationship to the senescence process of *Podospora*. *EMBO Journal*, **5**, 609–14.

Bierman, E.L. (1978). The effect of donor age on the *in vitro* life span of cultured human arterial smooth-muscle cells. *In Vitro*, **14**, 951–5.

Brown, W.R.A. & Bird, A.P. (1986). Long-range restriction site mapping of mammalian genomic DNA. *Nature*, **322**, 477–81.

Cann, R.L., Stoneking, M. & Wilson, A.C. (1987). Mitochondrial DNA and human evolution. *Nature*, **325**, 31–6.

Carle, G.F., Frank, M. & Olson, M.V. (1986). Electrophoretic separation of large DNA molecules by periodic inversion of the electric field. *Science*, **232**, 65–8.

Clayton, D.A. (1982). Replication of animal mitochondrial DNA. *Cell*, **28**, 693–705.

Drouin, J. (1980). Cloning of human mitochondrial DNA in *Escherichia coli*. *Journal of Molecular Biology*, **140**, 15–34.

Esser, K., Tudzynski, Z, Stahl, U. & Kuck, U. (1980). A model to explain senescence in the filamentous fungus *Podospora anserina* by the action of plasmid-like DNA. *Molecular and General Genetics*, **178**, 213–16.

Farrelly, F. & Butow, A. (1983). Rearranged mitochondrial genes in the yeast nuclear genome. *Nature*, **301**, 296–301.

Gellissen, G., Bradfield, J.Y., White, B.N. & Wyatt, G.R. (1983). Mitochondrial DNA sequences in the nuclear genome of a locust. *Nature*, **301**, 631–4.

Giles, R.E., Blanc, H., Cann, H. & Wallace, D.C. (1980). Maternal inheritance of human mitochondrial DNA. *Proceedings of the National Academy of Sciences, USA*, **77**, 6715–19.

Glaichenhaus, N., Leopold, P. & Cuzin, F. (1986). Increased levels of mitochondrial gene expression in rat fibroblast cells immortalized or transformed by viral and cellular oncogenes. *EMBO Journal*, **5**, 1261–5.

Hadler, H.I., Dimitrijevic, B. & Mahalingham, R. (1983). Mitochondrial DNA and nuclear DNA from normal rat liver have a common sequence. *Proceedings of the National Academy of Sciences, USA*, **80**, 6495–9.

Hayflick, L. (1965). The limited *in vitro* lifetime of human diploid cell strains. *Experimental Cell Research*, **37**, 614–36.

Holliday, R. (1984). The unsolved problem of cellular ageing. *Monographs in Developmental Biology*, **17**, 60–77.

(1986). Testing molecular theories of cellular ageing. In *Dimensions in Ageing*, ed. M. Bergener, M. Ermini & H.B. Stamelin, pp. 21–34, London, Academic Press.

Holliday, R. & Tarrant, G.M. (1972). Altered enzymes in ageing human fibroblasts. *Nature*, **238**, 26–30.

Huschtscha, L.I. & Holliday, R. (1983). Limited and unlimited growth of SV 40-transformed cells from human diploid MRC5 fibroblasts. *Journal of Cell Science*, **63**, 77–99.

Jacobs, H.T., Posakony, J.W., Grula, J.W., Roberts, J.W., Ji–Hu Xin, Britten, R.J. & Davidson, E.H. (1983). Mitochondrial DNA sequences in the nuclear genome of *Stronglyocentrotus purpuratus*, *Journal of Molecular Biology*, **165**, 609–32.

Jacobs, J.P., Jones, C.M. & Baillie, J.P. (1970). Characteristics of a human diploid cell designated MRC5. *Nature*, **227**, 168–70.

Jelinek, W.R. & Schmid, C.W. (1982). Repetitive sequences in eukaryotic DNA and their expression. *Annual Review of Biochemistry*, **51**, 813–44.

Kearsey, S.E., Flanagan, J.G. & Craig, I.W. (1980). Cloning of mouse mitochondrial DNA in *E.coli* affects bacterial viability. *Gene*, **12**, 249–55.

Kirkwood, T.B.L. & Holliday, R. (1975). Commitment to senescence: a model for the finite and infinite growth of diploid and transformed human fibroblasts in culture. *Journal of Theoretical Biology*, **53**, 481–96.

Kirkwood, T.B.L. & Holliday, R. (1979). The evolution of ageing and longevity. *Proceedings of the Royal Society, London*. Series B, **205**, 531–46.

Kirkwood, T.B.L., Holliday, R.& Rosenberger, R.F. (1984). Stability of the cellular translation apparatus. *International Revue of Cytology*, **92**, 93–132.

Kleckner, N. (1981). Transposable elements in prokaryotes. *Annual Review of Genetics*, **15**, 341–404.

Koll, F. (1986). Does nuclear integration of *mit* sequences occur during senescence in *Podospora*? *Nature*, **324**, 597–9.

Kristensen, T. & Prydz, H. (1986). The presence of intact mitochondrial DNA in HeLa cell nuclei. *Nucleic Acids Research*, **14**, 2597–609.

Krokan, H., Bjorklid, E. & Prydz, H. (1975). DNA synthesis in isolated Hela Cell nuclei. Optimalisation of the system and characterisation of the product. *Biochemistry*, **14**, 4227–32.

Kuck, U., Osiewcz, D., Schmidt, U., Kappelhoff, B., Schulte, E., Stahl, U. & Esser, K. (1985). The onset of senescence is affected by DNA rearrangements of a discontinuous mitochondrial gene in *Podospora anserina*. *Current Genetics*, **9**, 373–82.

Majors, J.E., Swanstrom, R., de Lorre, W.J., Payne, G.S., Hughes, S.H., Ortiz, S., Quintrell, N., Bishop, J.M. & Varmus, H.E. (1981). DNA intermediates in the replication of retroviruses are structurally (and perhaps functionally) related to transposable elements. In *Mobile Genetic Elements, Cold Spring Harbor Symposia of Quantitative Biology*, vol. 45, part 2, pp. 731–8. New York, Cold Spring Harbor Press.

Maniatis, T., Fritsch, E.F. & Sambrook, J. (1982). Molecular Cloning, New York, Cold Spring Harbor Laboratory.

Marcou, D. (1961). Notion de longevite et nature cytoplasmique chez quelque champignons. *Annale des Sciences Naturelles Botanique*. Series 12, **2**, 653–764.

Michel, F. & Lang, B.F. (1985). Mitochondrial class II introns encode proteins related to the reverse transcriptase of retroviruses. *Nature*, **316**, 641–3.

Miyata, T., Hayashida, H., Kikuno, R., Hasegawa, M., Kobayashi, M. & Koike, K. (1982). Molecular clock of silent substitutions: at least six-fold preponderance of silent changes in mitochondrial genes over those in nuclear genes. *Journal of Molecular Evolution*, **19**, 28–35.

Newbold, R.F., Overell, R.W. & Connell, J.R. (1982). Induction of immortality is an early event in malignant transformations of mammalian cells by carcinogens. *Nature*, **299**, 633–5.

Nomiyama, H., Tsuzuki, T., Wakasugi, S., Fukuda, M. & Shimada, K. (1984). Interruption of a human nuclear sequence homologous to mitochondrial DNA by a member of the KPN1. 1.8 kb family. *Nucleic Acids Research*, **12**, 5225–34.

Osiewacz, H.D. & Esser, K. (1984). The mitochondrial plasmid of *Podospora anserina* : A mobile intron of a mitochondrial gene. *Current Genetics*, **8**, 299–305.

Pederson, P.L., Greenawalt, J.W., Reynafarje, B., Hullihen, J., Decker, G.L., Soper, J.W. & Bustamente, E. (1978). Preparation and characterisation of mitochondria and submitochondrial particles of rat liver and liver derived tissues. In *Methods in Cell Biology*, vol. 20, ed. D.M. Prescott, pp. 412–83. New York, Academic Press.

Rohme, D. (1981). Evidence for a relationship between longevity of mammalian species and lifespans of normal fibroblasts *in vitro* and erythrocytes in vivo. *Proceedings of the National Academy of Sciences, USA*, **78**, 5009–13.

Schibler, U. & Weber, R. (1974). A new method for the isolation of undegraded nuclear and cytoplasmic RNA from liver of *Xenopus* larvae. *Analytical Biochemistry*, **58**, 225–30.

Schwarz, D.C. & Cantor, C.R. (1984). Separation of yeast chromosome-sized DNAs by pulsed field gradient gel electrophoresis. *Cell*, **37**, 67–75.

Shmookler Reis, R.J. & Goldstein, S. (1983). Mitochondrial DNA in mortal and immortal human cells. *Journal of Biological Chemistry*, **258**, 9078–85.

Tapper, D.P., Van Etten, R.A. & Clayton, D.A. (1983). Isolation of mammalian mitochondrial DNA and RNA and cloning of the mitochondrial genome. In *Methods of Enzymology*, vol. 97, ed. S. Fleisher & B. Fleisher, pp. 626–34, New York, The Academic Press.

Todaro, G.J. & Green, H. (1963). Quantitative studies of the growth of mouse embryo cells in culture and their development into established lines. *Journal of Cell Biology*, **17**, 299–313.

Tsuzuki, T., Nomiyama, H., Setoyama, C., Maeda, S. & Shimada, K. (1983). Presence of mitochondrial–DNA-like sequences in the human nuclear DNA. *Gene*, **25**, 223–9.

Vawter, L. & Brown, W.M. (1986). Nuclear and mitochondrial DNA comparisons reveal extreme rate variation in the molecular clock. *Science*, **234**, 194–6.

Waino, W.W. (1970). The mammalian mitochondrial respiratory chain, pp. 56–7, New York, Academic Press.

Wakasugi, S., Nomiyama, H., Fokuda, M., Tsuzuki, T. & Shimada, K. (1985). Insertion of a long *KPN*I family member within mitochondrial–DNA-like sequence present in the human nuclear genome. *Gene*, **36**, 281–8.

Wallace, D.C. (1982). Structure and evolution of organelle genomes. *Microbiological Reviews*, **46**, 208–40.

Wright, R.M. & Cummings, D.J. (1983). Integration of mitochondrial gene sequences within the nuclear genome during senescence in a fungus. *Nature*, **302**, 86–8.

Zamboni, L., Mishell, D.R., Bell, J.H. & Baca, M. (1966). Fine structure of the human ovum in the pronuclear stage. *Journal of Cell Biology*, **30**, 579–600.

W.R. MILLS AND B.J. BAUMGARTNER

DNA synthesis in isolated chloroplasts

Chloroplasts of higher plants contain closed circular chromosomes (approximately 120 to 150 kilobase pairs in size) which code for all of the plastid's ribosomal and transfer RNAs as well as for about 100 plastid proteins (Dyer, 1984). During leaf development, chloroplasts are formed by the binary fission of pre-existing plastids (Possingham, 1980; Boffey, 1985); thus, replication of chloroplast DNA (cpDNA) obviously must occur to maintain adequate levels of DNA per plastid. Early in leaf development (Lamppa, Elliot & Bendich, 1980; Scott & Possingham, 1983) cpDNA replication and chloroplast division occur simultaneously, while later, plastid division and cpDNA replication are not tightly linked. For example, Lamppa & Bendich (1979) showed that levels of cpDNA in pea chloroplasts increase about 10 fold as embryos mature into shoots which have expanding leaves. Subsequently, the DNA per plastid peaks then declines as plant development continues (from approximately 270 copies per chloroplast in 6–7-day old plants to about 100 copies per plastid in 12-day old plants). Similar types of developmental changes occur in spinach (Scott & Possingham, 1983), wheat (Boffey & Leech, 1982), and beet (Tymms, Scott & Possingham, 1983), where again cpDNA levels per plastid increase to a maximum then decline as tissues mature.

DNA replication has been shown to be a very complex process in the few prokaryotic viral and bacterial systems that have been well studied (Clayton, 1982). For example, Adams, Knowler & Leader (1986) list 18 gene products and 11 enzymatic activities which play a role in *E. coli* replication. The successful study of DNA synthesis has often resulted from the use of correlative approaches: one has been to isolate and characterise DNA molecules at various stages during replication and thereby gain insight into the mechanism of DNA replication (Clayton, 1982). A second has been to isolate and purify enzymes and other factors required for the initiation, elongation and segregation of new DNA daughter strands. With bacterial and viral systems, temperature sensitive mutants have greatly aided enzyme studies, since the investigator could utilise both biochemical and genetic analyses (Adams *et al.*, 1986). It is likely that multiple approaches will also be necessary to elucidate the replication of cpDNA and its control.

It is our view that isolated intact chloroplasts should be useful in studying chloroplast DNA synthesis, since intact plastids are likely to contain all the enzymes, templates, DNA binding proteins and other factors necessary for cpDNA replication and thus to allow correct initiation, elongation and segregation *in vitro*. A similar system has been valuable in examining chloroplast protein synthesis (Ellis, 1981). We have prepared chloroplasts from pea leaves and followed cpDNA synthesis under conditions that ensure that incorporation is occurring in intact organelles (Mills & Baumgartner, 1983). Although the intact plastid system has proved useful for several types of study, isolated chloroplasts have not been commonly utilised to examine cpDNA biosynthesis, possibly because of the difficulty in isolating intact chloroplasts from many organisms (Zimmermann & Weissbach, 1982) and because the isolated plastids showed relatively poor synthetic activity (Bohnert, Schmitt & Hermann, 1974) compared to intact pea chloroplasts.

Early studies on DNA synthesis in isolated chloroplasts

It has been approximately 20 years since the first reports on DNA synthesis in isolated chloroplasts appeared (Spencer & Whitfeld, 1967; Tewari & Wildman, 1967). Only a few additional studies have been published (Scott, Shah & Smillie, 1968; Spencer & Whitfield, 1969; Bohnert et al., 1974; Keller & Ho, 1981; Zimmermann & Weissbach, 1982; Mills & Baumgartner, 1983). In almost all of these, activity required the presence of $MgCl_2$ and all four deoxyribonucleoside triphosphates. In addition, when DNase was included in the reaction mixture it completely abolished the incorporation of labelled precursors into DNA. These data indicate that the observed activity was occurring in chloroplast fragments lacking complete envelopes rather than in intact organelles. Nevertheless, exogenous template was not required; and when the DNA synthesised by the plastid preparations was assayed by either density gradient centrifugation (Tewari & Wildman, 1967; Scott et al., 1968; Spencer & Whitfeld, 1969; Keller & Ho, 1981), hybridization analysis (Tewari & Wildman, 1967) or restriction enzyme digestion (Zimmermann & Weissbach, 1982) the label was present mainly in chloroplast DNA. Bohnert et al. (1974) were the first to utilise conditions favouring synthesis by intact chloroplasts. In this case radiolabelled deoxyribonucleosides (instead of deoxyribonucleoside triphosphates) were employed as the tracer and light was used as the energy source; in contrast to most other studies, labelled TTP was not readily incorporated into DNA. If synthesis is occurring mainly in intact chloroplasts this result might be expected, since nucleoside triphosphates are not thought to be carried across the chloroplast envelope at high rates. More recently (Mills & Baumgartner, 1983), it was confirmed that isolated plastids are able to use light as the sole energy source to drive [^{3}H]–thymidine incorporation into DNA, and additional evidence was obtained that only intact chloroplasts participate in this process. Based on these data a series of experiments were begun aimed at gaining information regarding the nature and regulation of chloroplast DNA biosynthesis.

Table 1. *Characteristics of light- and ATP-driven [^{3}H]–thymidine incorporation into DNA by isolated intact pea chloroplasts.*[a]

Treatment	[^{3}H]–thymidine incorporation (% of light control)	
	Light	Dark
Control 100 1		
MgCl$_2$ (10 mM)	33	14
ATP (10 mM)	109	30
ATP (10 mM) + MgCl$_2$ (10 mM)	253	149[b]
Osmotically lysed	4	1
Triton X–100 (1%)	3	9
DNase I (10 μg/ml)	81	99
DCMU (5 mg/ml)	3	ND[c]
NH$_4$Cl (10 mM)	1	ND
Ethidium bromide (500 μM)	1	4
N–Ethylmaleimide (8 mM)	2	4
Rifampicin (100 μg/ml)	38	43

[a]Chloroplasts were isolated from 9-day-old pea (*Pisum sativum L.* var Little Marvel) plants using a modification of the method of Mills & Joy (1980). Following Polytron homogenisation of leaf tissue (about 1 g fresh weight/5 ml medium) in ice cold homogenisation medium (300 mM sorbitol, 50 mM Tricine (pH 7.9), 10 mM NaH$_2$PO$_4$, 2 mM MgCl$_2$, 2 mM EDTA, 4 mM 2–mercaptoethanol and 0.1 per cent bovine serum albumin), the brei was filtered through 2 layers of Miracloth. After reducing nuclear contamination by spinning at 500 × g, plastids were pelleted through 40 per cent Percoll containing homogenisation medium. Chloroplasts were judged to be greater than 90 per cent intact using either phase contrast microscopy (Spencer & Unt, 1965) or the ferricyanide method of Lilley *et al.* (1975). Chlorophyll and protein were determined using the procedure of Wintermans & DeMots (1965) and Bradford (1976) respectively. Chloroplast numbers were estimated by haemocytometer counting. Intact plastids were assayed in the following medium: 300 mM sorbitol, 50 mM EPPS (pH 8.3) and 30 mM KCl in the light. For ATP-driven synthesis in darkness, samples additionally contained ATP and MgCl$_2$ at 10 mM each; in this case tubes were wrapped in aluminium foil. Typically, samples also contained 3.3 μCi [methyl–^{3}H]–thymidine and were incubated at 25 °C. After incubation, reactions were stopped by adding 10 per cent trichloroacetic acid and extracted using a modification of the method described earlier (Mills & Baumgartenr, 1983).
[b]All subsequent dark samples contained ATP and MgCl$_2$ at 10 mM each.
[c]Not determined

Light-driven DNA synthesis in intact chloroplasts

Chloroplasts isolated from 6–12-day old pea plants readily incorporated [^{3}H]–thymidine into acid-insoluble material using light as the sole energy source

(Table 1; Fig. 1), as little synthesis was seen in darkness. There was no requirement for ATP, $MgCl_2$ or deoxynucleotides (other than [³H]–thymidine). With these conditions, incorporation rates were 5–80 pmol/mg chl/h measured over a 40 min period; however, incorporation was linear for only a rather short period as the rates began to approach zero after 10–25 min. The addition of 10 mM ATP and 10 mM $MgCl_2$ (but not deoxynucleotides) had little effect on initial incorporation (Fig. 1) but markedly extended the labelling period. In isolated chloroplasts, not only light-driven DNA synthesis but also protein (Mills & Wilson, 1978), amino acid (Mills, Lea & Miflin, 1980) and RNA (Hartley & Ellis, 1973) synthesis proceed at rapid rates for only 10 to 30 minutes and begin to plateau thereafter. Our data suggest that loss of ATP during incubation may be an important factor (though almost certainly not the only one) contributing to this general loss of activity with time.

Several pieces of evidence suggest that light-driven synthesis is occurring mainly in intact organelles (Table 1; Fig. 1). First, since broken chloroplasts cannot

Fig. 1. Influence of light and MgATP on thymidine incorporation into DNA by isolated chloroplasts. Each sample contained (2 ml final volume) 325 μg chlorophyll and 45 μCi of [³H]–thymidine. Control tubes contained the following: 300 mM sorbitol, 50 mM EPPS (pH 8.3), 30 mM KCl, chloroplasts and radiolabelled thymidine. ATP and $MgCl_2$ (10 mM each) were present in the indicated samples. At specified intervals, 200 μl aliquots were removed, added to 800 μl 10 per cent TCA containing 10 mM sodium pyrophosphate and processed for scintillation counting. Dark samples were wrapped in aluminium foil.

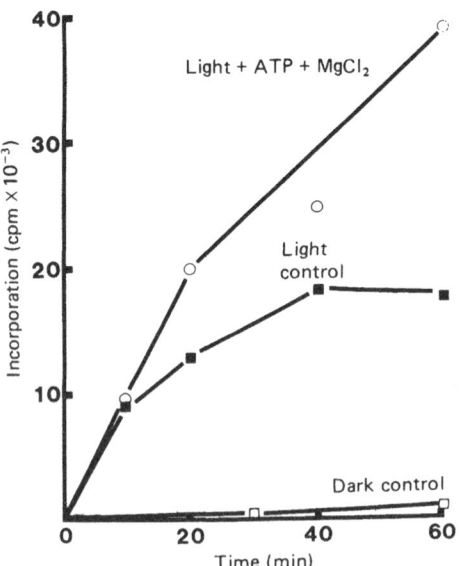

synthesise ATP in the absence of exogenous substrates (Ellis, 1981), light-dependence provides strong evidence that synthesis is by intact plastids. Second, ruptured chloroplasts exhibited little ability for DNA synthesis since both osmotically and triton X–100 disrupted organelles were inactive (Table 1). Since Triton X–100 solubilises chloroplasts but not nuclei or bacteria (Parenti & Margulies, 1967), the loss of activity in the presence of Triton X–100 further indicates that nuclear or bacterial contaminants were not significant. Third, DNase inclusion in the incubation medium only slightly reduced labelled thymidine incorporation (Table 1). This is expected if synthesis only occurs in plastids with intact outer membranes which would prevent exogenous enzymes from penetrating the organelle. Fourth, synthesis did not require added ribo- or deoxyribo-nucleosides or nucleotides other than [^{3}H]–thymidine. Although this contrasts with most previous studies (e.g. Zimmerman & Weissbach, 1982), where cpDNA synthesis was strongly dependent on added ribo- and deoxyribonucleotides, such a result might be anticipated since intact chloroplasts should retain their stromal constituents, including deoxynucleotides. Finally, DNA synthesis was strongly inhibited by photophosphorylation inhibitors; both the electron flow inhibitor DCMU and the uncoupler NH_4Cl nearly totally blocked labelled thymidine incorporation at the concentrations tested (Table 1).

DNA synthesised in isolated intact chloroplasts was analysed by restriction endonuclease and exonuclease III treatments followed by agarose gel electrophoresis. The ethidium bromide staining pattern of restriction enzyme digests (Fig. 2, lane A) was typical for pea cpDNA (McKown & Tewari, 1984). In addition, autoradiography of the gels revealed identical patterns, with each stained band being radioactively labelled (Fig. 2, lane B). This indicates that [^{3}H]–thymidine was incorporated into double stranded cpDNA and that labelling occurred throughout the entire genome. The presence of labelled closed circular DNA was examined by exonuclease III digestions. This enzyme degrades double stranded DNA with a 3'–hydroxyl end (i.e. linear or nicked circular molecules) but does not digest closed circular DNA molecules (Adams *et al.*, 1986). Autoradiography revealed a labelled gel band which was resistant to exonuclease III (data not shown). When gels were sliced into fractions and analysed by liquid scintillation counting, from 5–15 per cent of the label loaded on the gel was recovered in the resistant band. The activity of the exonuclease III was confirmed since it completely eliminated Eco R1 restriction fragments of cpDNA (gel not shown).

The data outlined above clearly indicate that cpDNA is being labelled in the isolated intact chloroplasts. They do not, however, allow us to conclude that replication type synthesis is occurring *in organello*. Despite this, several pieces of evidence suggest that the observed activity may reflect cpDNA replication. First, our experimental conditions ensure that incorporation is proceeding in intact chloroplasts (Table 1; Fig. 1; Mills & Baumgartner, 1983). Since they retain their membrane integrity, stromal enzymes, cofactors and substrates, their internal conditions are likely to mimic those found *in vivo* (Ellis, 1981) and thus favour cpDNA replication. Second, both restriction endonuclease and exonuclease III analysis indicate that the entire chloroplast genome becomes radiolabelled in isolated plastids and that whole supercoiled (closed circular) molecules are formed. Again, these results are consistent with replication type synthesis. Third, developmental changes in synthetic activity of the isolated plastids (see section on development below) are well correlated with changes in the numbers of cpDNA molecules *in vivo*. Finally, Gold *et al.*, (1987) concluded that a partially purified pea chloroplast extract (which is likely to be less complete than intact chloroplasts) exhibited replication rather than repair type activity. Nevertheless, additional studies will be necessary to establish clearly that replication is proceeding in the isolated plastids.

Fig. 2. Analysis of DNA synthesised by isolated intact pea chloroplasts in light. Samples were incubated as described in Table 1. Labelled DNA was isolated using the 'heat extraction' method of Baumgartner, Riley & Mills (1986). Radioactivity incorporated into cpDNA was analysed by *Sal* I or exonuclease III digestion and agarose gel electrophoresis (Maniatis, Fritsch & Sambrook, 1982). Following electrophoresis on 0.7 per cent agarose, gels were treated with Enhance (New England Nuclear), dried on filter paper and analysed by exposure to Kodak X–Omat film at −70 °C. The presence of closed circular DNA molecules was examined by ethidium bromide titrations on agarose gels (Johnson & Grossman, 1977) or by exonuclease III digestion studies (Adams *et al.*, 1986). DNA amounts were estimated by measuring the absorbance at 260 nm (Maniatis *et al.*, 1982) or by the diphenylamine colorimetric method (Burton, 1956). Purity of the DNA was estimated from A_{260}/A_{280} ratios. These ranged from 1.7–1.95 and are indicative of low protein contamination.

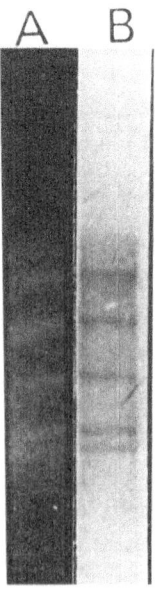

A   B

ATP-driven DNA synthesis by intact chloroplasts in darkness

Several chloroplast enzymes are subject to light/dark regulation via thiol/disulphide exchange (Cseke & Buchanan, 1986). Therefore, we asked the following question: are enzymes of cpDNA biosynthesis subject to light activation by the ferredoxin/thioredoxin system in a manner similar to enzymes of CO_2 photoassimilation? As one way to answer this question we sought to find conditions which allowed ATP-driven synthesis in intact isolated plastids in the dark.

Chloroplasts isolated from young (7–10-day old) pea shoots incorporated [^{3}H]–thymidine into acid precipitable material in darkness if both 10 mM ATP and 10 mM $MgCl_2$ (MgATP) were present (Table 1). In their absence, little assimilation was observed. The greatest activity was seen when ATP and $MgCl_2$ were either equimolar or when $MgCl_2$ was in slight excess (3–5 mM). If either $MgCl_2$ or ATP were substantially higher they were inhibitory (data not shown). The MgATP combination was usually 3–5 times as effective as ATP alone (Table 1). Marked [^{3}H]–thymidine incorporation in darkness did not require exogenous deoxyribonucleotides or nucleosides other than labelled thymidine. The rates of MgATP-driven thymidine incorporation ranged from 26–100 pmol/mg chl/h and are equal to (Mills & Baumgartner, 1983) or greater than (Table 1) light driven synthesis. The time course for dark (MgATP-driven) DNA synthesis is similar to the light plus MgATP incorporation seen in Fig. 1. Since the rates of MgATP-driven thymidine assimilation in the dark were similar to those observed for light-driven activity, the enzymes of cpDNA synthesis do not appear to be activated by light (Table 1). Added support for this view comes from dithiothreitol (DTT) preincubation studies. DTT is able to mimic light by reducing and activating certain chloroplast enzymes (Buchanan, 1986). DTT pretreatments had little influence on either light- or MgATP-driven (dark) synthesis (data not shown). These results are consistent with the suggestion that cpDNA synthesis enzymes are not modulated by the thioredoxin system and support the observation by Rose, Cran & Possingham (1975) that the rates of cpDNA synthesis remain relatively high in dark grown tissues.

As with light-driven synthesis, several lines of evidence suggest that MgATP-stimulated synthesis occurs primarily in intact plastids (Table 1). First, lysed chloroplasts showed little capacity to catalyse thymidine incorporation, as neither osmotically ruptured nor Triton X–100 disrupted plastids were active. As we discussed earlier, this also indicates that nuclear and bacterial contamination was not significant (Parenti & Margulies, 1967). Second, exogenous deoxynucleosides or deoxynucleotides were not required for synthesis; as discussed earlier, this result would only be expected if incorporation were occurring in intact organelles. Finally, DNase I had little effect when present in the incubation medium; thus, only intact chloroplasts appear to be synthetically active (Table 1).

Tables 1 and 2 illustrate the influence of several inhibitors on both light- and MgATP-driven thymidine incorporation. The DNA intercalator ethidium bromide and

Table 2. *Influence of DNA gyrase inhibitors on [³H]–thymidine incorporation into DNA by isolated intact chloroplasts.*

	(% of Control)[a,b]	
	Light-driven[a] [³H]–thymidine incorporation	MgATP-driven[b] [³H]–thymidine incorporation in darkness
Nalidixic acid		
10 μg/ml	99	99
100 μg/ml	98	82
1000 μg/ml	73	72
Oxolinic acid		
10 μg/ml	79	99
100 μg/ml	82	83
1000 μg/ml	38	70
Novobiocin		
10 μg/ml	73	94
100 μg/ml	66	87
1000 μg/ml	33	42
Coumermycin		
10 μg/ml	56	98
100 μg/ml	53	76
1000 μg/ml	30	65

[a]Samples were incubated for 30 minutes in light in the following medium: 300 mM sorbitol, 50 mM EPPS (pH 8.3), 30 mM KCl and 3.3 μCi [³H]–thymidine. Data represent averages of 3 experiments; rates of labelled thymidine incorporation ranged from 12.1 to 20 pmol/mg chl/h.
[b]Samples were incubated as described above for light-driven synthesis except that ATP and MgCl₂ were present in the medium (10 mM each) and that tubes were wrapped in aluminium foil. Data represent averages of four experiments; incorporation ranged from 14.3 to 16.8 pmol/mg chl/h.

the sulphydryl group inhibitor N–ethylmaleimide strongly inhibited synthesis (Table 1). These substances also inhibited purified cpDNA polymerase from pea (McKown & Tewari, 1984). The RNA polymerase inhibitor, rifampicin, which blocks initiation of replication in *E. coli* (Adams *et al.*, 1986) inhibited labelled thymidine incorporation (about 60 per cent inhibition) at 100 μg/ml (Table 1). This result is particularly interesting since chloroplast RNA polymerases are generally considered to be resistant to rifampicin (Greenberg *et al.*, 1985). Although these data suggest that

Table 3. *Influence of leaf age on [³H]–thymidine incorporation and CO₂ dependent O₂ evolution by isolated intact chloroplasts and on [³H]-TTP incorporation by chloroplast extracts.*

Assay	Activity[a,b] (% of 4th leaf)		
	2nd leaf	3rd leaf	4th leaf
[³H]–thymidine incorporation			
Light-driven	12	10	100
Dark ATP-driven	17	25	100
[³H]-TTP incorporation by chloroplast extracts	7	21	100
CO₂-dependent O₂ evolution	13	52	100

[a]Chloroplasts were isolated from leaves at the indicated age (stage) from 12-day-old pea plants.
[b]Mean activity in the fourth leaf was 15.5 and 16 pmol/mg chl/h in light and dark respectively for thymidine incorporation. 176 pmol/mg chl/h for TTP assimilation and 143 μmol/mg chl/h for CO_2-dependent O_2 evolution.

plastids may contain a rifampicin-sensitive RNA polymerase which catalyses a priming step in cpDNA replication, there is as yet no additional evidence to support this notion. The effect of DNA gyrase inhibitors on thymidine incorporation is shown in Table 2. DNA gyrase is a type II topoisomerase (Adams *et al.*, 1986) which is thought to be required for bacterial DNA replication since treatment of *E. coli* with either coumermycin or nalidixic acid immediately stops DNA synthesis. Gyrase has been postulated to remove positive superhelical turns induced into DNA during replication, to put negative supertwists into the DNA ahead of the replication fork thus making unwinding easier, or to facilitate decatenation of cyclic daughter DNA molecules (Adams *et al.*, 1986). When tested at 1000 μg/ml, each inhibitor reduced thymidine incorporation (Table 2). In general, coumermycin and novobiocin were more effective than nalidixic or oxolinic acids; and light-driven synthesis was inhibited more than MgATP-driven synthesis in the dark. These data suggest that chloroplasts contain a bacterial-like topoisomerase II which is necessary for cpDNA synthesis; however, the data must be viewed with caution, since rather high inhibitor concentrations were required (Castora & Simpson, 1979). Final proof must come from enzyme isolation and characterisation studies.

Effect of plant and tissue age on thymidine incorporation in isolated chloroplasts

Both light- and MgATP-driven thymidine incorporation are highly dependent upon the age of the leaf tissue from which the chloroplasts were isolated

(Table 3; Fig. 3). The highest rates of light-stimulated synthesis were observed in plastids isolated from young (6–8-day old) plants (Fig. 3; Mills & Baumgartner, 1983) with peak activity at day 8. Synthetic activity of older plants (greater than 10 days of age) was much lower. These data are well correlated with those of Lamppa & Bendich (1979) who showed that the level of cpDNA in intact pea leaves increased until day 8 or 9 then either stabilised or declined thereafter. MgATP-stimulated DNA synthesis in darkness was also strongly influenced by the age of the plants from which the chloroplasts were isolated (Fig. 3); again plastids from young shoots were 4–10 times more active than those from older plants. The similar results seen in light and darkness suggest that the variation in synthetic activity seen with age is due to changes in the activity of the DNA synthesising machinery *per se* rather than to variation in the activity of the photosynthetic electron transport system which is also known to fluctuate during development (Robinson & Wiskich, 1976; Holloway, Maclean & Scott 1983). Not surprisingly, the age of the leaves from which the plastids were prepared also affects DNA synthesis (Table 3). In this case,

Fig. 3. Influence of plant age on light- and MgATP-driven (dark) DNA biosynthesis in isolated intact pea chloroplasts. Chloroplasts were purified from leaves of pea plants of various ages and incubated as described in Table 1. Chlorophyll was determined for each plastid preparation and activity is expressed per mg of chlorophyll.

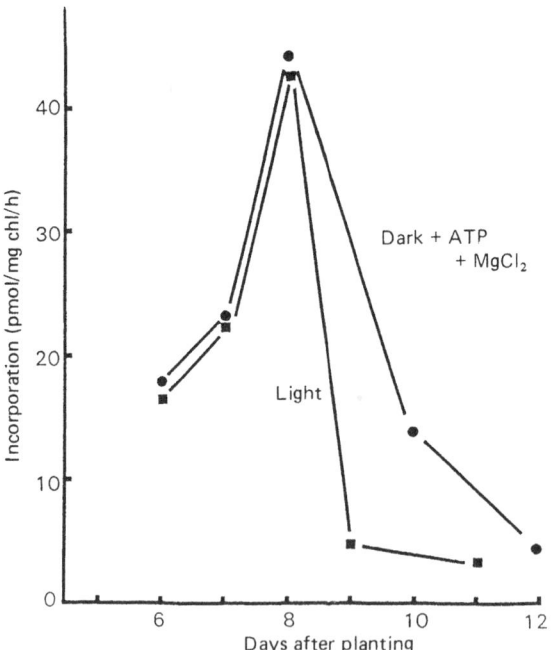

chloroplasts were isolated from the first (oldest), second and third (intermediate) and fourth (youngest) leaves of the same 12 day old plants (see Fig. 4). Activity in chloroplasts from the fourth (youngest) leaves was 4–10 times greater than that from the older leaves. Again, a similar pattern was seen for the light and MgATP-driven activity.

When conducting developmental studies with isolated chloroplasts, it is difficult to find an appropriate index for comparison. Rates of CO_2 photoassimilation and O_2 evolution have traditionally been expressed on a chlorophyll basis. However, it is well known that as leaf tissues mature the amount of chlorophyll per plastid increases severalfold (Leech, 1984). Thus, one might get a flawed view of developmental variation in synthetic activity by using chlorophyll as the sole index of comparison. To avoid this problem, we have also estimated protein levels and plastid numbers in the chloroplast preparations. Plastids isolated from the youngest tissues were 5–12 times more active than those from older tissues regardless of whether chlorophyll, protein or plastid number was used as the index (data not shown). Another reason for higher activity in plastids from young tissues could be increased percentage intactness in these samples. However, all preparations were purified by Percoll gradient centrifugation (Mills & Joy, 1980); therefore, plastid intactness was similar (approximately 90 per cent) for all samples.

Bouthyette & Jagendorf (1981) noted that chloroplasts from older (12–21 day-old) pea plants were less active in protein synthesis than those from younger (7–10-day

Fig. 4. Drawing of a 12-day old pea plant showing the developmental stages of leaves used in experiments described in Tables 3, 5 and 6. Leaf 1 is the oldest, while leaf 4 is the youngest.

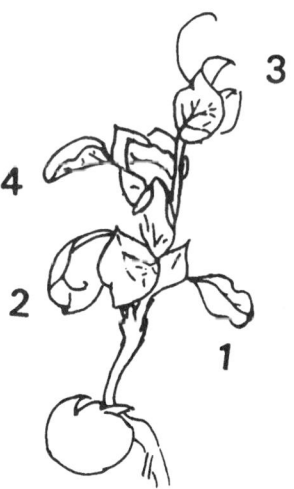

Table 4. *Influence of plant age, CaCl$_2$ and EGTA on light- and ATP-driven DNA biosynthesis in intact isolated chloroplasts.*

Age of Plants and Treatment	Isolation Conditions[a]			
	Minus EGTA[b]		Plus EGTA	
	Incubation Conditions[c]			
	Light	Dark	Light	Dark
		Thymidine incorporation (pmol/mg chl/h)		
7-day-old plants				
Control[c]	9.2 (100)[d]	11.6 (126)	3.6 (39)	11.6 (126)
CaCl$_2$ (10 mM)	1.2 (13)	4.5 (49)	0.7 (8)	3.9 (42)
EGTA (5 mM)	6.1 (66)	12.2 (133)	4.6 (50)	11.1 (121)
14-day-old plants				
Control	0.2 (3)	2.6 (28)	0.4 (4)	0.7 (7)
CaCl$_2$ (10 mM)	0.1 (1)	0.9 (10)	0.1 (1)	0.4 (4)
EGTA (5 mM)	0.4 (3)	2.5 (28)	0.2 (2)	1.1 (13)

[a]Chloroplasts were isolated and purified by Percoll gradient centrifugation as described by Mills & Joy (1980), except that 5 mM EGTA was included in the 'plus EGTA' isolation medium.
[b]EGTA = Ethyleneglycol–bis–(β–aminoethyl ether)N,N,N',N'–tetraacetic acid.
[c]Light controls were assayed in 300 mM sorbitol, 50 mM EPPS (pH 8.3) and 30 mM KCl. All samples contained 3.3 μCi [^{3}H]–thymidine and 50 μg chlorophyll. For dark incubations (tubes wrapped in aluminium foil) samples contained 10 mM ATP and 10 mM MgCl$_2$ in addition to the components listed above. All incubations were for 40 min.
[d]Numbers in parentheses are values expressed as percent of the 7-day light control.

old) plants. This was attributed to calcium inhibition of protein synthesis, because plastids from the older plants were found to contain elevated calcium levels and because the inhibition could be reversed by isolating the chloroplasts in a medium containing EGTA, a compound known to bind calcium tightly and specifically. To gain insight into the influence of calcium chloride and EGTA on labelled thymidine assimilation by isolated intact pea chloroplasts (Table 4), we have followed the influence of EGTA in the isolation medium, and both calcium chloride and EGTA in the incubation medium, on cpDNA synthesis. Calcium chloride (10 mM) inhibited both light- and MgATP-dependent (dark) thymidine incorporation, although light-stimulated synthesis was more sensitive (Table 4). However, inclusion of 5 mM

EGTA in the isolation or incubation medium did not restore the synthetic capacity in chloroplasts from older tissues. Therefore, though rising calcium levels in ageing pea leaf tissue may have an influence on DNA synthesis, it does not appear to be the only factor or even the key factor in mediating developmental variation in cpDNA biosynthesis.

[^{3}H]-TTP incorporation by chloroplast extracts

Here, we have reviewed several characteristics of intact chloroplast preparations which make them useful for studying cpDNA synthesis. However, there are also some disadvantages in employing intact plastids for developmental studies. For example, adenylate and phosphate uptake varies during greening of etioplasts to chloroplasts (Wellburn, 1984) with etiochloroplasts usually showing more rapid uptake than fully greened organelles. So an obvious question is this: is the variation in DNA synthesis by intact plastids simply due to changes in envelope permeability rather than to changes in the activity of enzymes of cpDNA replication? A second plastid characteristic which we know changes during development is the number of genome copies per plastid. We noted earlier that chloroplasts in young pea shoots have between 2 and 3 times as much DNA per plastid as those from older plants (Lamppa & Bendich, 1979). Therefore, one might anticipate higher DNA synthesis activity in young chloroplasts simply because of greater levels of template DNA. In order to circumvent these and other effects (e.g. developmental variation in stromal deoxynucleotide concentrations) we have also examined [^{3}H]-TTP incorporation by chloroplast extracts. Percoll purified chloroplasts were lysed in 2 per cent Triton X–100 then centrifuged at 18,000 × g for 10 min; the supernatant fraction was then used in DNA synthesis experiments. In agreement with McKown & Tewari (1984) and Gold *et al.* (1987), and in contrast to certain plastid RNA polymerases (Greenberg *et al.*, 1985), we found that enzymes of [^{3}H]-TTP incorporation were readily solubilised by Triton X–100, as about 90 per cent of the activity was recovered in the supernatant fraction (data not shown). With the incubation conditions used, TTP incorporation proceeds nearly linearly for over one hour (McKown & Tewari, 1984).

Table 5 illustrates some characteristics of [^{3}H]-TTP incorporation into DNA by pea chloroplast extracts. Activated calf thymus DNA was used as the control template since McKown & Tewari (1984) had previously shown it to be effective with pea chloroplast DNA polymerase. The plastid extract was capable of incorporating [^{3}H]-TTP without exogenous template (Table 5); however, when activated calf thymus was added at saturating levels (25 μg/ml) activity was stimulated 2–10 fold. Native calf thymus and isolated chloroplast DNAs were equally effective templates. By contrast the purified DNA polymerase from pea chloroplasts (McKown & Tewari, 1984) showed less activity with native calf thymus DNA and supercoiled pBR322 as templates. This indicates that the chloroplast extract contains nucleases (McKown & Tewari, 1984) and possibly topoisomerases (Saiedlecki, Zimmermann & Weissbach,

Table 5. *Characteristics of [³H]-TTP incorporation by pea chloroplast extracts*[a]

Treatment	Relative Activity
Control (Activated calf thymus DNA (25 μg/ml) template)	100
Minus exogenous DNA template	42
Native calf thymus DNA (25 μg/ml) template	86
Isolated pea chloroplast DNA (25 μg/ml) template	93
Poly rA:oligo dT (25 μg/ml) template	11
N-ethylmaleimide (1 mM)	17
Ethidium bromide (25 μM)	3
Aphidicolin (100 μM)	103
d₂CPT (1 μM)	99

[a]Reaction tubes (200 μl final volume) contained the following: 50 mM Tris–HCl (pH 8.0), 12 mM MgCl₂, 120 mM KCl, 1 μM each of dATP, dGTP, dCTP, TTP and 1 μCi [³H]-TTP, indicated DNA template and plastid extract. Samples were incubated for 1 hour. Radioactivity incorporated into DNA was determined as described in Mills & Baumgartner (1983). Mean activity in the activated calf thymus control was 240 pmol/mg chl/h.

1983) capable of relaxing closed circular cpDNA. Unlike spinach cpDNA polymerase, the pea extract did not use the synthetic template poly(rA):oligo(dT) (Table 5); however, the pea enzyme did resemble the spinach enzyme in its lack of sensitivity to dideoxycytidine triphosphate (Sala *et al.*, 1980). N-ethylmaleimide and ethidium bromide, which almost totally blocked thymidine assimilation by intact chloroplasts, also strongly inhibited TTP incorporated by the extracts (Table 4).

The highest rates of TTP incorporation were observed in plastid extracts from 6–9-day old plants: those from older tissues showed much less activity (Fig. 5). At saturating levels of DNA template, extracts from young tissues were 3–10 times more active than those from older tissues. In Fig. 5 protein is used as the index of comparison; however, similar patterns were found when either chlorophyll (Table 3) or plastid number were used as the indices. If an inhibitor is produced in plastids during leaf development, extracts from older plants should inhibit enzymes from younger tissues if they are mixed. When the less active extracts from older tissues were combined (1 : 1 ratio of protein) with samples from younger plants, activity was not lowered (data not shown). Thus, the loss of activity in plastid extracts from older plants does not appear to result from the production of an endogenous inhibitor. As with thymidine assimilation by intact plastids, leaf age again appears to have an important infuence on TTP incorporation by chloroplast extracts, since activity was approximately 4–10 times greater in samples from the youngest leaves (Tables 3 and 6).

Plastid extracts from young pea leaf tissues catalysed [^{3}H]-TTP incorporation at much higher rates than those from older tissues (Table 6, Fig. 5); this variation in activity is well correlated with changes in labelled thymidine assimilation into DNA by intact isolated chloroplasts (Table 3, Fig. 3). In contrast to intact organelles, soluble extracts are not subject to membrane permeability effects. Furthermore, since the assays were conducted with saturating template and deoxyribonucleotides,

Fig. 5. Influence of plant age on [^{3}H]-TTP incorporation by pea chloroplast extracts. Chloroplasts were purified from pea plants of varying ages as described in Table 1. The Percoll purified plastids were resuspended in extraction buffer (50 mM Tris–HCl (pH 8.0) and 2 per cent (v/v) Triton X–100). After 10 minutes incubation on ice, the samples were vortexed then centrifuged at 18,000 × g for 10 minutes. The resulting supernatant was decanted and used as the chloroplast extract. Samples (200 μl final volume) contained the following: 50 mM Tris–HCl (pH 8.0), 12 mM MgCl$_2$, 120 mM KCl, 1 μM each of dATP, dCTP, dGTP and TTP, 1 μCi [methyl–^{3}H]-TTP (specific activity 14 Ci/m mol), 5 μg of template DNA and chloroplast extract. Reactions were run at 30 °C for 30 to 60 minutes. Under these conditions, incorporation is nearly linear for over 1 hour (McKown & Tewari, 1984). Radiolabelled DNA was extracted by the method of Baumgartner *et al.* (1986); labelled TTP incorporation was determined by scintillation counting.

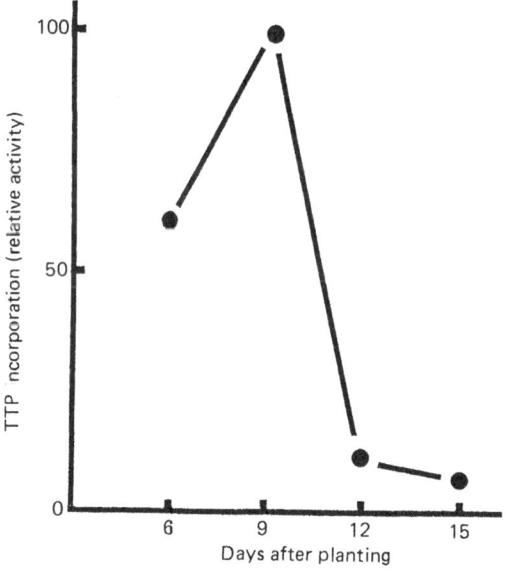

Table 6. *Influence of leaf age from which plastids were isolated on [³H]-TTP incorporation by pea chloroplast extracts*[a].

Leaf stage[b]	Polymerase activity (pmol/mg protein/h)
First leaf	0.59
Second leaf	0.64
Third leaf	1.89
Fourth leaf	8.84

[a]The assay mixture (200 μl final volume) contained 50 mM Tris–HCl (pH 8.0), 12 mM MgCl$_2$, 120 mM KCl, 1 μM each dATP, dGTP, dCTP, TTP 1 μCi [³H]-TTP, 5 μg of activated calf thymus DNA template and plastid extract. Radioactivity incorporated into DNA was determined as described in Mills & Baumgartner (1983).
[b]Chloroplasts were isolated from leaves at various stages of development from the same 12-day old plants (see Fig. 4). Note that the first leaves are the oldest while the fourth leaves are the youngest.

changes in substrate levels cannot account for the observed age-dependent fluctuations in catalytic activity. One explanation for the age variation is that synthetic capacity changes in response to developmental changes in the amounts of cpDNA polymerase and possibly other enzymes required for cpDNA replication. Variations in the activity of DNA polymerases during development are known in other organisms; for example, in cultured mouse cells the activity of DNA polymerase α changes dramatically with the rate of cell division (Adams *et al.*, 1986). Nevertheless, additional enzyme purification, characterisation and immunological studies will be required before we can fully assess the importance of any variation in the enzymes of cpDNA biosynthesis in controlling the replication of the chloroplast genome.

Summary

In summary, isolated intact chloroplasts purified from young pea leaves catalyse the light-dependent incorporation of labelled thymidine into cpDNA. Analysis of the labelled DNA clearly indicates that cpDNA is being produced. Intact plastids can also carry out ATP-driven cpDNA synthesis in darkness, provided that an appropriate amount of $MgCl_2$ is also present. Both light- and MgATP-driven synthesis are dependent on the age of leaf tissue from which the plastids are isolated, since only chloroplasts from young plants or the youngest leaves of older plants are highly active. The variation in synthetic activity in the isolated organelles appears to be physiologically relevant, since fluctuations in activity are well correlated with age-dependent changes in cpDNA *in vivo* (Lamppa & Bendich, 1979; Lamppa *et al.*, 1980).

Although our experimental results are consistent with replication type synthesis in the isolated intact chloroplasts, we cannot exclude the possibility of repair synthesis. Kolodner & Tewari (1975) used electron microscopy and denaturation mapping to show that maize and pea cpDNA replication initiates with the formation of two displacement loops which expand toward one another to form a Cairns replication intermediate. If initiation of cpDNA synthesis at specific sites could be shown in isolated chloroplasts, it would provide evidence for replication type synthesis there. Our attempts to partially synchronise cpDNA replication (which is necessary for studies of this type) have thus far been unsuccessful. Now we are attempting to analyse radiolabelled replicative intermediates produced *in organello*. We hope that these and other approaches will soon provide more useful information.

If we can establish that cpDNA replication is proceeding in the isolated chloroplasts, we feel they will be very useful in identifying origins of replication and studying the regulation of cpDNA synthesis. Even now the intact chloroplast system provides a simple means for specifically radiolabelling cpDNA. As we noted earlier, DNA replication is a complex process. Information regarding its mechanism or regulation has usually required the use of multiple experimental approaches (Adams *et al.*, 1986); the isolated chloroplast system is one of several which we feel will ultimately be of use in understanding chloroplast DNA replication.

Acknowledgements

This work was supported by the University Available Fund, the University of Houston–Clear Lake Organised Research Fund, a Sigma Xi Grant-in-Aid-of-Research and in part by National Science Foundation Grant PCM–8314328. We are grateful to S.F. Capo, R.C. Hopkins, M.R. Kasschau, J. Nye and S.P. Riley for helpful discussions, to J. Rheo Jr. for technical assistance and to B. Baumgartner for the artwork.

References

Adams, R.L.P., Kowler, J.T. & Leader, D.P. (1986). *The Biochemistry of the Nucleic Acids, Tenth Edition*. p.526., Chapman & Hall, London.

Baumgartner, B.J., Riley, S.P. & Mills, W.R. (1986). A rapid heat extraction method for the isolation of chloroplast DNA. *Plant Physiology*, **80**, S–10.

Boffey, S.A. (1985). The chloroplast division cycle and its relation to the cell division cycle. In *The Cell Division Cycle in Plants*. ed. J.A. Bryant & D. Francis. pp.233–46. Cambridge University Press.

Boffey, S.A. & Leech, R.M. (1982). Chloroplast DNA levels and the control of chloroplast division in light-grown wheat leaves. *Plant Physiology*, 69, 1387–91.

Bohnert, H.J., Schmitt, J.M. & Hermann, R.G. (1974). Structural and functional aspects of the plastome. III. DNA– and RNA–synthesis in isolated chloroplasts. *Portugaliae Acta Biologica*, 14, 71–90.

Bouthyette, P.–Y. & Jagendorf, A.T. (1981). Calcium inhibition of amino acid incorporation by pea chloroplasts and the question of loss of activity with age. In *Photosynthesis*. ed. G. Akoyunoglu, Vol. V. pp. 599–609. Balaban International Science Service, Philadelphia.

Bradford, M.M. (1976). A rapid and sensitive method for the quantification of microgram quantities of protein utilizing the principle of protein-dye binding. *Analytical Biochemistry*, 72, 248–54.

Buchanan, B.B. (1986). The ferredoxin/thioredoxin system. In *Thioredoxin and Glutaredoxin Systems: Structure and Function*. ed. A. Holmgren. Raven Press, New York.

Burton, K. (1956). A study of the conditions and mechanism of the diphenylamine reaction for the colorimetric estimation of deoxyribonucleic acid. *Biochemical Journal*, 62, 315–23.

Castora, F.J. & Simpson, M.V. (1979). Search for a DNA gyrase in mammalian mitochondria. *The Journal of Biological Chemistry*, 254, 11193–5.

Clayton, D.A. (1982). Replication of animal mitochondrial DNA. *Cell*, 28, 693–705.

Cseke, C. & Buchanan, B.B. (1986). Regulation of the formation and utilization of photosynthate in leaves. *Biochemica et Biophysica Acta*, 853, 43–63.

Dyer, T.A. (1984). The chloroplast genome: its nature and role in development. In *Chloroplast Biogenesis*. ed. N.R. Baker & J. Barber. pp. 23–69. Elsevier, Amsterdam.

Ellis, R.J. (1981). Chloroplast proteins: synthesis, transport and assembly. *Annual Review of Plant Physiology*, 32, 111–37.

Gold, B., Carrillo, N., Tewari, K.K. & Bogorad, L. (1987). Nucleotide sequence of a preferred maize chloroplast genome template for *in vitro* DNA synthesis. *Proceedings of the National Academy of Science, USA*, 84, 194–8.

Greenberg, B.M., Narita, J.O., DeLuca–Flaherty, C.R. & Hallick, R.B. (1985). Properties of chloroplast RNA polymerase. In *Molecular Biology of the Photosynthetic Apparatus*. Cold Spring Harbor Laboratory.

Hartley, M.R. & Ellis, R.J. (1973). Ribonucleic acid synthesis in chloroplasts. *Biochemical Journal*, 134, 249–62.

Holloday, P.J., Maclean, D.J. & Scott, K.J. (1983). Rate-limiting steps of electron transport in chloroplasts during ontogeny and senescence of barley. *Plant Physiology*, 72, 795–801.

Johnson, P.H. & Grossman, L.I. (1977). Electrophoresis of DNA in agarose gels. Optimizing separations of conformational isomers of double- and single-stranded DNAs. *Biochemistry*, 16, 4217–25.

Keller, S.J. & Ho, C. (1981). Chloroplast DNA replication in *Chlamydomonas reinhardtii*. *International Review of Cytology*, 69, 157–90.

Kolodner, R.D. & Tewari, K.K. (1975). Chloroplast DNA from higher plants replicates by both the Cairns and the rolling circle mechanism. *Nature*, 256, 708–11.

Lamppa, G.K. & Bendich, A.J. (1979). Changes in chloroplast DNA levels during development of pea (*Pisum sativum*). *Plant Physiology*, 64, 126–30.

Lamppa, G.K., Elliot, L.V. & Bendich, A.J. (1980). Changes in chloroplast number during pea leaf development. An analysis of a protoplast population. *Planta*, **148**, 437–43.

Leech, R.M. (1984). Chloroplast development in angiosperms: Current knowledge and future prospects. In *Chloroplast Biogenesis*. ed. N.R. Baker & J. Barber, pp. 1–21. Elsevier, Amsterdam.

Lilley, R.McC., Fitzgerald, M.P., Rientis, K.G. & Walker, D.A. (1975). Criteria of intactness and the photosynthetic activity of spinach chloroplast preparations. *New Phytologist*, **75**, 1–10.

Maniatis, T., Fritsch, E.F. & Sambrook, J. (1982). In *Molecular Cloning. A Laboratory Manual*. Cold Spring Harbor Laboratory.

McKown, R.L. & Tewari, K.K. (1984). Purification and properties of a pea chloroplast DNA polymerase. *Proceedings of the National Academy of Science, USA*, **81**, 2354–8.

Mills, W.R. & Baumgartner, B.J. (1983). Light-driven DNA biosynthesis in isolated pea chloroplasts. *Federation of European Biochemical Society Letters*, **163**, 124–7.

Mills, W.R. & Joy, K.W. (1980). A rapid method for isolation of purified physiologically active chloroplasts, used to study the intracellular distribution of amino acids in pea leaves. *Planta*, **148**, 75–83.

Mills, W.R., Lea, P.J. & Miflin, B.J. (1980). Photosynthetic formation of the aspartate family of amino acids in isolated chloroplasts. *Plant Physiology*, **65**, 1166–72.

Mills, W.R. & Wilson, K.G. (1978). Effects of lysine, threonine and methionine on light-driven protein synthesis in isolated pea (*Pisum sativum L.*) chloroplasts. *Planta*, **142**, 153–60.

Parenti, F. & Margulies, M. (1967). *In vitro* protein synthesis by plastids of *Phaseolus vulgaris*. I. Localization of activity in the chloroplasts of a chloroplast containing fraction from developing leaves. *Plant Physiology*, **42**, 1179–82.

Possingham, J.V. (1980). Plastid replication and development in the life cycle of higher plants. *Annual Review of Plant Physiology*, **31**, 113–29.

Robinson, S.P. & Wiskich, J.T. (1976). Stimulation of carbon dioxide fixation in isolated chloroplasts by catalytic amounts of adenine nucleotides. *Plant Physiology*, **58**, 156–62.

Rose, R.J., Cran, D.G. & Possingham, J.V. (1975). Changes in DNA synthesis during cell growth and chloroplast replication in greening spinach leaf disks. *Journal of Cell Science*, **17**, 27–41.

Sala, F., Amileni, A.R., Parisi, B.& Spadari, S. (1980). A γ-like DNA polymerase in spinach chloroplasts. *European Journal of Biochemistry*, **112**, 211–17.

Scott, N.S. & Possingham, J.V. (1983). Changes in chloroplast DNA levels during growth of spinach leaves. *Journal of Experimental Botany*, **34**, 1756–67.

Scott, N.S., Shah, V.C. & Smillie, R.M. (1968). Synthesis of chloroplast DNA in isolated chloroplasts. *Journal of Cell Biology*, **38**, 151–7.

Siedlecki, J., Zimmermann, W. & Weissbach, A. (1983). Characteristics of a prokaryotic topoisomerase I activity in chloroplast extracts from spinach. *Nucleic Acid Research*, **11**, 1523–38.

Spencer, D. & Whitfeld, P.R. (1967). DNA synthesis in isolated chloroplasts. *Biochemical and Biophysical Research Communications*, **28**, 538–42.

Spencer, D. & Whitfeld, P.R. (1969). The characteristics of spinach chloroplast DNA polymerase. 1969. *Archives of Biochemistry and Biophysics*, **132**, 477–88.

Spencer, D. & Unt, H. (1965). Biochemical and structural correlation in isolated spinach chloroplasts under isotonic and hypotonic conditions. *Australian Journal of Biological Science*, **18**, 197–210.

Tewari, K.K. & Wildman, S.G. (1967). DNA polymerase in isolated tobacco chloroplasts and nature of the polymerized product. *Proceedings of the National Academy of Sciences, USA*, **58**, 689–96.

Tymms, M.J., Scott, N.S. & Possingham, J.V. (1983). DNA content of *Beta vulgaris* chloroplasts during leaf cell expansion. *Plant Physiology*, **71**, 785–8.

Wellburn, A.R. (1984). Ultrastructural, respiratory and metabolic changes associated with chloroplast development. In *Chloroplast Biogenesis*. ed. N.R. Baker & J. Barber. pp. 1–21. Elsevier, Amsterdam.

Wintermans, J.E.G.M. & De Mots, A. (1965). Spectrophotometric characteristics of chlorophyll and their pheophytins in ethanol. *Biochemica et Biophysica Acta*, **109**, 448–53.

Zimmermann, W. & Weissbach, A. (1982). Deoxyribonucleic acid synthesis in isolated chloroplasts and chloroplast extracts of maize. *Biochemistry*, **21**, 3334–43.

J.A. BRYANT

DNA polymerases

DNA polymerases (more strictly, DNA-dependent DNA polymerases or DNA-dependent nucleotidyl transferases E.C. 2.7.7.7) are enzymes which carry out DNA synthesis. The process involves the successive insertion of deoxyribonucleoside monophosphates in a 5'→3' direction using parental strands as templates. The substrates for polymerisation are deoxyribonucleoside triphosphates; pyrophosphate is eliminated in the reaction and a phosphodiester linkage is formed between the remaining phosphate and the 3'–OH group of the previously inserted deoxyribonucleoside monophosphate (Fig. 1). Energetically, the reaction is driven by

Fig. 1. The polymerisation reaction catalysed by DNA polymerase.

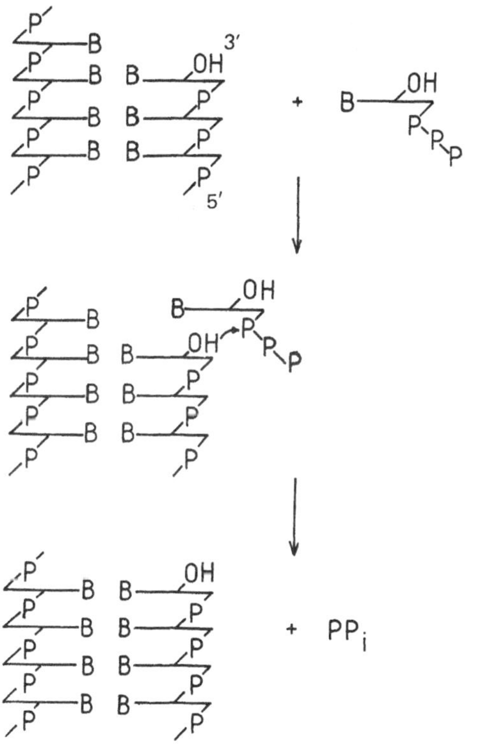

hydrolysis of the α–β phosphate ester linkage in the substrate, and, informationally the insertion of the correct deoxyribonucleotide is directed by its complement on the parental strand. This process can take place over a short stretch of DNA, as in DNA repair, or can lead to complete semi-conservative replication of the original DNA molecule.

The biochemistry of DNA-dependent DNA replication is universal, but the process takes place with parental DNA molecules of differing topologies, and there are also differences in the way that the replication process is organised. This is well illustrated by comparison of DNA replication in the nucleus with that in mitochondria or chloroplasts. Nuclear DNA is organised for replication as a series of replicons, each consisting functionally of an origin (from which replication proceeds bi-directionally) and two termini (Edenberg & Huberman, 1975; Van't Hof, 1975). As is evident from Fig. 2, the initiation of replication at an origin will result in the formation of two leading strands in which synthesis proceeds in a continuous manner, after a single priming event, and two lagging strands where synthesis is discontinuous, proceeding via the formation of Okazaki fragments each of which needs its own priming event. In the circular DNA molecules of chloroplasts and mitochondria, replication is initiated by the formation of a displacement loop or D-loop (Tewari, Kolodner & Dobkin, 1976; Clayton, 1982) caused by the synthesis, after priming, of a short piece of daughter strand DNA on only one parental strand (Fig. 3). In mitochondria, and probably also in chloroplasts, the D-loop can exist as a semi-permanent structure, although the short piece of daughter strand DNA is subject to turnover (Clayton, 1982). When full-scale replication does start, only one strand is initially copied. Second strand synthesis begins at a different origin of replication and is not initiated until first strand synthesis has proceeded some way round the circle. This mechanism does not involve the formation of Okazaki fragments and only two priming events are needed, one for each strand.

Fig. 2. Formation of leading (continuous) and lagging (discontinuous) strands after initiation at a replication origin in nuclear DNA.

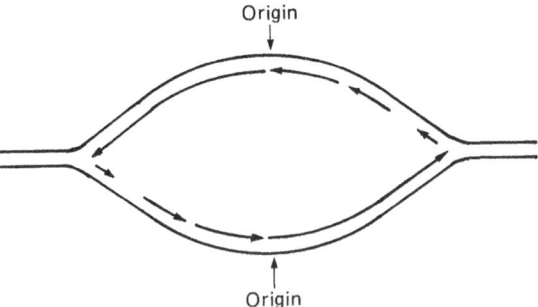

The difference between nuclei and mitochondria or chloroplasts also implies differences in the relationships between the enzymes involved. In particular, there are likely to be differences in the relative activities of DNA polymerase as compared with activities involved in synthesis and removal of RNA primers and in joining fragments (see Bryant, 1986; Fry & Loeb, 1986; Bryant & Dunham, 1988); differences in the properties of the DNA polymerases themselves are also likely.

Fig. 3. The D-loop mechanism for replication of organellar DNA. Dotted lines indicate the *continuous* synthesis of daughter strands. 1 and 2 indicate the origins of replication of each daughter strand.

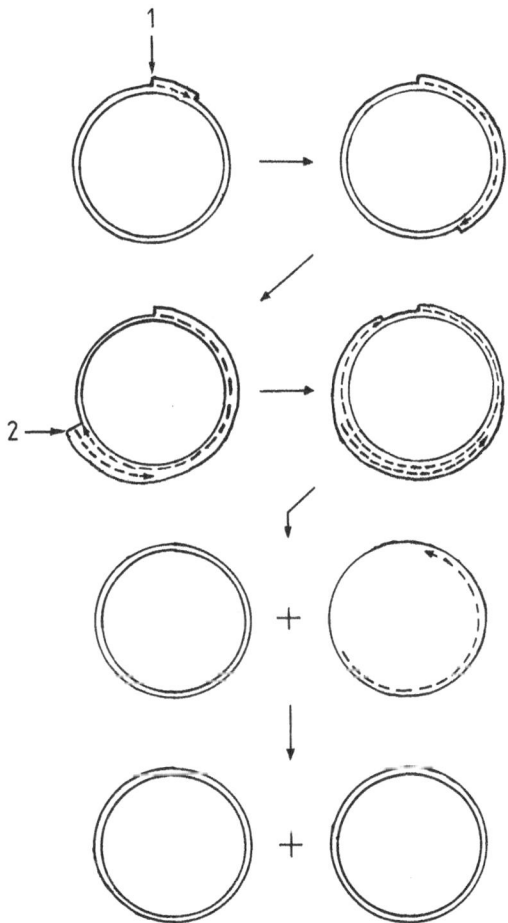

Animal cell DNA polymerases

Of all the DNA polymerases so far investigated, those of vertebrate animals have received by far the most attention. Cells of vertebrates (and in fact of advanced invertebrates such as insects) contain three well-defined different types of DNA polymerase (Table 1). These are called polymerase–α, polymerase–β and polymerase–γ. Polymerase–α and polymerase–β are very readily distinguished by their inhibitor responses: pol–α is inhibited by aphidicolin, phosphono–acetate, N–ethyl–maleimide and high concentrations of NaCl or KCl, but not by phosphate; pol–β is inhibited by phosphate but not by aphidicolin, phosphono–acetate, N–ethyl–maleimide or high concentrations of NaCl or KCl. DNA polymerase–γ is, in terms of activity, only a minor form, making up between 1 and 10 per cent of the total. Its obvious distinguishing characteristic is that, in the presence of Mn^{2+} it has a very marked preference for poly(rA).oligo(dT) as a template–primer system. Of these three, DNA polymerase–α is the enzyme which participates in the replication of nuclear DNA. The evidence for this is very wide-ranging. It has been reviewed exhaustively (Fry & Loeb, 1986) and is simply summarised here. First, there is a very good correlation between the activity of DNA polymerase–α and DNA replication in all animals in which the situation has been investigated. A typical set of data is shown in Fig. 4. Secondly, administration *in vivo* of aphidicolin, a specific inhibitor of pol–α, leads to inhibition of DNA replication (Spadari, Sala & Pedrali–Noy, 1982). Thirdly, a complex of proteins capable of DNA replication *in vitro* contains pol–α. Fourthly, a mouse cell line which is temperature–sensitive for DNA replication has a temperature–sensitive pol–α: reversion of the mutation abolishes the temperature–sensitivity both of replication and of pol–α (Eki *et al.*, 1986). In addition to its major role in the nuclear DNA replication, there is also evidence, based mainly on the effect of aphidicolin, that pol–α is involved in DNA repair, particularly in synthesis of long repair 'patches' (Fry & Loeb, 1986).

DNA polymerase has traditionally been regarded as a high mol wt enzyme, and it is relevant to our understanding of the enzyme's role to have an appreciation of exactly what is meant by high mol wt. In earlier studies of this enzyme, there was a good deal of uncertainty about mol wt, and the literature was a confusion of claims and counter-claims (see the review by Holmes *et al.*, 1977). The problem may be summarized as follows: DNA polymerase–α was apparently associated with a protein of mol wt between 100,000 and 200,000, although smaller proteins (50,000 to 75,000) also sometimes had polymerase–α activity. Under certain conditions (particularly low ionic strength), larger estimates (i.e. in excess of 200,000) were obtained, and these were always assumed to be some sort of non-specific aggregates of the enzyme protein. However, with the development of a technique for assaying DNA polymerase activity in polyacrylamide gels after electrophoresis in denaturing conditions (Spanos *et al.*, 1981), the situation has become much clearer. Current

Table 1. *DNA polymerases of vertebrates*[a]

Properties	Polymerase–α	Polymerase–β	Polymerase–γ
Mol wt			
Catalytic subunit	180–200,000	40,000	42–110,000 ?
Native enzyme	400–600,000	40,000	110–300,000 ?
Inhibition by			
N–ethyl–maleimide	Yes	No[b]	Yes
Aphidicolin	Yes	No	No
Phosphate	No	Yes	Stimulates
Phosphono–acetate	Yes	No	–
High K^+/Na^+ concentration	Yes	No	No
Requirement for K^+/Na^+	Low	High	High
Preferred divalent cation with normal template-primer systems	Mg^{2+}	Mg^{2+}	Mg^{2+}
Use of poly(rA).oligo(dT) as template-primer	No	Yes	Yes, preferred when Mn^{2+} is divalent cation
Tightly associated enzyme activities			
$3' \to 5'$ Exonuclease	Yes	No	No
Primase	Yes	No	No
Helicase	Yes	No	No
Template and primer recognition factors	Yes	No	No
Ap_4A binding	Yes	No	No
In vivo location	Nucleus	Nucleus (Chromatin)	Nucleus ? Mitochondria

[a]Data compiled from Fry & Loeb (1986).
[b]Pol–β from dogfish is inhibited by NEM (Philippe & Chevaillier, 1980).

views have been reviewed recently (Fry & Loeb, 1986) and are simply summarised here. Electrophoresis under denaturing conditions ensures that the different polypeptide subunits making up a complex protein are separated. Assay of enzyme activity therefore detects only the polypeptide carrying the catalytic site. For DNA polymerase–α in animals, the catalytic polypeptide has a mol wt of *c*. 180,000; but it is prone to proteolysis and fragments retaining activity, but with reduced mol wt (and particularly of *c*. 140,000 to 50,000), are all too easily obtained. These data thus establish the catalytic subunit of polymerase–α as having a mol wt of *c*. 180,000, although estimates as high as 200,000 have been obtained for some animals (Masaki, Tanabe & Yoshida, 1984).

In addition to the catalytic polypeptide (and its breakdown products), preparations of DNA polymerase–α which are apparently homogeneous under non-denaturing conditions, contain additional polypeptides which do not exhibit DNA polymerase activity. These may be up to 7 in number, and range in size from 30,000 to 70,000 Da (Hübscher & Stalder, 1985; Vishwanatha *et al.*, 1986). Thus the overall size of the mammalian polymerase is *at least* 400,000 Da. Further, it is all too easy to see how the earlier confusion arose, since different enzyme preparations may have shown differing extents of proteolysis of the catalytic subunit, and different extents of dissociation of the various subunits from each other.

The additional polypeptides include primase, helicase, 3'→5' exonuclease (proof-reading enzyme), template-primer recognition proteins and a protein which binds diadenosine tetraphosphate (a possible regulatory molecule). Polymerase–α therefore exists as a large complex which is able to carry out several of the activities involved in DNA replication (Hübscher & Stalder, 1985; Vishwanatha *et al.*, 1986). Intriguingly, there is evidence that it is only the population of pol–α involved in DNA replication which exists in this form; the enzyme population from cells not undergoing DNA replication appears to lack activities such as primase and helicase (Kozu, Seno & Yagura, 1986).

Fig. 4. The effect of partial hepatectomy (which leads to the onset of DNA replication and cell division) on DNA polymerase–α and DNA polymerase–β activities in rat liver.
Key: Filled blocks: pol–α; Open blocks: pol–β; O = control; 24 = 24 hours after partial hepatectomy; 48 = 48 hours after partial hepatectomy.
Based on Chang & Bollum (1972).

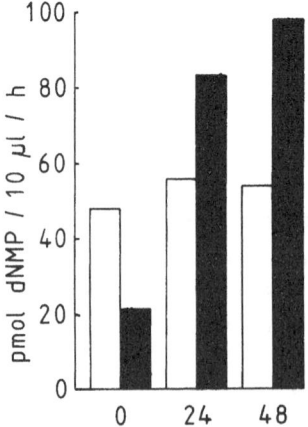

DNA polymerase–β is the simplest type of DNA polymerase found in animals, consisting of a single, small polypeptide chain. Its activity does not correlate with DNA replication (e.g. Fig. 4) and this, taken with the inability of cells to replicate nuclear DNA in the presence of aphidicolin (which does not inhibit pol–β) has been taken as evidence that it does not participate in nuclear DNA replication (Fry & Loeb, 1986). However, since pol–β is a nuclear enzyme (in fact, of the three polymerases, it is the most tightly bound in the nucleus), it is assumed to have some role in relation to nuclear DNA. Further, this role must be constant, since pol–β is as active in mature cells which have ceased replicating DNA as it is in dividing cells (e.g. Fig. 4). Use of aphidicolin in studies of *in vivo* and *in vitro* DNA replication shows that some types of repair, particularly involving only short 'patches', are resistant to the inhibitor, and pol–β is regarded as the enzyme which participates in this (Fry & Loeb, 1986).

DNA polymerase–γ is the most difficult enzyme to which to assign a role. Its dual location in both nucleus and mitochondria has been established in a series of very careful cell fractionation experiments (Fry & Loeb, 1986), but there is no information as to its role in the nucleus. Suggestions that it may participate in the late gap-filling phase in DNA replication or in some types of DNA repair have not been confirmed. Rather all the data suggest that the activities of pol–α and pol–β are completely sufficient for replication and repair of nuclear DNA. The role of the population of pol–γ in the mitochondria (which appears to be identical in properties to the nuclear pol–γ) is easy to guess at, i.e. replication of mitochondrial DNA, but harder to substantiate. The evidence for a role in mitochondrial DNA replication is as follows:

(i) mitochondrial DNA replication is not inhibited by aphidicolin (Spadari, Sala & Pedrali–Noy, 1982);

(ii) the activity of mitochondrial pol–γ correlates well with mitochondrial DNA replication and proliferation of mitochondria (Radsak & Schutz, 1978; Hardt *et al.*, 1980);

(iii) mitochondrial pol–γ is tightly bound to the inner mitochondrial membrane, the site of mitochondrial DNA replication (Adams & Kalf, 1980);

(iv) *in vitro*, with viral DNA templates, pol–γ is able to replicate DNA in an asymmetric, strand-displacing mode, as is required by the D-loop mechanism (Robertson, Bates & Stout, 1983).

Overall, therefore, the circumstantial evidence for replication of mitochondrial DNA by pol–γ is fairly strong, but direct evidence based on isolation of a replication complex from mitochondria, or on the effects of conditional mutations in pol–γ on mitochondrial DNA replication, is still lacking.

DNA polymerases of higher plants

The DNA polymerases of higher plants have received far less attention than their counterparts in higher animals. However, from the data available it is possible to

Table 2. *DNA polymerases of higher plants*[a]

Properties	Polymerase-α	Polymerase-β	Polymerase-γ
Mol wt			
Catalytic subunit	100–200,000	50,000	105–140,000
Native enzyme	400–600,000	50,000	105–140,000
Inhibition by			
N–ethyl–maleimide	Yes	Yes[b]	Yes
Aphidicolin	Yes	No	No
Phosphate	No	Yes	No
Phosphono–acetate	Yes	Yes[c]	–
High K^+/Na^+ concentration	Yes	No	No
Requirement for K^+/Na^+	Low	High	High
Preferred divalent cation with normal template-primer systems	Mg^{2+}	Mg^{2+}	?
Use of poly(rA).oligo(dT) as template-primer	No	No	Yes, preferred template with Mn^{2+} as divalent cation
Tightly associated enzyme activities			
$5' \rightarrow 3'$ Exonuclease	?	No	No
$3' \rightarrow 5'$ Exonuclease	?	No	?
Primase	Yes?	No	?
In vivo location	Nucleus	Nucleus (Chromatin)	Nucleus? Chloroplast? Mitochondrion?

[a]Based on data reviewed in Bryant (1986) and Bryant & Dunham (1988).
[b]Some pol–β preparations are resistant to NEM; those which are inhibited are inhibited less strongly than pol–α.
[c]Less strongly inhibited than pol–α.

construct a summary of the properties of these enzymes, as in Table 2. Before dealing with the three polymerases, two points need to be made. Firstly, the existence of pol–β in plants is still a matter of controversy. Some investigators have failed completely to detect a low molecular weight chromatin-bound polymerase, whilst others can detect such an activity with ease (see discussions in Litvak & Castroviejo, 1985; Bryant, 1986). However, the recent detection of a polymerase of this type, albeit at low levels, in a plant (wheat) in which earlier attempts to assay it had been unsuccessful (Litvak & Castroviejo, 1985) suggests that the presence of pol–β may be a general feature of higher plants. It should be pointed out, however, that the higher plant pol–β is sensitive to certain inhibitors (e.g. N–ethyl–maleimide) to which the animal type pol–β is resistant (compare Tables 1 and 2). Secondly, the number of plants in which pol–γ has been detected is small, but it has been found in every plant in which it has been looked for. In spinach, turnip and rice, the enzyme makes up a small proportion of the total polymerase activity, as in animals (Bryant, 1986; Dunham & Bryant, 1986), whilst in wheat, a pol–γ type enzyme is reported to make up *c.* 30 per cent of total polymerase activity (Castroviejo *et al.*, 1979).

There is extensive correlative evidence to link the higher plant pol–α with DNA replication (reviewed in Bryant, 1986) (Fig. 5); further, nuclear DNA replication in plants is inhibited by aphidicolin (see Table 2) just as in animals. These data are strongly indicative of a major role in DNA replication. Further evidence is beginning to accumulate to show that, as in animals, plant pol–α is a large, complex enzyme at the core of which is a large polymerase catalytic polypeptide (Bryant & Dunham, 1988). The identity of the other polypeptides in the complex is not known. Low levels of primase have been detected in pea pol–α (Bryant & Dunham, 1988), but in wheat primase activity is totally separate from pol–α (Graveline *et al.*, 1984).

In addition to role and complexity, plant pol–α also resembles the animal pol–α in its response to inhibitors (Table 2). Overall, the data point to an enzyme which is very similar to the pol–α of animals. However, despite this, there is apparently no antigenic cross-reaction between plant and animal pol–α (J.A. Bryant, S.G. Hughes & P.N. Fitchett, unpublished data).

There are too few data on plant pol–β to make generalisations about its role. However, in the two plants where detailed studies have been carried out, pea and soybean, there is no correlation between pol–β activity and DNA replication (Fig. 5 and Bryant & Dunham, 1988). There is, however, no direct evidence to link plant pol–β with DNA repair.

The situation regarding plant pol–γ is also not clear. By comparison with animals, three populations might be expected; a nuclear population of unknown function, a mitochondrial population and a chloroplastic population (since chloroplast DNA replication resembles mitochondrial DNA replication). If this were so, plant cells should contain very much more pol–γ than animal cells, since chloroplast DNA makes up a significant proportion of the total DNA (R.M. Leech, Th. Butterfass, J.V. Possingham, this volume) in contrast to the much lower amounts of mitochondrial DNA. However, only in one plant, wheat, does pol–γ appear to make up more than a few percent of the total, and there its activity seems to be correlated with nuclear DNA replication (Castroviejo *et al.*, 1979). In turnip, pol–γ fractionates into three peaks on DEAE–cellulose, and, on the sparsest of evidence, Dunham & Bryant (1986) have suggested that one of these may be chloroplastic. A very detailed characterisation of the chloroplastic DNA polymerase from spinach indicates that this is a γ–type polymerase and that its activity correlates with chloroplast DNA replication (Sala *et al.*, 1980). However, an equally detailed characterisation of the chloroplast DNA polymerase of pea (McKown & Tewari, 1984) revealed an enzyme which cannot clearly be equated with pol–γ (nor indeed with pol–α or pol–β).

The situation concerning the relationship between pol–γ and mitochondrial replication in plants is equally confusing. On the basis that incorporation of thymidine into spinach mitochondrial DNA (as assayed by autoradiography after *in vivo* labelling) is inhibited by ethidium bromide but not by aphidicolin, Sala *et al.* (1981) have suggested that pol–γ replicates mitochondrial DNA. This suggestion, however, presupposes that inhibition by DNA-intercalating agents (such as ethidium bromide) is diagnostic for pol–γ. The problem in assigning mitochondrial DNA replication to pol–γ is well illustrated by the data from wheat (Castroviejo *et al.*, 1979). Here the

Fig. 5. Relationship between DNA synthesis (labelling index), cell division (mitotic index) and the activities of pol–α and pol–β in roots of pea seedlings. Enzyme units are nmol dTMP incorporated per mg DNA per hour. Based on Bryant, Jenns & Francis (1981).

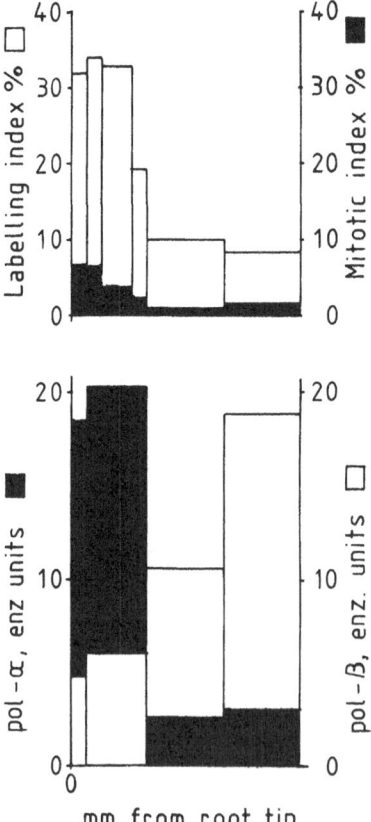

DNA polymerase from mitochondria only resembles pol–γ in that it will utilise poly(rA).oligo(dT) as a template-primer in crude enzyme preparations. However, highly purified enzyme preparations will not recognise the heteroduplex template-primer (Christophe *et al.*, 1981) although in respect of inhibitor responses, the wheat mitochondrial enzyme does resemble pol–γ. Further, as has already been noted, wheat also contains a highly active 'authentic' pol–γ which is presumed to be of nuclear origin. Crude DNA polymerase preparations from cauliflower recognise poly(rA).oligo(dT) as a template-primer (Fukasawa & Chou, 1980), but in view of the data from wheat it would be unwise to categorise this as pol–γ.

Overall, therefore, whilst the identity of the polymerase involved in nuclear DNA replication in plants is fairly clearly established (with the possible exception of wheat), the identity of the enzymes which replicate chloroplast and mitochondrial DNAs remain obscure.

DNA polymerases of lower eukaryotes

A survey of the major DNA polymerases from a wide range of unicellular eukaryotes indicates that all of these organisms so far investigated contain two major DNA polymerases (Bryant, 1986; Fry & Loeb, 1986). All have molecular weights of 100,000 or more, but, with the exception of yeast DNA polymerase I (see below), it is not clear whether these molecular weights relate to a single catalytic polypeptide chain or to a complex. Of the two polymerases, one is always much more closely correlated with DNA replication than the other, and is thus assumed to be involved in DNA replication. Support for this assumption comes from the finding that in the fungus, *Ustilago maydis*, mutations affecting polymerase I also affect DNA replication (Jeggo *et al.*, 1973). Further, the yeast pol–I occurs as part of a complex which includes DNA primase (Yagura *et al.*, 1983), and this complex is much more tightly bound in the nucleus during the S-phase than at other phases of the cell cycle (Tsuchiya *et al.*, 1984).

Investigation of the properties of these lower eukaryotic polymerases indicates that the putative replicative polymerases also resemble pol–α in being inhibited by aphidicolin and N–ethyl–maleimide. However, there are no really clear-cut differences between the two enzymes in any one organism (Bryant, 1986). Both enzymes are inhibited to a greater or lesser extent by N–ethyl–maleimide, and in yeast, both are inhibited by aphidicolin (Fry & Loeb, 1986). In some organisms, both enzymes exhibit the γ-like use of poly(rA).oligo(dT) as a template-primer; in others, one of them does, whilst in the remainder which have been examined, neither does so. A similarly confused situation exists in respect of the possession of exonuclease activity. It is probably, therefore, not wise to attempt to equate the two polymerases with pol–α and pol–γ, although the replicative enzyme resembles pol–α in being part of a complex and in being sensitive to aphidicolin. Finally, there is no firm evidence for the existence of a low molecular weight pol–β-like enzyme in unicellular eukaryotes, although one of the higher molecular weight enzymes may have some pol–β-like inhibitor responses in some organisms, and pol–β-like activities, albeit of higher molecular weight than mammalian pol–β, have been detected in the slime moulds *Dictyostelium* (Baril, Scheiner & Pederson, 1980) and *Physarum* (Holler *et al.*, 1987).

There have been surprisingly few investigations of mitochondrial and chloroplastic DNA polymerases in these organisms. In a study carried out as long ago as 1970, the yeast mitochondrial DNA polymerase was characterised by Wintersberger and Wintersberger, and in respect of its template-primer dependence and responses to a limited range of inhibitors, this enzyme is pol–γ-like. Chloroplast DNA polymerase has been isolated from *Euglena gracilis* (Keller, Beidenbach & Meyer, 1973; McLenna & Keir, 1975) and a 'cytoplasmic' polymerase in *Chlorella vulgaris* has been assigned, on rather slim evidence, to the chloroplast (Aoshima, Nishimura & Iwamura, 1983). There have not been any detailed enough characterisations to clarify these polymerases as α-like, β-like or γ-like.

Concluding remarks

Whilst the enzyme which catalyses replication of nuclear DNA in multicellular eukaryotes (and, to a lesser extent, in unicellular eukaryotes) is well characterised, we are still a long way from achieving a general understanding of the DNA polymerases of mitochondria and chloroplasts. Further, for replicative DNA polymerase (i.e. pol–α) in vertebrate nuclei, we are beginning to understand how the polymerase catalytic polypeptide interacts with other proteins in the pol–α complex, how the complex interacts with other enzymes involved in replication, such as topoisomerase and ligase, and finally how this array of enzymes interacts with the DNA itself. For chloroplasts and mitochondria, we have no such knowledge. We can guess that the comparative simplicity of organelle DNA means that the replication system will be less complex than in nuclei. Even so, the following enzymes will be needed, at the very least: topoisomerase (to convert from supercoiled to relaxed circular DNA), primase, DNA polymerase, proof-reading (i.e. $3' \rightarrow 5'$) exonuclease, primer-removing nuclease (e.g. RNase H) and ligase. A helicase may also be necessary; however, it is also possible that the relatively small size of the organelle DNA, the highly processive, strand-displacing mode of action of pol–γ and the concerted activity of topoisomerase may obviate the need for helicase. For several of these enzymes, there have been no reports at all of the presence in either chloroplasts or mitochondria, and it is thus clear that there is much work to be done. This will involve detailed enzymology and the development of active *in vitro* organelle DNA replication systems.

Acknowledgements

Recent work in my own laboratory has been supported by AFRC, SERC and Unilever Research. I am also grateful to Prof. Val Dunham of Western Kentucky University, for fruitful collaboration and much discussion in our efforts to unravel the mysteries of plant DNA replication.

References

Adams, W.J. & Kalf, G.F. (1980). DNA polymerase isolated from the mitochondrial chromosome appears to be identical to DNA polymerase–γ. *Biochemical and Biophysical Research Communications*, **95**, 1875–84.

Aoshima, J., Nishimura, T. & Iwamura, T. (1983). DNA polymerases of *Chlorella*. 1. Chloroplastic and nuclear DNA polymerases in synchronized algal cells. *Cell Strucure and Function*, **7**, 327–40.

Baril, E.F., Scheiner, C. & Pederson, T. (1980). A β-like DNA polymerase activity in the slime mold *Dictyostelium discoideum*. *Proceedings of the National Academy of Sciences, USA*, **77**, 3317–21.

Bryant, J.A. (1986). Enzymology of nuclear DNA replication in plants. *Critical Reviews in Plant Science*, **3**, 169–99.

Bryant, J.A. & Dunham, V.L. (1988). *DNA Replication and Repair in Plants*, Boca Raton, Fl., USA., CRC Press, in press.

Bryant, J.A., Jenns, S.M. & Francis, D. (1981). DNA polymerase activity and DNA synthesis in roots of pea (*Pisum sativum* L.) seedlings. *Phytochemistry*, **20**, 13–15.

Castroviejo, M., Tharaud, D., Tarrago–Litvak, L. & Litvak, S. (1979). Multiple deoxyribonucleic acid polymerases from quiescent wheat embryos. Purification and characterization of three enzymes from the soluble cytoplasm and one from purified mitochondria. *Biochemical Journal*, **181**, 183–91.

Chang, L.M.S. & Bollum, F.J. (1972). Variation of deoxyribonucleic acid polymerase activities during rat liver regeneration. *Journal of Biological Chemistry*, **247**, 7948–50.

Christophe, L., Tarrago–Litvak, L., Castroviejo, M. & Litvak, S. (1981). Mitochondrial DNA polymerase from wheat embryos. *Plant Science Letters*, **21**, 181–92.

Clayton, D. (1982). Replication of animal mitochondrial DNA. *Cell*, **28**, 693–705.

Dunham, V.L. & Bryant, J.A. (1986). DNA polymerases in healthy and cauliflower-mosaic-virus-infected turnip (*Brassica rapa*) plants. *Annals of Botany*, **57**, 81–9.

Edenberg, H.J. & Huberman, J.A. (1975). Eukaryotic chromosome replication. *Annual Reivew of Genetics*, **9**, 245–84.

Eki, T., Murakami, Y., Enomoto, T., Hanaoka, F. & Yamada, M. (1986). Characterization of DNA replication at a restrictive temperature in a mouse DNA temperature-sensitive mutant, ts FT 20 strain, containing heat-labile DNA polymerase–α activity. *Journal of Biological Chemistry*, **261**, 888–93.

Fry, M. & Loeb, L.A. (1986). *Animal cell DNA polymerases*. Boca Raton, Fl., USA.: CRC press.

Fukasawa, H. & Chou, M. (1980). Mitochondrial DNA polymerase from cauliflower inflorescence. *Japanese Journal of Genetics*, **55**, 441–5.

Graveline, J., Tarrago–Litvak, L., Castroviejo, M. & Litvak, S. (1984). DNA primase activity from wheat embryos. *Plant Molecular Biology*, **3**, 207 –15.

Hardt, N., De Kegel, D., Vanheule, L., Villani, G. & Spadari, S. (1980). DNA polymerase–γ, cytochrome C oxidase and mitochondrial integrity in rabbit spleen lymphocytes stimulated with concanavalin A. *Experimental Cell Research*, **127**, 269–76.

Holler, E., Fischer, H., Weber, C., Stopper, H., Steger, H. & Simek, H. (1987). A DNA polymerase with unusual properties from the slime mold *Physarum polycephalum. European Journal of Biochemistry*, **163**, 397–405.

Holmes, A.M., Hesslewood, I.P., Wickremasinghe, R.G. & Johnstone, I.R. (1977). The heterogeneity of deoxyribonucleic acid polymerase–α. *Biochemical Society Symposia*, **42**, 17–36.

Hübscher, U. & Stalder, H.P. (1985). Mammalian DNA helicase. *Nucleic Acids Research*, **13**, 5471–83.

Jeggo, P.A., Unrau, P., Banks, G.R. & Holliday, R. (1973). DNA polymerase mutants of *Ustilago maydis. Nature, New Biology*, **242**, 14–16.

Keller, S.J., Beidenbach, S.A. & Meyer, R.R. (1973). Partial purification of a chloroplast DNA polymerase from *Euglena gracilis. Biochemical and Biophysical Research Communications*, **50**, 620–8.

Kozu, T., Seno, T. & Yagura, T. (1986). Activity levels of mouse DNA polymerase–α-primase complex (DNA replicase) and DNA polymerase–α, free from primase activity in synchronized cells and a comparison of their catalytic properties. *European Journal of Biochemistry*, **157**, 251–9.

Litvak, S. & Castroviejo, M. (1985). Plant DNA polymerases. *Plant Molecular Biology*, **4**, 311–14.

Masaki, S., Tanabe, K. & Yoshida, S. (1984). Large peptides of 10S DNA polymerase from calf thymus: rapid isolation using monoclonal antibody and tryptic peptide mapping analysis. *Nucleic Acids Research*, **12**, 4455–68.

McKown, R.L. & Tewari, K.K. (1984). Purification and properties of a pea chloroplast DNA polymerase. *Proceedings of the National Academy of Science, USA.*, **81**, 2354–8.

McLennan, A.G. & Keir, H.M. (1975). Subcellular location and growth stage dependence of the DNA polymerases of *Euglena gracilis. Biochimica et Biophysica Acta*, **407**, 253–62.

Phillipe, M. & Chevaillier, P. (1980). Study of dogfish (*Scyliorhinus caniculus*) deoxyribonucleic acid polymerase–α and –β. *Biochemical Journal*, **189**, 635–9.

Radsak, K. & Schutz, E. (1978). Changes of mitochondrial DNA polymerase–γ-activity in synchronized mouse cell cultures. *European Journal of Biochemistry*, **89**, 3–10.

Robertson, A.T., Bates, R.C. & Stout, E.R. (1983). Purified DNA polymerase-γ replicates bovine parvovirus DNA to a unit-length product. *Biochemical and Biophysical Research Communications*, **117**, 580–6.

Sala, F., Amileni, A.R., Parisi, B. & Spadari, S. (1980). A γ-like DNA polymerase in spinach chloroplasts. *European Journal of Biochemistry*, **112**, 211–17.

Sala, F., Amileni, A.R., Parisi, B., Pedrali–Noy, G. & Spadari, S. (1981). Functional roles of the plant α-like and γ-like DNA polymerases. *Federation of European Biochemical Societies Letters*, **124**, 112–18.

Spanos, A., Sedgwick, S.G., Yarranton, G.T., Hübscher, U. & Banks, G.R. (1981). Detection of the catalytic activities of DNA polymerases and their associated exonuclease following SDS–polyacrylamide gel electrophoresis. *Nucleic Acids Research*, **9**, 1825–39.

Spadari, S., Sala, F. & Pedrali–Noy, G. (1982). Aphidicolin: a specific inhibitor of nuclear DNA replication in eukaryotes. *Trends in Biochemical Sciences*, **7**, 29–32.

Tewari, K.K., Kolodner, R.D. & Dobkin, W. (1976). Replication of circular chloroplast DNA. In *Genetics and Biogenesis of Chloroplasts and Mitochondria*, ed. T. Bucher, W., Neupert, W. Sebald & S. Werner, pp. 379–86. Amsterdam: Elsevier.

Tsuchiya, E., Kimura, K., Miyakawa, T. & Fukui, S. (1984). Characteristic alteration in the nuclear DNA polymerase activity during the cell division cycle of *Saccharomyces cerevisiae*. *Nucleic Acids Research*, **12**, 3143–54.

Van't Hof, J. (1975). DNA fiber replication in chromosomes of a higher plant (*Pisum sativum*). *Experimental Cell Reserch*, **93**, 95–104.

Vishwanatha, J., Coughlin, S.A., Wesolowski–Owen, M. & Baril, E.F. (1986). A multiprotein form of DNA polymerase–α from HeLa cells. Resolution of its associated catalytic activities. *Journal of Biological Chemistry*, **261**, 6619–28.

Wintersberger, U. & Wintersberger, E. (1970). Studies on deoxyribonucleic acid polymerases from yeast. 2. Partial purification and characterization of mitochondrial DNA polymerase from wild-type and respiration-deficient yeast cells. *European Journal of Biochemistry*, **13**, 20–7.

Yagura, T., Kozu, T., Seno, T., Saneyoshi, M., Hiraga, S. & Nagano, H. (1983). Novel form of DNA polymerase–α associated with DNA primase activity of vertebrates. *Journal of Biological Chemistry*, **258**, 13070–5.

INDEX